DAWN OF THE NEURON

DAWN OF THE NEURON

The Early Struggles to Trace the Origin of Nervous Systems

MICHEL ANCTIL

McGill-Queen's University Press
Montreal & Kingston · London · Chicago

ISBN 978-0-7735-4571-7 (cloth)
ISBN 978-0-7735-9732-7 (ePDF)
ISBN 978-0-7735-9733-4 (ePUB)

Legal deposit third quarter 2015
Bibliothèque nationale du Québec

Printed in Canada on acid-free paper

McGill-Queen's University Press acknowledges the support of
the Canada Council for the Arts for our publishing program.
We also acknowledge the financial support of the Government
of Canada through the Canada Book Fund for our publishing
activities.

Library and Archives Canada Cataloguing in Publication

Anctil, Michel, 1945–, author
Dawn of the neuron : the early struggles to trace the origin of
nervous systems / Michel Anctil.

Includes bibliographical references and index.
Issued in print and electronic formats.
ISBN 978-0-7735-4571-7 (bound). –
ISBN 978-0-7735-9732-7 (ePDF). –
ISBN 978-0-7735-9733-4 (ePUB)

1. Nervous system.   2. Neurophysiology – History.
3. Neuroanatomy – History.   4. Neurons – History.   I. Title.

QP353.A53 2015   611'.809   C2015-903831-6
                                  C2015-903832-4

Set in 9.5/13.5 Baskerville 10 Pro
Book design & typesetting by Garet Markvoort, zijn digital

CONTENTS

ACKNOWLEDGMENTS

Much of the information that has found its way into this book comes from scholarly publications, but invaluable information on the personalities depicted in the book came also from Internet sources. Visiting archival sources on the Internet rather than in their physical locations has increasingly become the norm for scholars in recent times. I am especially indebted in this regard to Donald R. Forsdyke, professor emeritus at Queen's University, Kingston, Ontario, for the material related to George Romanes that he collected and made available on his website. Other sources are available only at their physical locations, an example being the George Parker papers at the Harvard University Archives. Their reference desk at the Pusey Library was very helpful. Other material connected with George Parker was located at the Ernst Mayr Library of Harvard's Museum of Comparative Zoology, and my gratitude goes to Dana Fisher and Mary Sears, who greatly facilitated my research there. Being unable to travel to locations such as the Wellcome Library in London, England, for the collected papers of Edward Sharpey-Schafer, I was grateful to Professor Richard D. French of the University of Ottawa for providing information going back to his visit there for his own research in the 1970s. I am also indebted to Professor French for the inspiration that his papers on Romanes's jellyfish work and on Darwin and the physiologists gave me for the genesis of this book.

I thank the three anonymous reviewers of my manuscript, who provided numerous insightful comments and suggestions that greatly improved the scholarly quality of the book. For marked improvements in style, consistency, and narrative flow I am indebted to my copy editor, Jane McWhinney. And, last but not least, I am grateful to Jacqueline Mason, my editor at McGill-Queen's University Press, who believed in my project from the start, guided me, and encouraged me throughout the process in my moments of doubt.

PREFACE

When I began exploring the nervous systems of coelenterate animals in the early 1980s, I was motivated by the consensus at the time that such animals represented the earliest extant forms in possession of a nervous system. It was then thought that detailed knowledge of the organization and function of these nervous systems would provide a basis on which to construct theories on the earliest rudiment of a nervous system, which had presumably emerged in some ancestral form preceding coelenterates. However, like many of my colleagues engaged in this type of research, I tended to refrain from runaway speculations, restricting myself to short concluding statements about the possible evolutionary implications of my findings. The thought of constructing a theory of how nervous systems came into existence, although attractive at first, yielded to the humbling realization that such constructs could not have any sound foundation, as "footprints" of nervous system emergence are missing from the paleontological record.

But my fascination with this question persisted. Over the years I assembled, little by little, a collection of original papers and monographs by the pioneer zoologists who had examined coelenterate nervous systems, reaching back to the middle of the nineteenth century. In recent years I studied the texts by historians of biology that mentioned prominent figures in the field, and attempted to provide contexts to their works. And now, after the closure of my academic career, I have been able to dedicate time to the question more directly in its historical perspective.

As I went about structuring this book, it seemed inevitable that I would need to pursue two aims at once. On the one hand I envisaged a chronicle of the research activities on the topic, from the earliest efforts in the nineteenth century to those of the modern era, peppering them with cameo portraits of the pioneer zoologists who did the researches, insofar as details could be gleaned from the available documents. On the other, I had to thread my way through the chronology of scientific contributions to highlight the ideas of some of these pioneers on the origin of nervous systems, as demonstrated in their own work and that of others, and to show how these ideas evolved with time and with the ever-increasing sophistication of scientific methods.

This book traces through history the work of zoologists who were curious about the way nervous systems first came about, through their investigations of living metazoans that possess the most basic neural elements. It is not about how nervous systems evolved further from the template design seen in coelenterates, as in higher invertebrates and in vertebrate animals; that would require writing another book. I focus rather on research activities centred on animals presumed to possess the "likeness" of the earliest nervous systems, and on the drive on the part of researchers to generate hypotheses about the process by which nerve cells first entered the animal world. It is all about "origins," a quest as old as humanity, which begat the Bible and Darwin's *On the Origin of Species*. It is about the yearning to understand the origins of a structure that foreshadowed the organ responsible for much of the evolutionary success of higher animals – the brain.

It is now universally accepted that nerve cells are fundamental to the architecture and functioning of the brain. My first chapter narrates how cell theory led to this recognition. This clue, in turn, helped zoologists in pursuit of origins to turn their attention to nerve cells. But it is easy to forget that between the first observations of the brain as an organ in ancient times and the discovery of nerve cells, over two millennia had elapsed.

The awareness that the human brain is the seat of perception, motor control, and thought was recorded in antiquity, especially in the last few centuries BCE during the Hellenistic period of Greek Enlightenment. While the ancients, and later scholars of the Renaissance Period, came to recognize by dissection that animals also possess a brain, the significance of comparative anatomical descriptions did not sink in until Darwin's theory of descent by natural selection cast the subject in an evolutionary framework, as noted by Georg Striedter (2005) in his masterly essay on brain evolution. However, even then the focus was not so much the origin of brains as their evolution in a narrow sense; that is, the transition from the brains of lower vertebrates to those of mammals and the transition from the nonhuman primates to the human brain.

It is intriguing that Darwin himself studiously avoided raising the issue of the origin of brains in the original 1859 edition of his epochal book, *On the Origin of Species*. As Stephen Jacyna (2009) explained, it was not for lack of ruminating on the matter, as two of Darwin's notebooks from the 1830s contain ideas he had jotted down about "the relation between the mental powers of man and animals to 'brain.'" The reason for Darwin's reluctance, according to Jacyna, is that Darwin "realized that the thesis that species were not fixed but had evolved over time was sufficiently contentious without being mixed with the potentially incendiary claim that the human mind was in essence no different from the mental powers manifested by the lower animals and that all thought derived from the properties of the evolving brain." Darwin's conclusion that mammalian brains share a basic structure that evolved gradually toward man's cerebral development was expressed in later editions of *Origin of Species* and in *The Descent of Man*, but the question of how nerve cells initially arose in evolution was not part of the debate at that time.

Although Darwin was a skilled microscopist (Ford, 2009), he never used the instrument to examine cellular elements. Had he been interested in the origin of nervous systems rather than the origin of complex brains, he would have disqualified himself if only, as my second chapter shows, because the cellular level of organization of the "simplest" nervous systems required an expert familiarity with the best microscopes the era could offer. Ironically, it fell to a future anti-Darwinist, Louis Agassiz, to avail himself of the opportunity and to earn the distinction of reporting the first description of jellyfish nerve cells in 1850.

Promoters of Darwinism in Germany and England soon took on the mantle of coelenterate nervous system studies. In the third chapter we see how Ernst Haeckel, Darwin's bulldog in Germany and an influential evolutionist in his own right, dismissed Agassiz's contribution, only to come up with a facsimile description of the jellyfish nervous system. It turned out that zoologists Haeckel had mentored made more significant contributions to coelenterate neurobiology than he had done. After chronicling a halting start with Nikolaus Kleinenberg, whose zeal to uncover the "missing link" in the gestation of nerve cells led him to their misidentification in hydra, I devote the fourth chapter to the crucial contribution of the Hertwig brothers, Oskar and Richard, the formidable, trail-blazing team that brought the understanding of coelenterate neuroanatomy to a level unsurpassed for subsequent decades, and were the first to articulate a theory of nerve cell formation.

Almost simultaneously, George Romanes in England and Theodor Eimer in Germany put another foot in the door (beyond neuroanatomy), opening up the field of coelenterate behavioural physiology to glimpse possible scenarios for the evolutionary emergence of nervous systems. As the fifth chapter illustrates, Romanes was inspired by the close friendship he developed with his idol and intellectual father-figure, Charles Darwin, to make experiments on jellyfish that shed valuable light on the rise of nervous coordination in early evolution. His intellectual figure is contrasted with that of Eimer, a German competitor who had similar ambitions but had the misfortune of losing the publication race by a thread, forcing him to play second fiddle to Romanes and the Hertwig brothers in terms of legacy.

One of the most intriguing protagonists in this story is Edward Schäfer, the hero of the entire sixth chapter. Schäfer was drawn to the field through his friendship with Romanes and because Romanes found his skills as a histologist useful to his jellyfish research program. But lending a hand to Romanes led to a serendipitous event of major importance: the jellyfish nerve net Schäfer described allowed him to demonstrate for the first time the main tenet of the neuron doctrine: that nerve cells are discontinuous with each other. In that chapter we also see how his discovery failed to earn him proper recognition.

The work of intellectually towering figures is often followed by a host of more pedantic contributions that amplify the scientific knowledge base but fail to offer new paradigms. The seventh chapter showcases the motley crew of established scholars and doctoral students who to some extent devoted themselves to such work. In their surveys of the nervous systems of a greater variety of coelenterates, they enriched the field of study but, depending on their choice of methods and the intensity of their focus on the meaning of their contributions for the general question of the origin of nerve cells and nervous systems, they achieved varied success. This broadening of the field gives us the chance to look at the socio-cultural context of the work of these protagonists and to follow their respective fates through academic success or personal tragedy.

In America the search for coelenterate nervous systems started with Louis Agassiz at Harvard; and after a long hiatus in Europe, Harvard again hosted the pursuit in the figure of George Parker. As shown in the eighth chapter, Parker could be considered the intellectual heir of Romanes in both the approach and scope of his work. Parker stands out particularly through his strong commitment to a vast research program on the early evolution of the nervous system, which he articulated even before his career was launched. From his experiments on sponges and sea anemones, Parker proposed specific scenarios for the emergence of nerve cells by gradation. He introduced the concept of neuroid conduction, according to which an evolutionary stage of "pre-neurons" existed. In a strange twist, his views became tainted by the belief that the neuron doctrine applied to all animals possessing a nervous system except coelenterates. His synthesis of his views, *The Elementary Nervous System*, published in 1919, proved highly influential for decades to come. Yet, despite the stature of the man and his dominance of the field at the time, few disciples followed in his steps and he trained none.

As an American school of thought on the subject failed to materialize, the field once more crossed the Atlantic and landed in Germany. In the ninth chapter we meet the youthful figure of Emil Bozler, whose solid work on jellyfish, physiological and anatomical, helped propel the field in profitable directions. His neuroanatomy laid to rest, or so it seemed, the oft-revived notion that nerve cells are in protoplasmic continuity with each other and thus form syncytial nerve nets. This affirmation constituted another triumph for the neuron doctrine. The physiological work of Bozler and other Germans of the 1920s and 1930s helped advance our understanding of the nervous coordination behind coelenterate behaviour. As the chapter shows, this cohort of investigators, along with young colleagues in Great Britain, were also part of a new wave of interest in comparative physiology.

One representative of this new generation of comparative physiologists was the Cambridge zoologist Carl Pantin, to whom the tenth chapter is devoted. In contrast to his German colleagues, Pantin, like Parker, spent his entire career working on coelenterates (sea anemones). Pantin epitomizes the notion that a thoughtful and inventive handler of methods can accomplish scientific breakthroughs and generate new paradigms. This book is full of stories of investigators who did not heed this simple cautionary note. Through Pantin's experimental approach, a radically new understanding of the dynamics of nerve net function emerged, in which the concept of facilitation played a large part. In addition, Pantin's close attention to microscopical techniques allowed him and his students Elizabeth Batham and Elaine Robson to obtain striking images of the sea anemone nerve net.

The final chapter brings us to the modern era, the period after Pantin which from the 1950s saw the blossoming of exceptionally talented investigators such as G. Adrian Horridge and George Mackie. By this time electrophysiological tools had advanced in sophistication and allowed the monitoring of specific nerve cell activities in jellyfish, whose large cells made them suitable for such recordings. Horridge was thus able to observe that jellyfish nerve cells produced action potentials similar to those of higher animals, and Mackie discovered the phenomenon of epithelial conduction, whereby jellyfish epithelial cells can display the excitability and impulse conduction usually attributed to nerves. This discovery spurred Mackie to develop his own theory for the emergence of nerve cells. In this chapter we also see how electron microscopy and the development of chemical neuroanatomy helped identify synapses in the coelenterate nervous system. The book ends with a brief look at the current status of the field and the theories newly postulated by the present generation of investigators against the backdrop of the work and ideas of the old schools.

In this journey through time I have endeavoured to showcase the unending quest for a resolution to the issue of when, how, and in what form neurons first appeared in deep historical time; the interplay of personalities and societal circumstances that governed the prevalence of ideas and the filtering of observable facts; the jostling of schools of thought and cultural traditions that permeated the investigators' professional lives. It is a story of science and people.

DAWN OF THE NEURON

CHAPTER ONE

The Precursors

The time had now arrived when microscopists were beginning to
see actual cells in various animal tissues.

John R. Baker, 1948

At the heart of the story that will now unfold are the two most important theories contributing to our all-encompassing view of the biological world – cell theory and the theory of evolution. Without cell
theory, the hypothesis that nervous tissue is fundamentally made up of
individual cells would not have been articulated, and the subsequent
drive for microscopic observations of these cells, later called neurons,
would have been delayed for decades. And without the theory of evolution, as articulated by Darwin, it is arguable that the question of how
neurons and nervous systems arose and evolved would have remained
unasked for a long time.

Cell Theory and the Discovery of Nerve Cells

Cell theory gained currency once methods of observation allowed the
identification, however tentative, of the constitutive elements of plant
and animal tissues. These historical developments have been ably analysed in a monograph (Shepherd, 1991) and what follows is merely a
digest of the story. A critical step was of course the design and construction of the first microscopes. In their maiden form, microscopes were
crude assemblies of compound lens, tube, and eyepiece that allowed at
best a hazy and distorted view of the texture of biological samples. The
Londoner Robert Hooke in 1665 observed the array of "cells" in cork,
for instance, and the Dutchman Antonij van Leeuwenhoek described
the fine structure of the bee mouthparts around 1673. Not until four

decades later did Leeuwenhoek make his first observations of nerve tissue, a peripheral nerve of the cow (Hydén, 1967). It is perhaps an indictment of the poor optical quality of the early microscopes that Leeuwenhoek described the nerve as containing stacks of hollow tubes where myelinated axons reside. But such observations can also be seen as products of the intellectual era. The hollow tubes corresponded to a metaphor dating from antiquity about a spirit fluid flowing through tubes, in the manner of a circulatory system, to carry signals to the brain or to muscles.

If Leeuwenhoek stared unknowingly at axons, it took over a century for other investigators to revisit these nerve components with a fresh eye. With the advent of compound microscopes equipped with achromatic lenses around 1824 came the possibility of observing details of cell morphology with better accuracy at a greater magnification. The first to view nerve cells with these tools was the French medical doctor and microscopist Henri Dutrochet (1776–1847). A member of the French country gentry dispossessed by the French Revolution, Dutrochet became a military physician attached to the Napoleonic campaigns (Nezelof, 2003). After the Napoleonic era he settled on his mother's estate and was spurred to conduct scientific observations on plants and animals by his readings of Lazzaro Spallanzani (1729–1799), the Italian naturalist who first observed egg fertilization and animal regeneration, and by the work of the great Leeuwenhoek himself. Dutrochet set up a small laboratory on the estate and acquired the finest microscope he could lay his hands on.

In his first remarkable monograph, based primarily on his morphological observations and experiments on plants (Dutrochet, 1824), he not only described unequivocal cells, but marvelled that the small number of cell types that he saw could make up such a great diversity of tissues and organs. These results were published fifteen years before Theodor Schwann (1810–1882) advanced his own "cell theory" with botanist Mathias Schleiden (1804–1881). Their broad observations of animal and plant tissues led them to surmise that all biological structures originate from basic self-contained constituents called cells (Schwann, 1839). Even structures that in those years were considered counterintuitive to associate with cells – feathers, hair, teeth – were shown to derive from cells. Schwann never cited Dutrochet's earlier work. Schleiden was the first to describe the cell nucleus, which Dutrochet had overlooked. This failure takes the shine off Dutrochet's achievement, but it is excusable given the limits set by microscopes using incident rather than transmitted light and by the paucity of adequate dyes (carmine, haematoxylin, aniline dyes) for staining preparations at that time. Still, Dutrochet is a legitimate contender for the title of original exponent of cell theory. He also conducted experiments that unveiled the dynamics of water in plant cells, for which he coined the word *osmosis*; this discovery marks him as the pioneer of the discipline of cell physiology.

Dutrochet's writings suggest that he gained an accurate view of frog nerve axons as early as 1827, although he failed to observe or recognize neuronal cell bodies connected to these axons (Dutrochet, 1837). It is worth citing the relevant passage of his 1837 monograph, as it is instructive of the way Dutrochet's observational skills worked. The quoted passage follows a debate in which Dutrochet had contested the view of his good friend Pierre Prévost (1751–1839), a Swiss physicist and naturalist, who claimed that the nerve fibre was flat like a ribbon. Dutrochet was able to settle the dispute in 1827 by employing the latest microscope model equipped with achromatic lenses:

I had an opportunity at that time to put it to use in Geneva with Mr Prévost. With this microscope, with the aid of reflected light, we examined together the structure of the nerve fibre of the frog. We saw most distinctly that this nerve fibre is truly a cylinder whose walls are made up of clusters of globules. Mr Prévost agreed with me, that according to this decisive observation, the nerve fibre is not, as he had thought, made up of four elemental fibres composed in turn of cells placed in a line. This nerve fibre, I repeat, is a cylinder whose walls are formed of globules juxtaposed in a random fashion, as seen in figure ... 1. It is not known whether this cylinder is filled or whether it is tubular. (474–5)

By "globules" Dutrochet meant "cells," as he used that word for his descriptions of cells in other tissues. His illustration in his 1837 monograph suggests that what he saw as globules was probably the myelin sheath, a tubular outgrowth of the supporting cells of neural tissue (neuroglia) which wraps itself around, and acts as an electrical insulator of, the axon (myelinated axon).

This auspicious beginning was followed a decade later by the astute observations of a Polish-German biologist, Robert Remak (1815–1865). This remarkable investigator, who is credited with important discoveries such as the fact that cells form by the division of pre-existing cells and that there are three – not four as previously believed – germ layers in the early embryo (ectoderm, endoderm, and mesoderm), failed to gain the recognition he deserved because of his Jewish background (Schmiedebach, 1990). Throughout his career he had trouble getting professorial appointments and this curse followed his family for generations: his grandson died in Auschwitz in 1942. He was reduced to pursuing his microscopic research at home, financing it with income from his medical practice and tuition fees.

In his early twenties Remak made critical contributions to the microscopic anatomy of the nervous system. In two successive papers he was able to differentiate between what we now call myelinated axons ("primary bands within a thin-walled tube": Remak, 1836) and unmyelinated axons ("organic fibres": Remak, 1837). Although Remak suggested that these axons are connected to cell bodies, he offered no incontrovertible evidence to support such a claim. What he was convinced of is that the axon is not hollow as then thought – empty or fluid-filled – but packed with protoplasm. This view was slow to catch on; it took several years for it to be accepted. One prominent supporter of Remak's work was Theodor Schwann, the official flag-bearer of cell theory. In the late 1830s Schwann was able to show that Remak's "thin-walled tubes," the myelin sheath around the nerve fibre, merges with a cell body, henceforth named the Schwann cell.

It was Christian Gottfried Ehrenberg (1795–1876), a prolific German naturalist and zoologist credited with founding the discipline of micropaleontology (Siesser, 1981), who first documented seeing nerve cell bodies. He was a friend of Alexander von Humboldt, with whom he conducted scientific expeditions in remote parts of Europe and Asia. Ehrenberg became a microscopist out of a keen interest in unicellular organisms (protists). In that capacity he pioneered observations of vertebrate spinal ganglia (Ehrenberg, 1833) in which he fuzzily described what appeared to be ganglion cells, although he did not label them as such. He offered a more convincing description of nerve cells three years later (Ehrenberg, 1836). Apparently, while sampling ponds and other water basins for his microorganisms, he came across leeches and other invertebrates and seized the opportunity to examine their microscopic anatomy. In the cerebral ganglia of leeches he clearly saw and drew nerve cell bodies with their complement of cytoplasm, nucleus, and axon hillock (Fig. 1.7A in Bennett, 2001). In the same year another German, Gabriel Gustav Valentin (1810–1863), observed in the cerebellum strikingly similar cells, which he called "Kugeln," translatable as "corpuscles."

Valentin was a pupil of the famous Czech anatomist and physiologist Jan Evangelista Purkyně (1787–1869). Purkyně was no stranger to the microscope, and his skills were more developed than those of his contemporaries. He pioneered the use of new, more efficient chemical fixatives to preserve tissues and he introduced Canada balsam as an optical medium for immersing tissues between pieces of glass for microscopic observation. He also introduced the use of the microtome to slice tissues thinly, thereby further improving the optical quality of microscopic observations. Taking full advantage of these technical innovations, he made sterling observations of nerve cells in the cerebellar cortex (Purkyně, 1838). The drawings of these ganglion cells, which bear his name, clearly show the branching processes that became later known as dendrites.

From Darwin to Studies on the Evolution of Nervous Systems

Darwin's theory of evolution by natural selection came to the attention of the public over twenty years after nerve cells were first investigated with reliable microscopic techniques. While the publication of Darwin's seminal book (Darwin, 1859) ushered in a slew of research programs aimed at validating his hypothesis under the aegis of Thomas Henry Huxley in England and Ernst Haeckel in Germany, examination of nervous systems – and especially brains – in the light of Darwin's theory was contemplated with little urgency or relish. The reason is simple: Darwin's views were rebuked from the start by many critics from inside as well as outside the scientific community, so it was predictable that merely being open to considering the hypothesis of the human brain, the "organ of the soul," as a derived product of an ancestral animal brain was not only anathema but a threat to one's career. As a result, comparing nervous systems in an evolutionary perspective had a late start and proceeded cautiously at a crawl.

Georg Striedter (in Kaas, 2009) has provided a valuable sketch of the historical development of views on nervous system evolution in the context of pre- and post-Darwinian ideas about how diversity of forms is obtained and organized. Pre-Darwinians would ask, for instance: do invertebrate ventral nerve cords and the vertebrate spinal cord share a common plan, as understood by Geoffroy St Hilaire and his followers – that is, do they share variations stemming from a basic design (*Bauplan* or archetype)? Or is the construction of these neural structures dictated by common functional imperatives, irrespective of how distant from each other these taxa are in the fixed "scale of nature," as Georges Cuvier would argue? Once Darwin provided an evolutionary scenario naturalists could live with, then it became a simple matter of time before St Hilaire's "common plan" transmuted to the notion of forms connected to each other by common descent from an ancestral form – that is, homology in the modern sense. Similarly, organizing taxa in a hierarchical system graphically rendered by the rungs of a ladder (*scala naturae*) yielded to phylogenetic trees reflecting assemblages of taxa rooted in common ancestry.

But researchers' efforts to capitalize on opportunities opened up by Darwin's revolutionary ideas through comparative analyses of nervous systems were slow to take off. The first contribution of this kind came only in 1920 with the publication of a treatise on the comparative anatomy of nervous systems (Ariëns Kappers, 1920). This book, published in the Netherlands, was divided into two parts: the first, written by Peter Droogleever Fortuyn, covered invertebrate nervous systems and the second, written by Cornelius Ariëns Kappers, treated the vertebrates.

Droogleever Fortuyn had only a passing interest in this topic, so his treatment is no more than an exhaustive compendium of the literature on invertebrate nervous systems from sponges to protochordates. He contributed little original work of his own to the subject, and in later years he moved away from it, dabbling in such fields as anthropology, genetics, and even eugenics. Consequently, his enumerative approach was not conducive to phylogenetic insights from comparisons of invertebrate phyla. However, his thorough analysis of the available literature of the nineteenth and early twentieth centuries will be useful to us later here.

Ariëns Kappers, on the other hand, was a dedicated practitioner of comparative neuroanatomy who had made and continued to make significant contributions to the literature on vertebrate nervous systems. His 1920 monograph was later translated into English and augmented (Ariëns Kappers, 1929, Ariëns Kappers, Huber, et al., 1936). While the 1920 version dealing with vertebrates presents only a few phylogenetic insights, the later versions contained copious hints of evolutionary trends from fish to mammals for the different brain compartments and the spinal cord. Ariëns Kappers used his extensive knowledge of vertebrate brains and spinal cords to propose his theory of neurobiotaxis, according to which nerve fibres from different peripheral locations that need to be stimulated simultaneously become "attracted" to each other during development and converge in brain centres where they become connected. Examples of this would be connections of smell and taste inputs in the brain and learning of sensory experiences by association. Vertebrate brains, Ariëns Kappers suggested, evolved partly as a result of neurobiotactic processes. Although interesting at the time, his theory is nowadays discredited (see Striedter, 2009).

Another early practitioner of comparative neuroanatomy was the Swede Bertil Hanström (1891–1969), a professor at Lund University who published a monograph on invertebrate nervous systems (Hanström, 1928a). As the book was written in German, Hanström made a conscious effort to provide a summary in English in an American journal specializing in the subject (Hanström, 1928b). Perusing the book and the article make it clear that Hanström, in comparison to Droogleever Fortuyn, went far in providing phylogenetic hypotheses. He particularly developed the idea that the migration of invertebrate nerve cell bodies to the periphery of ganglia during the centralization of nervous systems was an evolutionary trend determined by chemotaxis and the metabolic needs of cell bodies, brought about by bringing the latter closer to oxygen supplies. As a result, Hanström argued, invertebrate nerve cells evolved from multipolar or bipolar, as seen in primitive nerve nets, to unipolar or pseudo-unipolar in the ganglia of higher invertebrates. In this way the cell body was devoid of dendrites because it was removed from the synaptic fields in the deeper neuropile, and nerve impulses consequently travelled from dendrites to axons without passing through the cell body. He further argued that the reverse trend occurred in vertebrates as a result of the greater development of blood vessels deeper in the body required to accommodate the metabolic needs of the nerve cell bodies.

While the legacy of Ariëns Kappers's treatise on vertebrate neuroanatomy continued with English translations and revisions into the 1960s, it was almost three decades before Hanström's equivalent contribution on invertebrate nervous systems found an heir. In 1965 a new comprehensive treatise on invertebrate nervous systems appeared which combined anatomical and physiological approaches. The monumental two-volume monograph by Theodor Bullock and Adrian Horridge (1965) was acknowledged as a great scholarly achievement and acted as a *de facto* mentor to an entire generation of invertebrate neurobiologists. The book inspired researchers to open new fields of investigation, but more significantly, it led to the development of seminal invertebrate models to investigate sensory-motor circuits and learning at the level of the individual neuron, challenges into which vertebrate neuroscientists had made very little headway. The pivotal role of large identifiable neurons in leeches, cockroaches, and sea slugs comes to mind.

From the Origin of Species *to the Origin of Neurons*

One may ask: have zoologists shown as much interest in unravelling how neurons emerged in early multicellular animals as they showed in comparing nervous systems among phyla after the publication of *On the Origin of Species*? Unfortunately, in answer to this question, one can only offer conjectures.

The publication record of that era suggests that zoologists went about acknowledging that the different systems or organs of animal bodies *are* there, so the task at hand was to deal with them through descriptions, comparisons, and deductions about phylogenetic trends. That these systems or organs must have had a beginning in the deepest of time seems to have eluded their grasp. The irony is that, although Darwin was himself more interested in how species arose than in how they compared, his successors did not follow suit when addressing systems or organs. Zoologists became engrossed in probing how nervous systems are modified, but neglected to ask how neural structures started in the first place.

I am not suggesting that all zoologists of the period followed that pattern. This book will indeed show that a few exercised their curiosity in the direction of origins. But the scholar is hard-pressed to unearth the relevant literature, as it is often buried in specialized journals. In the past, the issue of the origin of nerve cells and their assembly was rarely discussed in scientific forums, and even then, only among neurobiologists. The only instance when the issue came before a wider readership occurred recently, interestingly, in an article by Greg Miller published on the occasion of the two-hundredth anniversary of Darwin's birth and the concurrent 150th anniversary of the publication of *On the Origin of Species* (Miller, 2009). The article discussed comprehensively and in lay language the current status of the field: hypotheses on the emergence of neurons from other cells, how these hypotheses fit into the grand scheme of early animal evolution, the role of head development in shaping neural circuitries, and how genomic information supports the view that a wide array of signalling molecules currently associated with synapses were deployed before neurons made their entry in the animal world.

In view of this newfound interest in the origin of nervous systems, it is perhaps timely to examine in turn the historical origin and early development of studies on this subject. In undertaking this task, I delve into the historical background that led to the pioneering studies on the origins and early evolution of nervous systems in the second half of the nineteenth century and early twentieth century. Interest in understanding where neurons came from and what they stood for arose in parallel with the great strides in unravelling the basic concepts of neuronal structure and function during the same period. I illustrate how some of these early studies contributed support for the neuron concept even before Ramón y Cajal articulated the concept that won him the Nobel Prize. And finally I show that, albeit with notable exceptions, Continental Europe and Anglo-Saxon pioneers in the field went about their research with different approaches reflecting scientific-cultural divides, the Europeans emphasizing anatomical studies and Anglo-Saxons experimental investigations. Throughout, I offer socio-cultural contexts to the scientific events and experimentation that took place, and vignettes of the protagonists with a view to unmasking the motives that drove them.

Let us begin our narrative.

CHAPTER TWO

Louis Agassiz and the First Description of Jellyfish Neurons

I do not know which in this organism is the most wonderful, – the apparent simplicity of the whole structure; or the diversified indications of active life; or the complications and variety that the simplest elements of structure exhibit.

Louis Agassiz, 1850

To the modern investigator, searching for the current manifestation of the earliest nervous systems means training an inquisitive eye on the animals presenting the simplest organization. For naturalists working in the mid-nineteenth century the same intuition held. They observed sponges, jellyfish, sea anemones, and comb-jellies, and saw that despite their deceiving "vegetable" appearance they were capable of sensing their environment and of producing coordinated motor activity. The graceful pulsating swim of jellyfish and comb-jellies stands out in this regard, in contrast to sedentary forms like sponges and sea anemones. The great naturalist Louis Agassiz was entranced by their beauty and sensitive to the lessons to be learned from watching them:

> It is indeed a wonderful sight, to see a little animal not larger than a hazel-nut, as transparent as crystal, as soft as jelly, as perishable as an air-bubble, run actively through as dense a medium as water, pause at times and stretch its tentacles, and now dart suddenly in one direction or another, turn round upon itself, and move suddenly in the opposite direction, describe spirals like a bird of prey rising in the air, or shoot in a straight line like an arrow, and perform all these movements with as much grace and

precision, and elongate and contract its tentacles, throw them at its prey, and secure, in that way, its food, with as much certainty, as could a larger animal provided with flesh and bones, teeth and claws, and all the different soft and hard parts which we consider generally as indispensable requisites for energetic action; though these little creatures are, strictly speaking, nothing more than a little mass of cellular gelatinous tissue.

The study of such animals is therefore of high physiological importance, as it will enlarge our views of animal functions, and give more precision to our ideas of cellular life; and the more so, because in this, as well as in other Naked-eyed Medusae, we can satisfy ourselves with the greatest ease, that the different organs that perform here different functions are entirely and exclusively composed of cells; not in the same sense as it can be said of the body of higher organisms, but strictly so, – the cells here not undergoing any extensive metamorphosis by which they are transformed into distinct tissues of different structures, though derived from cells. (Agassiz, 1850)

In this passage Agassiz reveals himself as an intellectual heir to his mentor Georges Cuvier in that he too organizes animals by levels of complexity in the pre-Darwinian mold and in viewing structures as subservient to their function. And a mere eleven years after cell theory was deftly promoted by Theodor Schwann and Matthius Schleiden, Agassiz had taken the step of acknowledging that even the simplest of multicellular animals were constituted of cells, although little tissue specialization was detectable. But should nerve cells be included among the other cell types he observed? Before we engage in answering this question, it is opportune to trace the path that led Agassiz to his discovery.

A Naturalist for All Seasons

It seems ironic that the first zoologist to observe one of the earliest nervous systems, a topic suited for evolutionists, was a deeply religious man who later fiercely opposed Darwinism. But there was more to Agassiz than this. After all, he is considered perhaps one of the greatest scientists of nineteenth-century America. He founded Harvard's Museum of Comparative Zoology (Winsor, 1991), was a founding member of the American Association for the Advancement of Science, and was the first to create educational facilities in the field that later led to the mushrooming of seaside institutions such as the celebrated Marine Biological Laboratory and the Woods Hole Oceanographic Institution in Woods Hole, Massachusetts.

Edward Lurie's hailed biography of Agassiz is full of details and insights that have informed this narrative (Lurie, 1960). Jean-Louis Rodolphe Agassiz (1807–1873) was born in the French-speaking region of Switzerland to Protestant parents, and his father was a well-regarded pastor in his community. From his childhood Louis, as he was soon known, took a keen interest in what the local Swiss natural landscape offered to the curious collector of fauna. At fifteen he already was a proficient amateur naturalist and had written out a plan for his future that saw him as the foremost field naturalist and writer of his time, following self-prescribed stints of academic training in prestigious institutions of learning in Germany and France. That he devised such a plan testifies to his precocious maturity, exceptional intelligence, and ambitious drive.

Agassiz managed to overcome the predictable opposition of his father to these grand schemes. As a man deeply rooted in his canton and seeing the best prospects for a son in business, religion, or medicine, Louis's father was fairly representative of his day and place. Louis did not challenge his father openly, but he deployed the limitless stores of charm and persuasion that were to define his character throughout his life. He coaxed local notables and family friends to impress on the pastor the wisdom of letting a brilliant mind flourish and make its mark in the world. Besides, as an insurance policy for his financial future, pressure was brought to bear on Louis to complete a medical degree in parallel to studies of natural history. Despite his parents' continued misgivings about their son's unrelenting devotion to natural sciences, Louis managed to study under the best professors of Europe in embryology, zoology, and paleontology – Friedrich Tiedemann, Friedrich Leuckart, and Heinrich Georg Bronn in Heidelberg; Ignaz Döllinger and Lorenz Oken in Munich; Georges Cuvier in Paris – and still successfully complete his perfunctory medical degree. His drive to become the top naturalist of Europe was unwavering as he push his dream forward with industry and leadership.

By 1832 Agassiz's dedication to science and his winning ways had impressed two top luminaries of Europe, Georges Cuvier and Alexander von Humboldt, who decided to do their utmost to pave the way toward the young man's goals. Cuvier turned his fossil fish collections over to Agassiz, an event which, among others, led to the monumental and influential work *Recherches sur les poissons fossiles*, published in 1843 while Agassiz held the position of director of the Musée d'histoire naturelle de Neuchatel in his home country. Humboldt acted as his mentor, using his numerous connections to solicit funds for Louis's research and domestic needs and generally supporting his protégé's career. It was also in Neuchatel that Agassiz produced another work of great historical importance, *Études sur les glaciers* (1840), in which he set out his observations on the Swiss Alps and elsewhere, and proposed the theory of glaciation and the ice ages that impressed geologists like Charles Lyell and naturalists like Charles Darwin, and still have some currency today. In addition to these scholarly achievements, Agassiz always found time to indulge in his favorite hobby: to give lectures and lead expeditions to nature trails as a means of educating the general public about the wonders of God's creation.

It took a disastrous turn in Agassiz's fortune to take him away from continental Europe and eventually give him access to coastal surroundings where jellyfish abound. The very qualities that defined his character – the grandeur and scope of his projects, the ability to enlist fervent admirers in these pursuits by dints of enthusiasm, charm, and leadership – proved fatal when blended with his inability to cope with conflicts among his team of assistants or to manage the expenses incurred by the "scientific factory," as his assistant zoologist Karl Vogt called the workplace in Neuchatel. Agassiz's financial woes came to a head in the wake of expenses incurred by his owning a printing press, ostensibly to accelerate publication of his works, and of money mismanagement by his secretary and manager, Edward Desor. To put it plainly, Agassiz faced bankruptcy. But he also faced failure in his domestic life. In 1833 he had married Cécile Braun, the sister of a German fellow student from his university days in Heidelberg and Munich. Three children came from this union, and the first ten years of marriage seem to have passed smoothly. However, between 1842 and 1845 Cécile became increasingly dissatisfied with the narrow provincial life of Neuchatel, with Louis's total entrapment in his scientific work to the detriment of his family, and with her husband's inability to discern what she saw as the shadiness of Desor's character and the financial ruin it portended. Just as his professional problems were reaching crisis point, Cécile left him and returned to Germany with the children.

Agassiz had little time to dwell on despondency or guilt; he was soon offered generous funds to tour the United States and study its natural history. Excited by the prospect of new projects, he feverishly made preparations and, after a few frustrating delays, arrived in Boston in October 1846. He was feted by Boston's elite, who saw in the renowned European a much-needed catalyst for the nascent American scientific enterprise. In return, Agassiz seized the opportunity to erase his debts by accepting as many speaking engagements as possible in Boston, New York, and other major centres. His legendary charm worked wonders with the American public. He used it to greatest advantage, however, to enlist prominent Bostonians in "moving heaven and earth" to advance his scientific explorations of the American landscape. He was soon won over by the hospitality of his hosts and their diligent catering to his every whim, so the prospect of making the United States his permanent home gradually took hold. The decision was cemented by the creation of the Lawrence School of Science at Harvard College, named after its sponsor and prominent Massachusetts cotton manufacturer Abbott Lawrence. In the fall of 1847 Agassiz was appointed its professor of geology and zoology.

The first year of his professorship kept him very busy, what with co-authoring *Principles of Zoology*, continuing speaking engagements and preparing lectures, and leading an expedition to the Great Lakes. But learning of his wife Cécile's death in August 1848 started a spiral of grief that lasted until shortly before he was to report the presence of nerve cells in jellyfish. Cécile had always had a frail constitution, and her death from tuberculosis was not a surprising outcome. Yet it revived feelings of remorse about his conduct with his family that heightened his sense of loss. His assistant, Desor, who had moved to Boston along with a few others of his Neuchatel staff after his Harvard appointment, chose this moment to spread rumours behind Agassiz's back that the latter "had consistently neglected the wants of his family, thus contributing to Cécile's failing health and ultimate death" (Lurie, 1960). In addition, Desor had developed illusions of grandeur and believed that his own scientific contributions were overshadowed by the Agassiz name. If this was not sufficient, he accused his employer of having an affair with his Irish servant. Agassiz was finding late what his wife and others had warned him of for years – that Desor was disloyal and malicious. Desor's accusations polarized Boston society and the academic community, and so the dispute went to arbitration. It dragged on for several months, threatening to undermine Agassiz's reputation. Finally, in February 1849, all charges against him were determined to be unfounded and were dismissed. Nevertheless, the emotional reverberations of these distressing matters, especially for a man who relied heavily on his charisma in social intercourse, were probably still felt three months later, when Agassiz made an oral presentation of his findings on the structure and habits of medusae to the American Academy of Arts and Science.

A Not-So-Trumpeted Discovery

In the full monograph expanding on his verbal report to the American Academy (Agassiz, 1850), Agassiz humbly acknowledged his limited expertise in the field he was entering: "A newcomer into the field of these researches, after having spent the best of my earlier years in other labors, and happy to find still something to glean where Péron and Lesueur, Cuvier, Eschsholtz, Ehrenberg, Milne-Edwards, Merton, Brandt, Lesson, Sars, Loren, Steenstrup, von Siebold, Dujardin, Will, Edward Forbes, Sir John Dalyell, and others, have reaped full harvest, I have not the pretension to offer anything very complete upon this subject."

But, being driven by curiosity in everything he pursued, Agassiz was willing not only to learn from Nature's book but also to master the zoological literature. He had had little exposure to living invertebrates, as his publications to date had focused on vertebrates and he had long been ensconced in the Swiss mountains. This was to change in the summer of 1847 when he was invited on board the Coast Survey steamer *Bibb*, which happened to survey the harbour and Bay of Boston under the command of Captain Charles Henry Davis (later an admiral) (Agassiz, 1885). From the deck Agassiz wrote enthusiastically to a friend: "I learn more here in a day than in months from books or dried specimens. Captain Davis is kindness itself. Everything I can

2.1 · Louis Agassiz at his work desk. Courtesy NOAA
(image theb3381, NOAA's People Collection).

wish for is at my disposal so far as it is possible." As a result, Agassiz
was able to collect large numbers of marine invertebrates, especially
echinoderms and jellies. He pursued these excursions during the following two summers and used the accumulated materials to write his
monograph (Agassiz, 1850), his first major original research work since
settling in the United States.

The pioneer contribution of Agassiz to cnidarian neurobiology went
largely unnoticed for well over a hundred years, however, until the academic detective work of George Mackie (Mackie, 1989, 2004). Himself
a zoologist of great distinction who has made lasting contributions to
jellyfish neurobiology, Mackie discerned in Agassiz's 1850 monograph

the birthplace of the description of nerves in the most "simple" animals after sponges. The following account of this discovery relies largely on Mackie's adroit and enlightening papers.

Agassiz went to work on the materials collected from the *Bibb* cruises with great zeal and professionalism. He soon found that fresh jellies (jellyfish and comb-jellies) yielded the secrets of their tissue organization more readily under the microscope than pickled materials. The macerating fluids at his disposal in the late 1840s, while adequate for higher invertebrates, proved inadequate for the transparent, fragile jellies. The coarse collecting equipment of the *Bibb* was probably such that many of the jellies arrived on deck in various degrees of freshness, if not in a moribund state. By tinkering with them, Agassiz was able to peel off the decaying superficial epithelial layer (called ectoderm today), thus revealing the epithelial grade of their histological organization and making the underlying nerves and muscles more easily detectable. His tinkering was carried into his microscope work also. Around 1846 Agassiz had purchased an Oberhäuser compound microscope from the firm of Georges Oberhäuser in Paris. What allowed him to observe the tiny cells and slim nerves of jellyfish was the use of the most powerful objective the microscope could offer. According to Mackie, this objective could theoretically resolve structures half a micron thick. However, as a result of the optical distortion experienced with some of Oberhäuser's lenses and Agassiz's bold move to immerse the lens in water because of the short working distance at his disposal (specially designed immersion lenses were unavailable at the time), the resolving power of his microscope was not optimal for his purpose. Nevertheless, he managed, in Mackie's words, "to give a substantially correct account of a major component of the nervous system in any coelenterate" (Mackie, 1989).

Even before Agassiz enters into detailed descriptions of nervous system elements, he offers something tantamount to a blanket statement on the presence of a nervous system in jellyfish and its nature in relation to the nervous system of "higher animal forms":

There is unquestionably a nervous system in Medusae, but this nervous system does not form large central masses to which all the activity of the body is referred, or from which it emanates.

There is no regular communication by nervous threads between
the centre and periphery, and all intervening parts; and the
nervous substance does not consist of heterogeneous elements,
of nervous globules and nervous threads, presenting the various
states of complication and combination, and the internal structural differences, which we notice in the vertebrated animals, or
even in the Mollusca and Articulata. (Agassiz, 1850)

This statement is of interest because Agassiz here appears already
to be contrasting the decentralized, radiate cnidarian nervous system,
just examined for the first time, with the centralized nervous system
of more highly organized animals, which includes brains, nerve cords,
ganglia, and branching nerves for communication between the centre
and periphery of the bilateral body. This very contrast was highlighted
in various ways by his successors in zoology textbooks up to the late
twentieth century. But Agassiz was unable to pinpoint exactly what
substituted for the lack of centralized elements in jellyfish, as microscopic observation of the fine nerve cells forming the two-dimensional
nerve nets of cnidarians eluded him. And as Mackie himself was to
show later (Mackie, 1971), some cnidarians, and especially jellyfish, do
display varying degrees of nervous system centralization, thereby refuting views expressed first by Agassiz, and later by many others, as
we shall see.

Agassiz's basic observations centred on one jellyfish species, *Hippocrene* (now *Bougainvillea*) *superciliaris* thanks to its ease of maintenance
and longevity under the microscope, but he also gathered additional
or complementary observations from a species of the genus *Sarsia*. As
he succinctly put it: "the nervous system consists of a simple cord, of
a string of ovate cells, forming a ring around the lower margin of the
animal [outer nerve ring], extending from one eye-speck to the other,
following the circular chymiferous tube, and also its vertical branches
[radial nerve plexus], round the upper portion of which they form another circle [upper nerve ring]. The substance of this nervous system,
however, is throughout cellular, and strictly so, and the cells are ovate.
There is no appearance in any of its parts of true fibres."

In addition to the principal components of the nervous system described above, Agassiz noticed a swelling of the outer nerve ring as it
enters the base of the four tentacle bulbs:

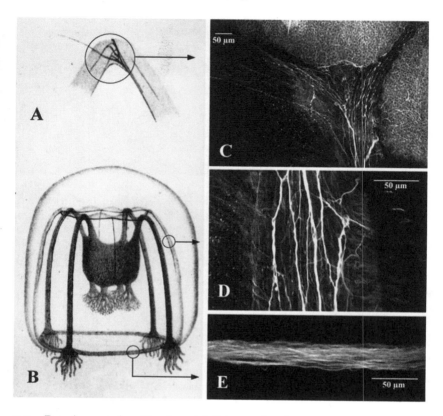

2.2 · Drawings rendered by Agassiz's lithographer Auguste Sonrel of *Bougainvillea superciliaris* (A, B) and photomicrographs (C–E) of the main neural structures, as indicated by arrows, from whole mounts of *B. principis* immunolabelled by George Mackie with anti-tubulin antibody. (A) Agassiz's "plexuses under the arches of the radiating tubes." (B) A general view of the nervous structures. (C) Apex of radial nerve plexus where nerves branch off laterally to form the upper nerve ring. (D) Enlarged view of radial nerve plexus. (E) Outer nerve ring. From Mackie (2004). Courtesy Springer-Verlag.

As for the bulb when seen with the naked eye, it appears simply as a dark speck, from which the tentacles issue. Under a moderate power ... it appears like a crescent-shaped projecting mass, thickest towards its base, thinner towards its prominent margin ... The substance, also, of the bulb presents then a regular arrangement, the pigment-cells being disposed in conical masses, with their points turned towards the black dots, while their wider bases

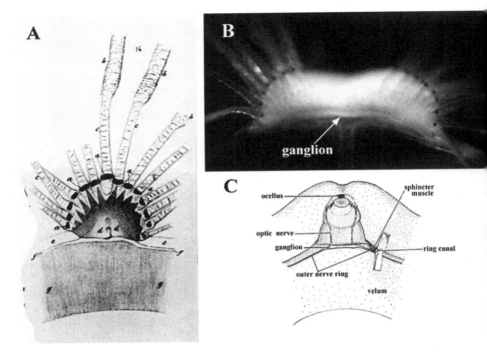

2.3 · (A) Tentacle bulb of *B. superciliaris* as prepared by Auguste Sonrel, showing eye-specks (a), distal portion of tentacles (b), proximal portion of tentacles (c), ganglion (d), chymiferous tube (e), nerve ring (f), velum (g). (B) Photograph by George Mackie of *B. principis* tentacle bulb corresponding to A. (C) Drawing of a corresponding view of tentacle bulb in *Sarsia*. From Mackie (1971, 2004). Courtesy Springer-Verlag.

unite in a semicircular dark mass ... Seen from below and fully stretched out, the triangular ganglion is still more clearly seen, and the radiating pigment-cones most distinctly noticed. The black specks form a curved series arising from the lower margin of the bulb at the base of the shorter outer tentacles ... From these observations it must be plain that the bulb is not hollow, but that it consists of a mass of dense cells arranged in a particular way ... But one thing is plain ... that the pigment-cones of the bulb point to the centre of the eye-specks, which shows a close connection between these dark dots and the centre of the bulb where the nervous ganglion is seated. (Agassiz, 1850)

Mackie's own observations with modern microscopic and immuno-histochemical techniques (Mackie, 2004) confirmed the majority of Agassiz's descriptions (Figs. 2.2, 2.3). Agassiz had indeed seen the main elements of the outer nerve ring at the bell margin, including the ganglionic swellings where nerve inputs from the tentacle bulbs and ocelli (eye-specks) converge. He had noticed the radial nerves running up to the peduncle and manubrium, where nerve fibres diverge laterally to form an upper nerve ring. Agassiz also described inter-radial nerve tracts running down from the upper nerve ring, but Mackie was unable to confirm their existence. While Agassiz saw all these structures in a coarse manner, the intrinsic optical limitations of his lenses and their immersion in water did not allow him actually to detect the low-diameter constituent axons of the jellyfish nerves (Mackie, 2004).

But Agassiz's discoveries are not even recorded in Bullock and Horridge's comprehensive 1965 treatise on invertebrate nervous systems, of which over seventy pages are dedicated to cnidarians and ctenophores. Why is that? According to Mackie (1989, 2004), it may have resulted from the prejudices of the time and from Agassiz's own wavering over his original findings. Colleagues read Agassiz's paper through a haze of skepticism; their own knowledge of jellyfish and other cnidarians made it inconceivable that these animals could possess a nervous system, as the only nervous systems they knew (in more complex, bilateral animals) could not possibly fit into the radial body of a jellyfish. As Mackie (2004) put it: "Contemporary workers found it hard to conceive of a nervous system that had no center, but consisted simply of strands or cords of nervous tissue as Agassiz described it." And then the barrage of criticism from Agassiz's colleagues, some of whom will come into this story further on, managed to shake his confidence in his own observations, to the point that he publicly retracted them in a monograph over a decade later. On reading this passage, one cannot help suspecting that Agassiz went to great lengths to find fault with his own perceptual abilities:

> As to a nervous system, it has not been possible to detect the least signs of a structure indicating its presence. When the innermost wall is seen in profile, along the radiating tubes and at the four intermediate points, its thickness resembles a thin cord, which might easily be mistaken for a nervous thread. The most intimate

structure, the cells of the innermost wall, along the radiating tubes, do not differ from those on each side; all are alike excessively, and round. When the animal is contracted in the manner described above, the innermost wall, at its eight points of adherence, comes strongly into profile, and, on this account, the nerve-like appearance of its thickness is more apparent than at any other part; but when the disk is uncontracted, and the innermost wall presses uniformly against the whole surface of the middle one, it is possible to observe this same appearance anywhere between these eight points. Looking at the disk from above, the innermost wall, where it bends downwards to become the outer wall of the proboscis, resembles, in profile, a quadrangularly-disposed cord, surrounding the inner wall of the proboscis like a nervous ring. At the junction of the transverse partition with the edge of the disk, the inner wall bends on itself at right angles, and there, again, when looking across the edge of this angle, its thickness appears like a nervous ring, running along the inner edge of the circular tube. The statement, in my paper on *Sarsia*, Mem. Amer. Acad. of Sci. and Arts, Vol. IV pp. 246 and 247, that these Acalephs have a specialized nervous system, was based upon these appearances. (Agassiz, 1862)

That Agassiz, a proud man, should indulge in so much self-censorship is baffling precisely because, as Mackie (2004) remarks, "what strikes the present-day reader of Agassiz's account is his professionalism. Methods and results are described in great detail, conclusions are expressed cautiously, alternative explanations are considered, earlier literature is scrupulously cited, and great pains are taken to make the illustrations as realistic as possible." So, with such solid scientific credentials brought to bear on the subject, why should Agassiz have considered that he erred so thoroughly in identifying nerves in his material? We have no clear answer, but there are ample grounds to suspect that much of the original work that went into *Contributions to the Natural History of the United States* was done by his students and assistants, although they received almost no credit for it from Agassiz (Winsor, 1991). In this case, it is possible that Agassiz was induced to retract his earlier observations of nerves because his own scientific staff made the original observations on new material and they failed to see nerves in those jellyfish, thereby invalidating the observations of their mentor with the old specimens.

The Legacy of Agassiz's Discovery

When scrutinizing Agassiz's 1850 text, it is striking how his descriptions of the various anatomical elements – epithelial, muscular, digestive, nervous – are compartmentalized and sealed off from one another. Nowhere in the paper did Agassiz ask what the neurons he was describing were for. While he describes at length the motor behaviour and acknowledges the "power of feeling" of jellyfish, he makes no attempt to speculate on a possible functional relationship between nerves, muscles, and the various activities of the jellyfish. Agassiz appeared to be more interested in the cellular composition of the structures than in their functional significance, a surprising outcome for a pupil of Cuvier. As a result, he left no legacy other than the original discovery of nerves in a cnidarian, but even that found no taker among his contemporary and future colleagues.

And of course Agassiz was not prepared to look more deeply into the potential significance of finding nerves in an animal belonging to the lower rung of the scale of nature, beyond the ready argument that it is "the product of an original and continuous intervention by the Supreme Being" (Lurie, 1960). When he was writing his 1850 paper, he had already read Lamarck's *Philosophie zoologique* (1809) and Chambers's *Vestiges of the Natural History of Creation* (1844), two pre-Darwinian conceptions of the idea of organic evolution, which he opposed vehemently. And when he later criticized and dismissed Darwin's *On the Origin of Species by Means of Natural Selection* (1859), even prominent American scientists such as Asa Gray and James Dwight Dana, who had "deified" Agassiz during his early years in the United States (Lurie, 1960), now found him inadequately equipped to respond to the new intellectual ferment in the natural sciences. Consequently, it was left to later generations to assess the evolutionary significance, not of his own finding as it turned out, but of Mackie's verified observation that there are indeed nervous systems in jellyfish and other cnidarians.

CHAPTER THREE

German Zoologists and the Rediscovery of Nerves in Jellyfish

Its transparent body is shaped like a bell, and moves through the water by regular contraction and expansion, like the lung in breathing. Where the clapper of the bell should be, we find a stomach with a mouth for eating, hanging down from the curved upper part. At the edge of the curved surface are many long fibrils that close on the approaching prey and paralyse it by their sting. Then it thrusts it into its mouth and swallows the object into its stomach.

Wilhelm Bölsche and Joseph McCabe, 1906

Between 1859 and 1872 several German investigators entered the field of cnidarian zoology. And several of them claimed first descriptions of nerves in these animals. The field appeared wide open again, thanks in part to Agassiz's retraction of his 1850 finding. Other issues played a part in the confused jostling that ensued, a jostling centred on the formidable personality of Ernst Haeckel. A close look at the stories of the protagonists will help us disentangle this confusion and apportion merit on claims where appropriate.

Ernst Haeckel Elbows His Way to the Prize

Ernst Haeckel (1834–1919) was one of the most eminent biologists of the nineteenth century. He was also restless, controversial, and larger than life. Born Ernst Heinrich Philipp August Haeckel in Potsdam, Prussia, he vigorously pursued his ambition to become not only a first-rate and prodigiously productive zoologist but also an intellectual force to reckon with on the nineteenth-century European cultural scene, as his recent biographers have persuasively demonstrated (Di

3.1 · Ernst Haeckel as he appeared in *Popular Science Monthly*,
vol. 6, Nov. 1874, 108–9.

Gregorio, 2005; Richards, 2008). From his position as professor of
zoology at the University of Jena, Haeckel developed far-reaching research projects in field locations around the world, which led to count-
less publications and the naming and description of numerous new
species. Along with Thomas Huxley in England Haeckel became the
chief champion – some would say bully – of Darwinism in Germany,
and wrote several books expounding on the evidence for the theory of
evolution and mapping phylogenetic relationships. He is the author of
the controversial recapitulation theory, according to which ontogeny
recapitulates phylogeny; that is, the embryonic development of a more
evolutionarily advanced form is a replay of its adult ancestral forms as
they evolved. This theory has since been discredited.

Haeckel's acquaintance with jellyfish began when he was twenty years old and a student of the great naturalist and physiologist Johannes Müller at the University of Berlin. As Haeckel's hagiographer, Wilhelm Bölsche, relates:

> His artistic eye was caught with their beauty, as it was afterwards with the radiolaria. "Never shall I forget," he says, "the delight with which, as a student of twenty years I gazed on the first *Tiara* and *Irene* [species of medusae], and the first *Chrysaora* and *Cyanea*, and endeavoured to reproduce their beautiful forms and colours." His predilection for the medusae never disappeared. At Nice in 1856 he met them again in the Mediterranean ... At Naples and Messina he completed his mastery of them. When he had done with the radiolaria for the time after publishing the great monograph of 1862, the next task that loomed up on his horizon was the need for a "monograph on the medusae." It would be a long time, however, before he could complete the work in any fullness. A work of Agassiz that purported to do it, but, in his opinion, only confused the subject – he disliked both the Agassizs, father [Louis] and son [Alexander], and the father became one of his bitterest opponents on the Darwinian question – gave him a negative impulse to the study. (Bölsche, 1900)

It is apparent that jellyfish gradually became a serious object of study for Haeckel and also that, just as he felt ready to write a definitive monograph on them, the prior contribution of Agassiz rose publicly to frustrate him, the more so because Haeckel disliked intensely what Agassiz stood for, especially with regard to Darwinism. Haeckel had recently entered into an extensive correspondence with Darwin that led to his first major philosophical work to extol his own brand of evolutionism, *General Morphology*. The preparation of this work served in a way to distract Haeckel from giving full attention to the jellyfish monograph, so he decided instead to deal with one jellyfish family after another and publish by instalments. As a result, the monograph that assembled all his disparate jellyfish contributions appeared only in 1879.

There is a personal back-story to the specific work that included Haeckel's descriptions of the jellyfish nervous system that is strangely evocative of the emotional turmoil that preceded Agassiz's own work.

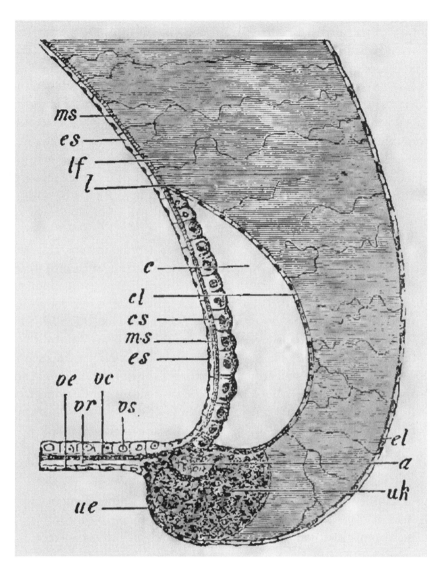

3.2 · Section through the margin of the umbrella of the jellyfish *Glossocodon
eurybia*. a, nerve ring; cl, subumbrella; el, mantle; uk, cartilage of the ring; vs,
velum. From Haeckel (1865a).

Haeckel had married his cousin Anna Sethe on 18 August 1862. The marriage had a beneficial effect on Haeckel. As Bölsche remarked, "The happiness of his home life, with a highly gifted woman who shared all his ideas with the freshness of youth, began to chain the restless wanderer with pleasant bonds to his place." But the happiness was short-lived. Anna caught typhoid fever and died in February 1864. Haeckel, only thirty, was grief-stricken and soon realized that only by throwing himself headlong into work would he find relief. So in the spring of 1864, still deep in mourning, he left for Nice where he collected jellyfish and worked on their structure (Di Gregorio, 2005; Richards, 2008). The results of these investigations were published the following year (Haeckel, 1865a).

Before undertaking to describe the jellyfish nervous system, Haeckel deals with the previous literature on the topic. He wastes little time in dismissing Agassiz's claim to have identified nerves, citing sources other than himself to level charges that Agassiz confused other structures for nerves. There are two reasons to suspect that Haeckel was disingenuous in rejecting Agassiz's claim. The first is his bitter dislike of the man, as alluded to in the passage above from Bölsche. This sentiment is abundantly evident in the venomous tone of the following assessment of Agassiz in a book by Haeckel, not meant to offer vignettes of his contemporary colleagues, but rather to assess the current status of ideas on evolution:

> Louis Agassiz principally owed his exceptional and wholly
> predominant situation among American naturalists, not to the
> scientific value of his own work, but to the marvelous talent he
> had of appropriating to himself the work of others, to the rare
> mercantile aptitude which he knew how to deploy to make large
> amounts of capital conspire to make his ideas become real, and
> finally to the prodigious spirit of organization which allowed him
> to create collections, museums and the most grandiose institutes.
> Louis Agassiz was the most ingenious and most active swindler
> who ever worked in the field of natural history (Haeckel, 1875,
> cited by Winsor, 1991)

A man with this brand of vindictive spirit could not be expected to assess the value and merit of Agassiz's contribution with impartiality.

The second reason to question Haeckel's attitude is found in his description of the jellyfish nervous system, which is eerily similar to that of Agassiz (see Fig. 3.2):

> Moreover, in *Cunina* as in *Geryonia*, there is a fine nervous ring on the margin of the umbrella, contiguous internally to the insertion of the velum, superiorly to the lower margin of the annular vessel, externally to the cartilaginous ring ... The most essential anatomical difference between the Geryonidae and Aeginidae is to be found in the position and structure of organs of sense (marginal vesicles), which are certainly very different in the two families (and also as regards their intimate structure). In the Aeginidae the sensory vesicles are situated freely on the outer margin of the umbrella, and are elevated upon short peduncles. In the Geryonidae, on the contrary, they are enclosed in the gelatinous mass which forms the lowest margin of the mantle, and each vesicle is seated here upon a ganglion-like enlargement of the nervous ring. (Haeckel, 1865b)

Even after making allowances for differences in wording and disparities between West Atlantic and Mediterranean jellyfish species, it is clear that Haeckel merely re-describes the lower nerve ring and ganglionic thickenings associated with sensory organs that Agassiz had previously noticed.

Haeckel dismissed with equal dispatch the description of jellyfish nerves by a fellow German naturalist, Fritz Müller. Müller was quite different from Haeckel in character and outlook. Son of a minister, he became an atheist and an advocate of free love. He obtained a doctoral degree from the University of Berlin, and then decided to study medicine. But he could not practise medicine because he refused to proffer the graduation oath, which contained references to God! Müller left Germany in the wake of the revolution of 1848, which spread throughout Europe and which he supported. In 1852 he settled in Brazil, where he remained for the rest of his life and carried out numerous investigations of local fauna, including jellyfish.

In 1859 Müller published a description of two jellyfish species in which he claimed to have identified nerves (Müller, 1859). This was nine years after Agassiz's contribution and six years before Haeckel's.

In this article Müller took stock of the same elements of the jellyfish nervous system that Haeckel was to describe later, except that Müller's text is less clear and some passages are confusing. Haeckel seems to have used Müller's clumsy way of expressing himself to quarrel with the scientific validity of his contribution. According to Haeckel, what Müller identifies as the lower nerve ring is in fact the "cartilaginous" ring. Both Müller and Haeckel had failed to recognize the upper nerve ring, although Haeckel saw the radial nerve plexus that links with the upper nerve ring.

Haeckel made use of improved microscopic techniques for his study. What he could not make out by playing with light refraction on fresh or moribund material he complemented by observing vertical histological sections through the bell of the jellyfish. But despite these improvements, his understanding of the jellyfish nervous system was no better than Agassiz's. A careful scrutiny of the texts supports my contention that Haeckel, deliberately or unconsciously, knocked down the competition in order to promote himself as the original discoverer of the jellyfish nervous system. But Haeckel's strong personality prevailed over his fellow Darwinian bulldog Thomas Huxley, who came to accept Haeckel's version and in turn influence the young Romanes, the subject of a later chapter (postscript in Romanes, 1876). It remained for Haeckel's students, the Hertwig brothers, to provide a complete and detailed account of the jellyfish nervous system, as we shall soon see.

Nicolaus Kleinenberg and the Misconstrued Nervous System of Hydra

So far in this story, jellyfish have stolen the limelight. But another cnidarian, the diminutive, freshwater hydra, was at least as well known by naturalists and had been the object of investigations since the eighteenth century, thanks to its easy culture and indoor maintenance. It was just a question of time before someone took up the challenge of looking for nerve cells in hydra. That zoologist was Nicolaus Kleinenberg, an assistant of Haeckel, to whom he dedicated the study of hydra in which the question of the existence of nerve cells was discussed.

Nicolaus Kleinenberg (1842–1897) was born in the Latvian city of Liepaja. He studied medicine at the University of Dorpat but soon shifted his interest to botany and attended the lectures of the co-champion of cell theory, Mathias Jacob Schleiden. Soon after beginning graduate

3.3 · Nicolaus Kleinenberg in 1873, as painted by Hans von
Marées. Photograph by James Steakley. Courtesy Royal
Library of Denmark, with permission.

studies in botany at the University of Jena, Kleinenberg fell under the
influence of Haeckel and acquired a position as his assistant. It was
during this assistantship (1869–70) that Kleinenberg prepared his dissertation on the development of hydra, in which the problem of nerve
cells is raised. He was granted his doctorate in 1871, and his dissertation was published the following year (Kleinenberg, 1872). In 1873 he
joined Anton Dohrn's team in Naples, where the latter had just opened
the Stazione Zoologica (this famous marine station will make further
appearances later in our story). However, Kleinenberg had a difficult
personality that prevented him from getting along with people, even
with Haeckel, whom he rejected because of his mentor's ideologically

charged Darwinism. By 1875 he had resigned his assistantship in Naples. Dohrn obtained a professorship for him at the University of Messina in Sicily, and he remained in Italy until his untimely death.

In his published dissertation Kleinenberg turns his attention to the nervous system in the anatomical descriptions of the adult hydra, ahead of the larger section on the embryological development of hydra. On reading his discussion it becomes clear that Kleinenberg, contrary to his predecessors working on jellyfish, did not look at nerve elements as isolated entities. He was bent on considering nerve cells as part of a functional unit with muscles. This preference reflected his general rejection of the phylogenetic outlook of Haeckel when analysing biological systems and his embrace of the physiological perspective. He actively looked for a "primitive" analogue of the frog neuromuscular system in hydra. Aware of the basic epithelial organization of hydra and all cnidarians, which is constituted of the external ectoderm and internal endoderm, he focused his search for nerve elements among the cell types in these epithelia.

In what we now know is the epithelio-muscular cell, Kleinenberg saw the ancestor of the neuromuscular unit of vertebrates. Whereas the neuromuscular unit is composed of a motor neuron axon terminating on the surface of a muscle cell, in hydra the two-cell components are functionally integrated into one single cell with sensory capability, which Kleinenberg calls a "neuromuscular cell." In his view the two-cell motoneuron/muscle cell innervation unit of more complex animals derived from the ancestral neuromuscular cell. He thought that an all-in-one sensory-motor processing unit suited an epithelial animal like hydra. Kleinenberg overlooked the nerve cells of hydra, probably because the fixative he used was adequate for preserving the epithelia but not the fine nerve cells present in hydra. Therefore, the opportunity to find a nervous system in a cnidarian other than jellyfish was missed. Nevertheless, Kleinenberg's mistake was to prove seminal for theories on the origin of nerve cells that emerged in the modern age.

CHAPTER FOUR

The Hertwig Brothers Open a New Era

The Hertwigs were in the vanguard of specialized academic biology that characterized the expanding universities of Imperial
Germany.

Paul J. Weindling, 1991

The main legacy of Ernst Haeckel, as far as the description of the cnidarian nervous system is concerned, is not so much his own contribution as the fact that he delivered the Hertwig brothers to the scientific
community. Along with Theodor Eimer (of whom more later) the
brothers presented the first comprehensive exposition of the cnidarian
nervous system and helped propel the investigations of the evolutionary origin of nervous systems.

The Making of Golden Boys

Oskar (1849–1922) and Richard (1850–1937) Hertwig were born to a
wealthy family in Friedberg, Hesse, in the nascent Imperial Germany.
Their father was a businessman and a member of the upper German
bourgeoisie. Just a few years after their birth the family moved to
Mühlhausen in Thuringia, where Oskar and Richard completed their
pre-college education in 1868 (Weindling, 1991). Health problems that
Oskar suffered in his youth slowed down his academic progress. As
a result he slipped back into the classes of his younger brother, and
the siblings remained together even throughout their education. As
members of the comfortable Protestant middle class, they had every
opportunity for a university education. Their father had studied under
the famous chemist Justus Liebig, but in his days the lack of good
prospects for a scientific career had pushed him toward commerce. It

4.1 · Oskar (*left*) and Richard Hertwig (*right*) as they looked late in their careers. Left, from Weindling (1991), Fig. 12; right, from an anonymous 1894 portrait. Courtesy of Paul J. Weindling (left) and Museum of Natural History Berlin, Historical collection of pictures and writings, Zool Mus. B 1/5 (right).

is possible that the father projected his frustrated ambition onto his sons and therefore actively encouraged them to pursue scientific careers (Weindling, 1991).

Several considerations went into the selection of the University of Jena for their studies. Its proximity to their home was attractive, but Jena's reputation for the high quality of its medical and scientific curriculum and the high calibre of its scientific staff carried the day. The climate of intellectual freedom in Jena, fostered over the years by luminaries such as Goethe, Lorenz Oken, and Friedrich Schelling, had nurtured the emergence of the *Naturphilosophie* that had stormed Europe and later led to an early acceptance of Darwinism. After all, Haeckel taught there, and the comparative anatomist Carl Gegenbaur, if anything another Darwinist, was considered the biggest asset of the university and a scholar highly reputed all over Europe. Just as the brothers attended the same classes in high school, despite the age difference of a year and a half, so they pursued their university studies side by side, sharing the science and medical curricula. They also took advantage of a uniquely German academic feature of the era – the possibility of attending classes in other universities for a well-rounded education. In this way they travelled to Zurich and Bonn where they learned from top scholars in organic chemistry and experimental physiology, mineralogy, and archaeology (Weindling, 1991). More significant for this narrative, in Bonn they profited greatly from Max Schultze's expertise on the latest microscopic techniques as applied to biology.

Also in Bonn the brothers passed the state medical examination. But their studies were interrupted by the Franco-Prussian war of 1870, which they spent respectively as surgeon major in the army (Oskar) and medical orderly in a field hospital (Richard) (Smola, 2004). When they resumed their studies in Jena, the Hertwigs came under the influence of Haeckel who, seeing their intellectual potential, managed to loosen the grip of their medical education and interest them in his own sphere of activity.

For the Hertwigs, Haeckel's approach created a bridge between medicine and zoology. The Hertwig brothers were not interested in the zoology prevailing at that time, which was dominated by systematic thinking; under Haeckel's influence, however, they became curious about the evolutionary history of human organs and human physiological processes. This of course called for studies of phylogenetically lower creatures, and the Hertwigs thus became zoologists under Haeckel's influence while rejecting most of what was done by contemporary zoologists ... He got the Hertwig brothers interested in marine biology and, in 1871, took them with him to the Dalmatian coast, a beautiful area which was then part of the Austrian Empire and belongs to Croatia now. Here the Hertwigs immediately set up a place to work in an old monastery. After that the Hertwigs went there frequently, and managed well despite the primitive conditions. They acquired extensive knowledge of marine fauna, and this later allowed them to find the species they needed in their research. (Smola, 2004)

It was at that time that Haeckel started regarding the two brothers as his intellectual heirs, his *Goldsöhnen* (golden boys) (Weindling, 1991). Their early forays to coastal outdoors (Helgoland, Corsica,

Villefranche, Naples), as well as guidance by their mentors, gave them scope to select research topics of interest and led each to write his doctoral dissertation on lower animals: Oskar on his startling discovery of the sea urchin fertilization process at the cellular level; Richard on the lymph nodes of the sturgeon's heart. The dissertation was one step toward obtaining the *Habilitation*, designed to test the teaching and research abilities expected from prospective professors. Once the Hertwigs became "habilitated" in 1875, it was time to look for academic positions. Both brothers were ambitious and hungry for an opportunity to develop an illustrious academic career. Haeckel in turn was eager to find appointments for his golden boys and, with the enlisted help of Gegenbaur and others, he managed to secure positions for them in Jena on the first rung in the ladder of German academic posts, that of *Privatdozent*, equivalent to a lecturer whose salary is derived solely from teaching classes. Additional appointments were found for them, to keep them in Jena: Oskar as comparative anatomist in the medical faculty and Richard as zoologist in the faculty of philosophy (Weindling, 1991). In these roles as fledgling academics, the Hertwig brothers embarked on their eventful voyage of nervous system discovery.

A Landmark Scientific Contribution

In influencing the brothers' decision to tackle the thorny issue of the jellyfish nervous system, Haeckel played a compelling role in two ways. First, he knew that his own 1865 histological work on the jellyfish nervous system had only touched the surface. And he had reason to encourage his protégés to embrace a more ambitious, large-scale study. Second, the Hertwigs had fallen under the spell of Haeckel's Darwin-inspired descent theory, which suggested that budding elements of nervous tissue such as are found in early human embryos were present in adult animals like jellyfish and evolved toward the brains of higher animal forms. As Ulrich Smola (2004) explains, "Of all the highly developed features of humans, the brain was then and still is the one with the most secrets. It was a totally new idea at the time that an organ like this, with all its mental and physical capacities, could be the result of historical evolution." So it was understandable that most zoologists felt squeamish about undertaking such a daunting task. Paul Weindling ably summarized the issue:

The development of the nervous and sensory system, which the Hertwigs investigated in medusae, was an issue of critical importance for Darwinians. Darwin himself confessed that the problem of the eye gave him "a cold shudder" (Darwin and Darwin, 1888). His followers set about the solution of this problem by study of "primitive forms" of sensory systems. Whether there was a sensory and nervous system at all in medusae had been controversial. Both Gegenbaur in 1856 [*sic*] and Haeckel in 1865 had made pioneering histological study of medusae. The physiological experiments of the Darwinist George John Romanes [of whom more later] had suggested the probability of a central nervous system, capable of originating impulses, leading to the inference of a very fine plexus of nerve fibres ... The cellular investigations of the Hertwigs on medusae thus filled an important gap in the theory of descent. (Weindling, 1991)

Encouraged by their predecessors' tantalizing, if piecemeal, observations of jellyfish nerve cells, and by reading Romanes's report, which was intuitively suggestive of sensory and motor capabilities in jellyfish (Romanes, 1876), the Hertwig brothers travelled to Messina, Sicily, in the winter of 1877. Messina was then considered the "Mecca of the German Privatdozent" (Groeben, 1985) because young investigators could find an astonishing trove of marine animals almost literally brought to their laps by unusual upwelling currents in these waters. They could rely on steady supplies of nineteen jellyfish species for the histological processing required to reveal nerve cells (Hertwig, 1878a). The Hertwigs made significant headway because they paid more attention to methodology than their predecessors (Agassiz, Müller, Haeckel). As Weindling (1991) observed, "they were able to use far more sophisticated techniques than Haeckel, and overcome problems of investigating the enormous gelatinous mass." The brothers understood that to avoid ambiguous nerve cell identifications in a diffusely arranged tissue such as that of gelatinous jellyfish, chemical reagents ought to stain nerve cells and fibres preferentially against the backdrop of other epithelial cells. Relying at first on advice from Schultze, they tested chromic acid but soon found that it caused cells to lose their shape. They turned to osmic acid impregnation in an attempt to preserve cell integrity and nuclei, but it resulted in the coagulation of cytoplasm and connective tissue. They finally opted for a mixture of osmic acid and acetic acid that resulted in better impregnation of nerve and other cells while largely preserving tissue integrity. The maceration worked, thanks to the loosening action of acetic acid on the extracellular matrix that binds cells into a tissue, thus spreading the nerve cells out under the pressure of the coverslip to facilitate viewing.

The Hertwigs' close attention to methodology was unusual among their colleagues working on the histology of coelenterates, who were content merely to rely on "routine" histological methods. An illustrative example of contemporary work in this regard is that of the German-born but Vienna-based zoologist Carl Friedrich Wilhelm Claus (1835–1899). The jellyfish work of this opponent of Darwin, published in 1877, contains muddled or confused descriptions of nerve and sensory cells backed up by poor illustrations that reflect his uninspired use of histological methods.

A mere few months after their investigations in Messina the brothers published a preliminary report on their findings, but without illustrations (Hertwig, 1877). They began their introduction by emphasizing that the phylogenesis of nervous systems and sense organs was a hitherto neglected topic in the field of comparative anatomy. The purpose of their investigation, they wrote, was to redress this situation. By using the word "phylogenesis," borrowed from Haeckel and meant as the sequence of events in the evolution of a group of animals or of an organ system in an animal group, they signalled that they were consciously pursuing their investigation within an evolutionary framework, in a clear departure from the outlook of their predecessors. They looked at coelenterates as the most simply constructed group of animals in which "ganglion cells" and sensory cells may have developed. Their mention of ganglion cells was symptomatic of the time; Agassiz and Haeckel also used these words. The observational experiences of these nineteenth-century investigators had been limited to higher animals in which nerve cells and fibres were condensed into ganglions and nerves, and nerve cell assemblages centralized into brains. So they brought that mindset to their first inquiries on cnidarians, assuming at the outset that some form of nerve condensation and centralization should exist in these forms; hence, the biased use of the phrase "ganglion cells."

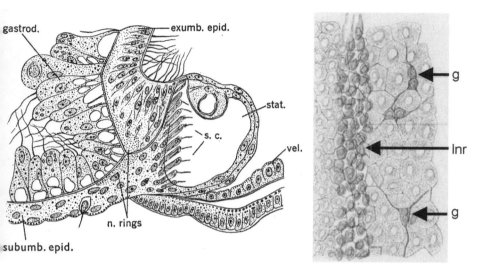

4.2 · (*Left*) Section through the bell margin of *Aequorea*, showing the exumbrellar epidermis (exumb. epid.), gastrodermis (gastrod.), upper and lower nerve rings (n. rings), sensory nerve cells (s.c.), subumbrellar epidermis (subumb. epid.), statocyst (stat.) and velum (vel.). Adapted from Hertwig and Hertwig (1878) by Bullock and Horridge (1965). Courtesy of Professor G. Adrian Horridge, with permission. (*Right*) Section at a different angle showing the lower nerve ring (lnr) with its band of cells standing between the velar epithelium and the subumbrellar epithelium over which lie bipolar or tripolar ganglion cells (g). From Hertwig and Hertwig (1878a), Plate 2, Fig. 3.

The Hertwigs' monograph of the following year (Hertwig and Hertwig, 1878) is a remarkable achievement. They provided a lucid appraisal of the state of the field at the time, an exhaustive and critical review of the relevant scientific literature, and carefully worded and extremely detailed descriptions of the nervous and surrounding elements along with caveats about technical pitfalls. In addition, they ventured into compelling discussions of the significance of their observations with regard to cell theory, the physiology and embryonic development of these organisms, and phylogenetic or evolutionary considerations. Ten plates of beautifully rendered illustrations were appended. The following passages highlight the main results and conclusions reported in this monograph.

The brothers first establish on firmer ground the presence of the lower and upper nerve rings, and they show that nerve cells with multiple processes, which they call "ganglion cells," connect with each other to form a loose network (nerve net) and with the nerve rings (Fig. 4.2). All these nerve elements originate in the ectoderm, the outer epithelial layer, as opposed to the inner layer, the endoderm. The lower nerve ring lies between the velar and subumbrellar epithelia and its nerve fibres lie deeper inside the ectoderm than the ganglion cells (Fig. 4.3). The lower nerve ring is constituted by numerous densely packed oval nerve cell bodies with a process at each end (bipolar arrangement), and these processes, running parallel, form the nerve fibre bundle of the ring (Fig. 4.4). Processes from ganglion cells are seen to connect with elements of the nerve ring, on the one hand, and with the muscle fibres in the velum on the other hand (Fig. 4.4), thus creating communication channels for the impulses travelling in the nerve ring to be propagated and delivered to the epithelio-muscular cells for motor responses. The manner in which the cells connect is not explicitly ascertained by the Hertwigs except to suggest that protoplasmic continuities between the nerve cells are involved. The microscopes used by the brothers lacked the resolution necessary to discern the boundaries of nerve terminals, and it is evident, particularly in illustrations such as figure 4.4, that impressions of continuities between cells are urged on the reader.

The upper nerve ring is situated across the mesogleal channel from the lower nerve ring. The Hertwigs emphasized the differences between the two nerve rings. The upper nerve ring is looser in its organization, containing fewer nerve cell bodies and nerve fibres of lower diameter

4.3 (*Opposite, above*) · Cross-section view of ectodermal epithelium with underlay of ganglion cells (g) and nerve fibres of the lower nerve ring. From Hertwig and Hertwig (1878a), Plate 2, Fig. 14.

4.4 (*Opposite, middle*) · Tangential section through the superficial side of the ectodermal epithelium, showing the lower nerve ring (lnr), ganglion cells (g) and the epithelio-muscular cells of the velum (vm). From Hertwig and Hertwig (1878a), Plate 5, Fig. 7.

4.5 (*Opposite, below*) · View of a segment of the upper nerve ring (unr) with ganglion cells (g) and sensory cells (sc). Osmic-acetic acid maceration. From Hertwig and Hertwig (1878a), Plate 5, Fig. 10.

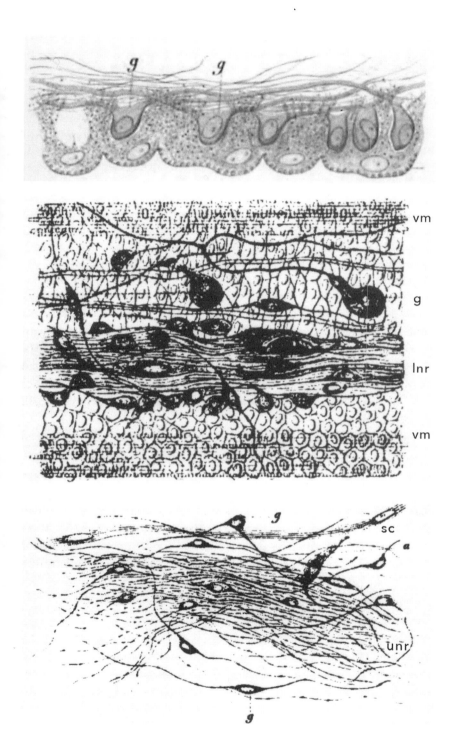

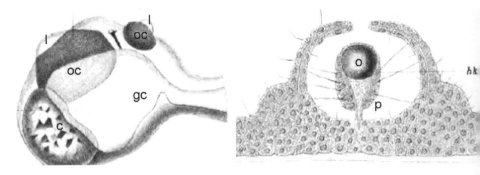

4.6 · (*Left*) Eye organ with two ocelli (oc) and lenses (l), cuticle (c), and gastrovascular cavity (gc). From Hertwig and Hertwig (1878), Plate 9, Fig. 9. (*Right*) "Hearing organ" (statocyst) with vesicle opened to the outside. Note the otolith (o), the stem (hk), and the pit (p). From Hertwig and Hertwig (1878a), Plate 3, Fig. 12.

than in the lower nerve ring (Fig. 4.5). Even the density of associated ganglion cells is lower in the upper nerve ring. Whereas the lower nerve ring receives sensory fibres from the "hearing organ" (later identified as the balance organ or statocyst), the upper nerve ring receives fibres from scattered sensory cells in the epithelium (Fig. 4.5) and from ocelli (small eye organs).

The Hertwigs were thus able to show that a diffuse, decentralized network of nerve cells, in the form of the subepidermal nerve net, co-exists in jellyfish with a beginning of centralization in the form of nerve rings. But even more remarkable was to find already present in jelly-fish the development of sensory organs, perhaps the only organs in an animal whose organization is based on simple epithelia. Eye structures of different complexity were described, from simple ocelli to eye organs that included multiple ocelli, each equipped with light-concentrating lenses and other accessories such as pigmentation and a supporting cuticle (Fig. 4.6). In addition, they described sensory structures, the "hearing organs," assumed to mediate, if not hearing, at least detection of vibration. The brothers were aware that these organs likely served also as organs of balance, as they included otoliths, mineral concretions that move according to the tilting of the animal in the water column, thereby pressing on specific sensory cells within the organ (Fig. 4.6).

The significance of their new discoveries was immediately apparent to these sibling investigators, and they were not shy about discussing the developmental and phylogenetic implications with regard to the origin of nervous systems. Weindling provides a very useful digest of the Hertwigs' thoughts on the subject found in their 1878 monograph:

They confirmed, firstly, the ectodermal origins of the ganglion cells and nerve fibres. Secondly, that the end organs [terminals] of the sensory nerves had been formed from sensory cells, which had originally covered the free surface of the body. Thirdly, that the end organs of the motor nerves were ectodermal muscle cells on the surface of the body. They rejected on a priori grounds all explanations of the origins of the nervous system as a secondary association of preformed parts; this was because of their fundamental belief that the whole and the parts were reciprocally adapted in the course of development. To the Hertwigs a ganglion cell was inconceivable without its connection to muscle or sensory cells. On this "organicist" principle they criticized the *Neuromuskeltheorie* of Nicolaus Kleinenberg that cells in *Hydra* combined muscular and sensory functions differentiated in higher animals. The theory was supported by Claus, Allman, van Beneden, Gegenbaur and Haeckel, who saw these cells as the primary germs of two-organ systems. For reasons based on developmental theory and physiology the Hertwigs objected to the theory and replaced the term *Neuromuskelzelle* by *Epithelmuskelzelle* [epithelio-muscular cell], to denote the phylogenetic significance of its position in the ectoderm. (Weindling, 1991)

In their monograph the Hertwigs seemed to imply that epithelio-muscular cells existed because there is no mesoderm layer in jellyfish from which dedicated muscle cells can differentiate, as in higher animals. Every cell with a function was epithelial, either as ectodermal or endodermal layers. Therefore, sensory, ganglion, and epithelio-muscular cells constituted the most primitive of nervous systems, set in an epithelium (ectoderm). How, in the Hertwigs' view, did this come about? They believed that in the beginning all ectodermal cells had been equal in type and interconnected to varying degrees by "protoplasmic continuities," thus forming what they called a *Zellverband*. This concept was likely fuzzy in the brothers' minds, but today it evokes the gap junctions known to exist between epithelial cells and even between some nerve cells in jellyfish, and which allow electrical impulses to propagate freely between such cells (see chapter 11). According to the Hertwigs, the nervous system arose as a result of division of labour within this *Zellverband*. From there further developments ensued during metazoan evolution.

> The theory was the basis for an account of the phylogenetic development of the nervous system. The Hertwigs suggested that in all Metazoa, the ectoderm, from which the animal nervous system derived, consisted of a single layer of undifferentiated cells; these were brought into contact by protoplasm. Once the ectoderm lost its primitive single-layered nature, the ganglia [ganglion cells] separated from the surface epithelium and migrated deeper – as in the *Hydra*. Division of labor caused differentiation of the central and peripheral nervous system from the ectoderm, in association with formation of the mesoderm. Cell groups formed localized sensory areas in the skin surface. These became the sense organs. (Weindling, 1991)

The brothers also discussed the functional roles of what they called *Centralorgane*, by which they meant the nerve rings and sense organs, as these structures to them represented examples of nerve cell and fibre concentration. In this they were helped by examining Romanes's report of his physiological investigation of jellyfish (Romanes, 1876), of which they had recently been made aware, but after they had already completed their research work. Romanes, to whom we will return later on, had tried to deduce the location of a "central organ" controlling locomotion by sectioning different parts of the jellyfish body and watching the effect of surgery on motor coordination. The Hertwigs asserted that the location of the two nerve rings and the sensory organs (eyes and statocyst) in their study coincided with the experimentally deduced central organs of Romanes. On the basis of Romanes's findings and their own, the brothers concluded that the lower nerve ring had primarily a motor function for swimming, whereas the upper nerve ring had for the most part a role in sensory integration.

The Hertwigs Strike Again: The Nervous System of Sea Anemones

The enthusiasm of youth and the lure of professional advancement impelled the brothers to maintain their already impressive rate of scientific production. They now chose to examine sea anemones, close relatives of jellyfish. In contrast to jellyfish, which are pelagic and free swimmers, sea anemones share with corals and sea pens a sedentary life by virtue of their attachment to sea bottoms. This important difference of lifestyle determined a departure of body organization from that of jellyfish. The Hertwigs wanted to know if this departure had repercussions on the tissue arrangement of ectodermal and endodermal layers, and more important, if the sedentary lifestyle of sea anemones made fewer demands on the organization of the nervous system than the lifestyle of more active jellyfish. At the start the brothers had less information to help them set a course of action, as sea anemones had so far been little investigated, even less so their nervous system. So they decided to investigate the general anatomy and histology of these animals in addition to the nervous system.

In March 1879 they returned to Messina to collect specimens, and then they moved to the Stazione Zoologica in Naples, where they could keep sea anemones alive in aquaria, process them, and prepare them for microscopic observations. The zoological station was the brainchild of Anton Dohrn (1840–1909) who, like the Hertwigs, had studied under Haeckel and in 1868 had become, if only for a short while, a *Privatdozent* in Jena. In his search for a favourable location to establish a zoological station, he joined the *Privatdozenten* who spent time in Messina in 1868–69. But Messina proved inadequate for Dohrn's vision of a zoological station that would serve scientists, as well as the public interest. Naples was a better fit, as Christiane Groeben explains:

The growing interest in exploring life at sea, the need for marine organisms for research in morphology and embryology, Dohrn's own marine experiences, his championing of Darwinism, and his need to prove himself – all these currents converged into the creation of the Naples Zoological Station. Dohrn's attention had turned to Naples because he wanted to connect the Station with an aquarium open to the public, the entrance fees thus providing the means to pay a permanent assistant. Dohrn therefore had to choose a large city that attracted many tourists. Naples at that time was still one of the largest and most attractive cities in Europe with more than 500,000 inhabitants and about 30,000 tourists a year. (Groeben, 1985)

To cover the operating costs, Dohrn devised the ingenious formula of the "table system," which was later adopted by other stations of its kind, notably the Marine Biological Laboratory in Woods Hole, Massachusetts. Having the use of a table gave a researcher access to a laboratory space, animal supplies, chemicals, and the use of various facilities and the library. Tables were allotted to an institution or a private individual in return for an annual fee, and a university or an academy could nominate scientists to use its table at particular times. It so happened that the Hertwigs used the Berlin Academy table between 11 April and 11 May 1879 for their work on sea anemones. On the basis of a short visit in 1876, the Hertwigs thought highly of the Naples station, but this current move did not meet with Haeckel's approval. A serious rift had developed between Dohrn and Haeckel, mainly over a clash of personality and because Dohrn "became increasingly skeptical of Haeckel's Darwinism" (Weindling, 1991). Both were headstrong individuals, and particularly in the case of Haeckel, you concurred with his ideas or else you became his enemy. Not surprisingly, in Naples the brothers felt the coldness of the scientists in Dohrn's inner circle, although they managed to enlist the technical assistance of visiting scholars to the station, including Nicolaus Kleinenberg, the author of the monograph on the *Hydra* discussed earlier.

Being an industrious pair, the siblings published the results of their investigation as a monograph in the very year of their fieldwork (Hertwig and Hertwig, 1879). Publication as journal articles came later – results and illustrations in the final 1879 issue of the *Jenaische Zeitschrift für Naturwissenschaft*, and the general discussion in early 1880 (Hertwig and Hertwig, 1879–80). Before describing their results, the brothers, now both promoted to the rank of professor, took unusual care to write out their methodology over many pages: the choices they made regarding anaesthesia and chemical fixation of specimens; processing to tissue embedding and histological section preparation; section staining and slide mounting medium; the scientific justification of these choices and the advice collected from various colleagues to arrive at some of these

choices. It was a painstaking exercise they had not undertaken when reporting on their jellyfish findings, but one they felt was necessitated by
the nature of their material. Sea anemones are fleshier, tougher animals
than transparent and fragile jellyfish. They tend to contract tightly and
produce copious amounts of mucous when handled. The brothers responded to this challenge by paying more attention to seeking out the
most suitable histological methods to address it and be rewarded with
good yields of quality observations under the microscope.

Smola relates the ingenious way in which the brothers went about
solving one of the technical problems of handling sea anemones:

Even today it is entertaining to read of the patience, skill and
inventiveness that he applied in preparing objects for the microscope. In one of his publications, for example, Richard explained
the problems of fixing actinia. These first had to be sedated, and
unfortunately this always seemed to make the animals contract,
regardless of whether one used curare, potassium cyanide, chloroform or even morphine. The solution of this problem ultimately
came from Salvatore [Lo Bianco], a helper responsible for fixing
animals at the marine station in Naples who had the idea of
exposing animals to tobacco fumes. This worked surprisingly
well, but it is up to the reader of the manuscript to figure out how
many pipes and cigars Salvatore had to smoke to obtain fumes
that were dense enough. (Smola, 2004)

Their histological approach indeed led to wonderful observations
never before recorded. First of all, the brothers were able to see clearly
the organization of the body column on the outside, culminating above
in the oral disc and its crown of tentacles. Inside the body wall they
found the central gut (pharynx), from which radiate the mesenteries
that run to varying extents toward the body wall, thereby dividing the
internal space into several gastro-vascular cavities. By cross-sectioning
the body walls, the Hertwigs saw that sea anemones are three-layered
organisms, consisting of two epithelial layers (ectoderm and endoderm) and a middle support layer that they likened to the mesoderm
of higher animals, but a nascent mesoderm. On close inspection they
spotted two important morphological novelties that separate sea anemones from all other coelenterates.

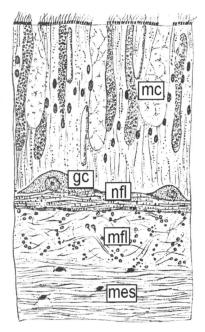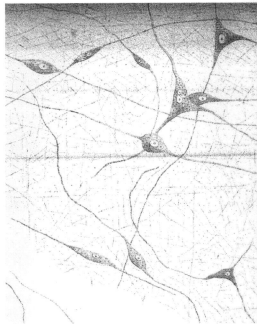

4.7 · The ectodermal nerve net of a sea anemone. (*Left*) Cross-section
through the epithelium of the oral disk. Note the "ganglion cells" (gc) and
the nerve fibre layer (nfl). Mucous cells (mc), the muscle fibre layer (mfl)
and the middle layer (mesoglea, mes) are also seen. Adapted from Hertwig
and Hertwig (1879) by Bullock and Horridge (1965). Courtesy of Professor
G. Adrian Horridge, with permission. (*Right*) Sagittal section through the
ganglion cell sublayer showing the net of intersecting bipolar and tripolar
ganglion cell processes. From Hertwig and Hertwig (1879), Plate 4, Fig. 6.

First, they noticed that the endoderm in sea anemones takes an exceptionally large share of the body, the like of which is not seen in any
other coelenterates. Not only does it include the septa, mesentery filaments, acontia, and sexual organs, but it also incorporates the greater
part of the body's muscles and part of the nervous system. It struck
the brothers as a shocking difference from the organization of jellyfish, in which the ectoderm is the dominant layer where all the nervous system and most of the muscles reside, whereas the middle layer is
negligible. Second, they noticed a great similarity between the histological elements of the ectoderm and the endoderm (cellular compos-

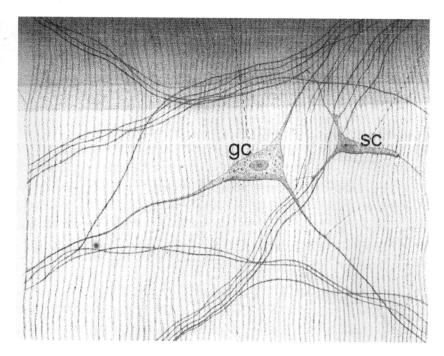

4.8 · The endodermal nerve net of a septum of a sea anemone lying over the muscle bands. Note the sensory cell (sc) and the ganglion cell (gc) bodies. From Hertwig and Hertwig (1879), Plate 6, Fig. 6.

ition), which is another substantial difference from jellyfish. They saw this continuity of cellular composition even in the innermost structures such as the pharyngeal wall and mesenteric filaments.

In the ectoderm they found the "ganglion cells" and their processes in the nerve fibre sublayer nestled between the superficial, columnar epithelial cells and the deeper muscle fibre sublayer (Fig. 4.7, left). By sectioning through the ectoderm in a different orientation they were able to show how the ganglion cell processes form an intertwined plexus (Fig. 4.7, right). This nerve net sits on the muscle fibre layer. A similar arrangement was observed in the endoderm (Fig. 4.8). In both ectoderm and endoderm, they insisted, sensory, ganglion, and muscle cells are linked by nerve fibres into a "unified system," which they called the "neuromuscular system." In this system, sensory cells bring input from the outside world directly to ganglion cells and to muscle fibres, so that both ganglion and sensory cells can make contact with muscle, thereby participating in the motor response of sea anemones.

The Hertwigs also showed that the ganglion cell and nerve fibre layer of the ectoderm was more developed in the oral disc – the mouth area of sea anemones – than in other parts such as the column and tentacles. In their view, this disc constitutes a kind of "central organ" for sea anemones, associated with an increased mass of muscle fibres, but less developed than the marginal nerve rings of jellyfish. The brothers were also able to gain fresh insight into the organization of muscle cells. They followed epithelial cell bodies through their more or less elongated neck toward deeper sublayers until they expand in fibrillar end-feet constituting the contractile muscle bands. They provided the definitive word on how epithelio-muscular cells are organized in these animals. The existence of a clearly defined nervous system alongside these epithelio-muscular cells dealt a deathblow to Kleinenberg's neuromuscular cell theory (see chapter 3).

The Far-reaching Legacy of the Hertwig Brothers

The years of the Hertwigs' sibling partnership must count among the most productive and paradigm-changing periods in nineteenth-century biology. Although they appeared to form a good team, however, the partners had different personalities. Oskar was described as somewhat aloof, whereas Richard was considered an extrovert. Because the latter mixed more easily with people, he was faster to make connections and early on sped ahead of Oskar in landing attractive professorships. Oskar was the better writer of the two, but throughout his career he did his best to avoid speaking engagements and attendance at scientific meetings (Weindling, 1991). Both were very class-conscious, but Oskar was particularly stuffy about his social position as a university professor, to the point that it smacked of snobbery to some. Others regarded him as pedantic, remote, unfriendly, or puritanical – pejorative qualifiers in the scientific circles the brothers moved in. Although Richard had an easy-going and sunnier personality, he was lumped with his brother during their partnership years by those who sought to berate them.

The situation came to a head during their last stay at the Naples Zoological Station in 1884 when Carl Vogt, the ex-collaborator of Agassiz, spearheaded a lampooning and spoofing campaign against them (Weindling, 1991). The insult to the Hertwigs, who were branded by the internationalist scientists at Naples as representing a new generation of blinkered, careerist, patriotic German professors (Weindling, 1991), created a clamour that drew strong protests from government offices where the Hertwigs found favour. Their personality differences as well as their opposing views on Darwinism – Oskar distancing himself from Haeckel and eventually rejecting Darwin's process of natural selection, Richard keeping true to both Haeckel and natural selection – were bound to create a rift between them, and the appointment of Oskar to the University of Berlin in 1887 irreversibly finalized their separation. Richard, in the meantime, had moved to Munich, where he spent the rest of his career and where, in contrast to Oskar, his personality attracted a number of students who in turn became highly successful and influential biologists, among them Theodor Boveri, Richard Goldschmidt, and Nobel Prize winner Karl von Frisch.

The few years the Hertwigs spent jointly investigating coelenterates were probably those when their complementary skills played out best. As a result of their harmonious working relationship at that time, they produced a solid body of work that was satisfying for the new and original discoveries they made, and also for the theoretical ramifications that influenced biological thought in decades to come. Their original investigations on coelenterate nervous systems became a landmark that inspired future workers in the field, not only because they represented a leap forward in understanding how the nervous system of these primitive animals is organized, but also because the Hertwigs were the first to see clearly the key phylogenetic position that coelenterates held for insights on how nervous systems first arose in the animal kingdom.

While acknowledging the importance of coelenterates, however, the Hertwigs looked also beyond nervous systems. Their observations of overall coelenterate tissue organization had raised issues in their minds about current views on germ layers, views that were stirring lively debates among zoologists of the era. In fact, their coelenterate work posed such pregnant questions that the brothers promptly vented their musings on the subject in a short monograph that tailgated their first jellyfish contribution (Hertwig, 1878b). The heart of the matter, as far as the brothers were concerned, was twofold. First, they took issue with the popular view that each germ layer was a fixed, independent morphological entity from which specific sets of differentiated tissues and organs consistently arose. They used the example of the jellyfish "sex organs" (gonads), which in some species develop from the ectoderm and in others from the endoderm, to make their point that the destiny of tissues and organ is determined by cells, not germ layers, and that the germ cell's location is not obligatorily assigned to one germ layer. And later, from their observations on sea anemones, they concluded that the nervous system could originate from both the ectoderm and endoderm. This was one element of their "germ layer theory."

The second point raised by the Hertwigs in this 1878 monograph (Hertwig and Hertwig, 1878b) concerned the origin of the third embryonic germ layer in addition to the ectoderm and endoderm – the mesoderm. The question of its origin was fiercely debated and the Hertwigs wrote that it constituted the Achilles' heel of developmental biology at the time. Coelenterates were believed to possess only ectoderm and endoderm, and therefore were referred to as diploblasts as opposed to more advanced animals called triploblasts, which possessed a mesoderm. Their mentor Haeckel had posited that his hypothetical primeval form of a multicellular animal, the "gastrea," was the ancestor of the gastrula stage of the embryos of more evolved animals, in which only two germ layers, ectoderm and endoderm, exist. This hypothesis implied that the emergence of the third layer, the mesoderm, was a later development. While studying coelenterate nervous systems in the context of the epithelial layers that appeared to mimic ectoderm and endoderm, epithets that had persisted in the coelenterate literature since Thomas Huxley (1849) and George Allman (1853), it occurred to the brothers that coelenterates may have occupied the stage where the evolutionary development of the mesoderm was played out.

In jellyfish they pinpointed three structures that in their opinion could be construed as steps toward the formation of a mesoderm. First, there is an acellular gelatinous layer (*Gallerte*) that glues the ectoderm and endoderm together, which is secreted by endodermal cells, according to the brothers. (This gelatinous mass is today referred to as the mesoglea.) Second, the Hertwigs observed what they interpreted as an intrusion of endodermal cells into the gelatinous mass to form connective fibres. In this way the mesoderm would initially be a by-product of the histological differentiation of the endoderm. And third, they described an infolding of ectodermal epithelio-muscular cells that squeezed the muscular parts deeper into the gelatinous mass and sep-arated them from the epithelial parts by means of a thin basal lamina. This the Hertwigs saw as the origin of mesodermal muscles.

This early burst of conceptual insight was complemented by the brothers' study of the histology of sea anemones. Interestingly, when the Hertwigs laid out their findings on sea anemones in monograph form, the publication was intended as the first instalment of a series of monographs on the germ layer theory (Hertwig and Hertwig, 1879). The only structure that potentially qualified as mesoderm in sea anem-ones took the shape of a gelatinous mass in which stellate cells of endo-dermal origin are scattered. Nevertheless, this observation inspired them to attempt a clarification of what should be considered *bona fide* mesoderm, in a scientific article that became their articulation of the coelom theory (Hertwig and Hertwig, 1882). In this paper the Hert-wigs argued that gelatinous mass with embedded cells such as is seen in sea anemones is akin to "mesenchyme," but not true mesoderm. True mesoderm, they asserted, is formed by epithelial invagination of the endoderm, as occurs in some jellyfish, and the formation of the body cavity of more evolved animals – the coelom – is a consequence of this process. Haeckel, ever the skilled communicator, explained the Hert-wigs' concept in his book *The Evolution of Man*: "a couple of pouches or sacs are formed by folding inwards at the primitive mouth, between the two primary germinal layers; as these pouches detach from the primitive gut, a pair of coelom-sacs (right and left) are formed; the coalescence of these produces a simple body-cavity" (Haeckel, 1897). Thus the coelom, he asserted, was not formed by creating a hollow in a solid tissue mass as previously believed, but was the natural outcome of pouch formation.

The Hertwigs got a lot of mileage from their coelenterate work. It allowed them to tackle big issues that preoccupied many of the prom-inent biologists of the time. This was particularly true of Oskar, who became a major proponent of ideas on dual evolutionary and develop-mental processes that are now the bread and butter of the recently emerged discipline called evo-devo (Hossfeld, 2003). It propelled the brothers to the leadership of their disciplines and afforded them polit-ical leverage to wrestle plum academic chairs from competitors. Not a bad harvest reaped from the seeds of the lowly coelenterate.

CHAPTER FIVE

The Experimental Approach to Jellyfish Neurobiology: George Romanes and Theodor Eimer

> With all his talents George Romanes had a simple lovable nature. He never concealed anything he might have in his mind, and even displayed in conversation an amount of self-appreciation of his own work and of his intellectual powers, which would have been characterized as vanity in most people, but in Romanes gave no offence, since it was a natural and unconscious part of his character.
>
> Edward Sharpey-Schafer, 1927

Up to this point, our story has followed the trail blazed by pioneers who asked themselves whether there *was* a nervous system in coelenterates and, in the affirmative, whether its nature differed radically or not from that of higher animals. They were interested only in tracing the physical nerve cells and fibres by anatomical/histological observations, and the Hertwig brothers alone showed genuine interest in the physiological implications of their observations, although they stopped short of performing experimental investigations themselves. Had the Hertwigs pursued experimental work, it is reasonable to assume that their experiments would have been designed to reveal the functional role of the different nerve and sensory elements they had already described. But if one started from the premise that the existence of nerve cells in these animals was yet to be demonstrated, then the objective of experimental manipulation would be to generate evidence that somehow an underlying nervous system is at work. The latter was the path taken by a different breed of biologists, represented by George Romanes and Theodor Eimer.

George Romanes: A Young Maverick of Great Promise

Unlike the Hertwigs, who appear to have been bred into the profession of biology, George John Romanes (1848–1894) had a background that seemed unpromising in this regard. He was born in Kingston, Ontario, Canada, on a date that makes him a contemporary of Oskar Hertwig by a year. He was the son of Reverend George Romanes, an Anglican minister originally from Scotland, who had moved to Ontario and was appointed professor of Greek at Kingston's Queen's College in 1847, only five years after its opening. The little we know of Romanes's early life we owe to documents archived at Queen's University, Kingston, thanks to the dedication of Professor Donald Forsdyke. Romanes's widow edited a collection of his correspondence along with tidbits of biographical sketches (Romanes and Romanes, 1896), but we are advised "not to rely too heavily on all the evidence that she gives about her husband, despite the fact that in many cases this is the only source available for Romanes' life" (Barnes, 1998). The usual suspect here is the urge of a wife to project the image of her husband she wishes to preserve, so there always lurks in one's mind the hidden agenda of the intimate or relative. Nevertheless, we shall try here to navigate those treacherous shoals.

Reverend Romanes, according to his correspondence quoted in an article (Ringereide, 1979), appears to have left Scotland for Canada in 1833 under the impression that an appointment to a church in the colony would improve his prospects of being offered a charge at home. So his commitment to grow roots in Canada did not appear very strong. He took charge of a congregation of the Church of Scotland in Smith Falls, Ontario, and in 1835 he married Isabella Gair Smith, the daughter of a fellow minister living nearby. When the Church of Scotland opened its first theological college in Kingston (Ringereide, 1979), Reverend Romanes must have felt his appointment there as professor of classical literature was a promotion and an alleviation of ministerial responsibilities. Upon hearing that he had inherited a considerable fortune in the old country, he wasted little time in breaking his thin ties with the colony. His son was two years old when the reverend moved his family to England. The family soon made its home near London's Regent's Park, where George grew up very comfortably and happily as a result of his father's newfound wealth. His good nature and deep religious feelings were hallmarks of his childhood, as the following passage suggests:

> He was said to be a kind and patient brother to all his siblings and, indeed, got on particularly well with his elder sister. He enjoyed shooting at his mother's family home in Ross-shire, Scotland [Dunskaith], and music became his "most perfect passion."
>
> Christianity was a strong and potent force in George's early years. His father was an Anglican, whilst his mother attended Presbyterian services. Neither of them apparently enforced their own particular views on the rest of the family, and it transpired that George and his siblings tended towards the English Church. Since religion played such a significant part in his life, it is little wonder that throughout his childhood and early adulthood, George grew up with the intention of following his father into Holy Orders. (Barnes, 1998)

A glance at his education history gives the impression that young George was at once indulged and somewhat neglected by his parents. A severe attack of measles stopped his formal school attendance, and "his education continued in a desultory way at home" (Ringereide, 1979). What he learned about the world beyond his English setting came through trips to Europe with his parents. Germany was a favourite, and Heidelberg the most frequently visited town, and as a result George became conversant in German – an asset for any aspiring scholar at the time – and he nursed a lifelong love for poetry and music. He learned plenty about the natural world from his summers spent at his mother's Scottish family estate, Dunskaith, where he enjoyed hunting and hiking excursions. When he reached the age of seventeen it finally dawned on his parents that their son was ill prepared to enter university and that the loose reins of their parenting needed to be tightened. So he was entrusted to a tutor who had but two years to bring his charge up to the required academic standards. The effort paid off; George's academic credentials were deemed sufficient to be granted admission to Gonville and Caius College, Cambridge, in July 1867. The young scholar seemed to fit naturally into campus life:

For two years he pursued an ordinary degree, reading mathematics, whilst remaining intent on becoming a clergyman ... Cambridge life seems to have been happy, and he participated in several college activities. In October of 1867, he was elected to the Gonville and Caius Boat Club, and between 1867 and 1871, he was also a member of the Gonville and Caius Debating Society. He made many friends, one of whom was Proby Cautley who later became the Rector of Quainton, near Aylesbury. It was in these early years at Cambridge he and Cautley would spend many hours discussing their calling. (Barnes, 1998)

Romanes's immersion in things religious turned to militancy when he was drawn to the evangelical movement that had seized Cambridge campuses at the time. For some it represented a form of Christian fundamentalism. According to Oliver Barclay (1977, cited by Barnes, 1998), "Cambridge University was a typically middle-class institution that was outwardly very religious but only superficially so in many quarters ... Most students at the time were wealthy and assured of an inheritance. Their university life was easy-going, sociable and sport- rather than work-dominated." Thus evangelicalism, expressed by the daily prayer meeting, could be construed as a reaction to the hedonistic model of campus life. In Barnes's opinion, it would have been very difficult for Romanes to escape the evangelical influence during his first two student years.

But something happened that deterred Romanes from pursuing a call to the ministry. A step that proved critical was his decision, helped to a degree by peer pressure, to read the natural sciences tripos (the Cambridge version of the Honours degree) toward the end of his second year. He had only eighteen months to study for the tripos, but he became so engrossed in the task that he cleared all the exams decently, if not with flying colours (Class B). Interestingly, that same year (1870) Francis Darwin, attending his famous father's alma mater, graduated with a Class A. Suddenly Romanes became aware that he was in his element; comparative anatomy, zoology, and the history and philosophy of science appealed to his nature. By graduation in January 1871 he had done away with any lingering thoughts of taking holy orders, partly because "his family had dissuaded him from that calling"

5.1 · George J. Romanes late in his life. This
is the Elliott and Fry portrait used as the
frontispiece in Romanes and Romanes (1896).

(Turner, 1974), but probably also because the tripos had awakened another calling in him – to nurture his growing interest in the natural
sciences and especially the discipline of physiology. As Ethel Romanes
asserts, "science entirely fascinated him; his first plunge into real scientific work opened to him a new life, gave him the first sense of power
and of capacity" (Romanes and Romanes, 1896).

The source of this new empowerment was Michael Foster (1836–
1907), who in 1870 had just taken the position of prælector at Cambridge, with a mission to develop the discipline of physiology there.
As Gerald Geison explains in his masterly historical analysis (1978),
Cambridge, like many other British universities, lagged far behind
French and especially German universities in the physiological sciences. Edward Sharpey-Schafer, who will play a significant role in our
story later, emphasized Britain's deficiency:

In the middle part of the nineteenth century Great Britain was far
behind France and Germany in the development of Physiology.
We had no pure physiologists and it was considered that any
surgeon or physician was competent to teach the science. Indeed,
long after this, the subject was in many medical schools left in
the hands of a member of the hospital staff, usually a young man,
who carried on clinical teaching as well. Hence, during a period
of time when other experimental sciences were rapidly progress-
ing, Physiology in this country could show no names worthy
to be mentioned with those of [François] Magendie, [Claude]
Bernard, [Johannes] Müller, [Hermann von] Helmholtz or [Carl]
Ludwig, to mention but a few of the brilliant physiologists of
France and Germany. (Sharpey-Schafer, 1927)

And yet discoveries garnered from physiological investigations,
aided by elegant and rigorous experimental protocols, had demonstrated their power to contribute to the understanding of pathologies
as well as normal functions. But the British scientific elite remained
unimpressed until visionary individuals and mentors such as Thomas
Huxley and William Sharpey, then considered the pioneer of physiology
in England such as it was, clinched Foster's appointment. As Sharpey-
Schafer (1927) recalls: "Sharpey then persuaded Michael Foster, a
young man of about thirty, who had received his medical education
at University College and Hospital and was engaged with his father in
general practice in Huntingdon, to give up medicine and come to University College [London] with the title of Professor of Practical Physiology. This appointment of Michael Foster proved a decisive factor in
the history of Physiology in Great Britain." It was decisive insofar as
it led to Foster's subsequent posting to Cambridge three years later.
As Foster moved to Trinity College, Cambridge, he was determined
to develop a full research program in physiology that would rival the
laboratories of Carl Ludwig and Emil Du Bois-Reymont in Germany.
　Foster soon began to create a research school around a theme dear to
him, to which he had devoted time and on which he had published prior
to his Cambridge appointment – the problem of the heartbeat (Geison,
1978). He attracted a number of bright students who could be counted
on to design and implement critical experiments to solve this problem. Among them were John N. Langley and Walter H. Gaskell, who

later became distinguished professors of physiology at Cambridge. Another was George Romanes, who entered Foster's circle late in 1870 or early in 1871. He was at first instructed in physiology by Foster, who must not have failed to fill in his student on the heartbeat question. When Foster first visited this problem in 1864, the prevailing view was that the rhythmicity of the heart's contractions was controlled in all animals by the nervous system. If the heart was isolated from the body or cut in pieces and the isolated heart or pieces continued to contract rhythmically, it was because nervous tissue was still embedded in the pieces and capable of exercising control.

But Foster had debunked this argument on two grounds. First, he did not detect nerve elements in the isolated hearts of his model animals (snails and frogs), although neither could he affirm their absence. And second, but more important, the embryonic heart contracts rhythmically even though the nervous system has not sufficiently developed to be functional at that stage. It is thus conceivable that Foster at some point alerted Romanes to the rhythmic contractions of the jellyfish bell which effect swimming, as an experimental model likely to throw some light on the heartbeat problem; or alternatively, that Romanes, having already watched jellyfish swimming, was induced to propose it as a model to Foster after the latter explained his views on the heartbeat. The latter view is supported by Romanes's contemporary Sharpey-Schafer (1927): "the subject seems to have suggested itself to him whilst convalescing from typhoid at Nigg on the Cromarty Firth [Scotland], where his family had a summer residence, and where the opportunities for such observations were considerable."

But the influence of Foster on Romanes manifested itself unequivocally in another sphere: Foster had embraced Darwinism wholeheartedly. Romanes had read none of Darwin's works and Foster urged him to close this big gap in his scientific culture. Romanes was immediately fascinated by what he read. In Ethel Romanes's words, "it is impossible to overrate the extraordinary effect they [Darwin's books] had on the young man's mind." However exciting life at Cambridge and the discovery of Darwinism may have been for Romanes, it was momentarily eclipsed in the spring of 1872 when he suffered a severe attack of typhoid fever that left him incapacitated for almost a year. On returning to Cambridge in 1873, he abandoned any ambition of an academic career on account of his fragile health. He looked forward instead to

devoting his life "leisurely" to scientific research in the way of a gentle-
man, not unlike several British naturalists of the time who enjoyed personal wealth, not the least of whom his new idol, Charles Darwin. Late
in 1873 Romanes published an article in the journal *Nature* on "Permanent variation of colour in fish" which attracted Darwin's attention and
spawned a lifelong correspondence and friendship between the two. As
Ethel Romanes explains: "From that time [1874] began an unbroken
friendship, marked on one side by absolute worship, reverence, and
affection, on the other by an almost fatherly kindness and a wonderful
interest in the younger's man work and in his career. That first meeting
[at Darwin's Down House] was a real epoch in Mr. Romanes' life. Mr.
Darwin met him, as he often used to tell, with outstretched hands, a
bright smile, and a 'How glad I am that you are so young!'"

Romanes's enthusiasm was channelled beyond Darwin as a person;
it spilled over to the Darwinian theory of evolution. Now he wanted
to conduct his research work in the context of evolutionary thought.
Armed with this theoretical framework, which was shared by Foster,
Romanes saw the importance of investigating the evolution of physiological systems. As Richard French has persuasively argued (French,
1970a), the intellectual intercourse among Romanes, Darwin, and
Foster resulted in Romanes's experimenting on jellyfish with a dual
objective: to shed light both on the underlying mechanism of the heartbeat and on the evolutionary emergence of the nervous system. In the
same year that he completed his master's degree at Cambridge (1874),
Romanes went to University College in the laboratory of John Burdon-
Sanderson, who had just succeeded his mentor William Sharpey as
professor of physiology there. When not in London, he used the summers of 1874 and 1875 in Scotland to work on jellyfish.

Jellyfish Experiments and the Ghost of the Nervous System

Ethel Romanes, whom Romanes married in 1879 and who therefore was
not privy to the "jellyfish years," wrote: "At Dunskaith a little laboratory
was fitted up in an adjoining cottage and here during the summer Mr.
Romanes worked constantly for some years, diversifying his labours
by shooting." At the seaside estate of his mother's family, Romanes collected jellyfish when weather permitted and began his experimental
work. What was his research plan? Strangely, the first article published

on the subject, his *Preliminary observations on the locomotor system of medusae* (Romanes, 1876), is silent on the goals of the research and on his methods for conducting it, and there is no proper introduction to the topic. His predecessors, as we saw in earlier chapters, had accustomed readers to greater transparency. Only in the synthesis of his papers in his 1885 monograph (Romanes, 1885) did he retrospectively state his purpose. Ethel Romanes, in *Life and Letters*, summarized his purpose better than her husband's belated effort:

> He set himself to try to discover whether or not the rudiments
> of a nervous system existed in these creatures. Agassiz had
> maintained it did, others considered his deductions premature,
> and Huxley, in his "Classification of Animals," summed up the
> much-debated question by saying that "no nervous system has yet
> been discovered in Medusae."
> Microscopically, it had already been shown that in some forms
> of Medusae there are present certain fine fibres running along
> the margin of the swimming bell, from their appearance said to
> be nerves, but in no case had it been shown that they functioned
> as such. Thus it was to solve this question, whether or not a
> nervous system, known to be present in all animals higher in the
> zoological scale, makes its first appearance in the Medusae, that
> Mr. Romanes entered upon a long series of physiological experi-
> ments, first on the group of small "naked-eyed" Medusae, and
> then on the larger "covered-eyed" form, the latter containing the
> common jelly-fish. (Romanes and Romanes, 1896)

It is clear from the first pages of his *Preliminary Observations* that Romanes launched his investigation on the assumption that the evidence of nerves was inconclusive, and that therefore, only by dint of experiment and deduction could he reach a conclusion as to whether nerves he could not see must nevertheless be present in order to account for the motor behaviour of the jellyfish. He had already dismissed Agassiz's observations and, although he regarded Haeckel's 1865 paper with less skepticism because he considered the German a highly reputable histologist, Romanes was not swayed by the histological evidence. His own attempt to find nerves was unsuccessful – a predictable outcome given his lack of expertise in histology and "the histological intractability of the gelatinous tissue of the jellyfish swimming bell" (French, 1970a). After all, even the Hertwigs, who were better-trained anatomists and could tap the expertise of several colleagues, had to surmount technical obstacles when dealing with coelenterates (see chapter 3). Romanes was initially not as blessed as the Hertwigs in the matter of resources for histology, but what he lacked in that domain was more than made up for by his experimental acumen.

Before we delve further into Romanes's investigations, it is perhaps timely to examine what his definition of nerves was – and how in his view one could recognize their existence. As Bullock and Horridge (1965) remarked in retrospect: "we can see the importance of the most exacting standards in evaluating evidence concerning the identification of nerve cells and their relations in coelenterates – and indeed in other animals and parts of animals where nervous tissue is diffusely distributed among other tissues." Dealing with animals that lacked organs that could be used as signposts to track suspected nerve cells was, to say the least, disorienting for the early investigators. What could they rely on? Well, they had to amass as much circumstantial evidence as possible that converged enough to reach a threshold of credibility, of persuasive power. As mentioned above, by Romanes's standards the contributions of Agassiz and even Haeckel had failed to reach that threshold. It did not help that for Romanes and many of his contemporaries their only experience of nerves was what they had seen in higher animals: nerve cells and fibres assembled tightly and neatly into tracts, fascicles, ganglia, and brains. This is the baggage they brought to the question of nerve cells in coelenterates, and it is conceivable that Romanes's skepticism about the claims of his predecessors had to do with his *a priori* expectations.

For anatomists like the Hertwigs the standards to go by had been clear: "the cells and fibres should be anatomically consistent with the identification of nerve cells, for example by staining like nerve cells in higher animals and also not staining in ways characteristic of nonnervous structures, such as collagen" (Bullock and Horridge, citing Pantin, 1952). But to a physiologist like Romanes, functional considerations were imperative in formulating an operational definition. He also kept an open mind as to the nature of the excitable tissue: at the evolutionary stage of coelenterates, why not consider the possibility that a distinctive nervous system has not yet developed, that some unformed "protoplasm" or dual-function muscle was the source of excitability and the conduit for spreading the excitation? Romanes's own words seem deliberately to entertain a measure of ambiguity on this question: "By the word 'nerves' here I mean certain physiologically differentiated tracts of tissue which either stimulation or section prove to perform the function of conveying impressions to a distance; and by 'physiological character of the contractile tissues' I mean the character of these tissues in respect of the degree in which the nervous element shows itself to be physiologically differentiated from the muscular element" (Romanes, 1877).

Strangely, a similar definition was offered by his mentor Michael Foster at about the same time, but Foster went further, entering territories Romanes feared to tread by indulging in bold and prescient theorizing on the origin of nerve cells:

In its simplest and probably earliest form, a nerve is nothing more than a thin strand of irritable protoplasm, forming a means of vital communication between a sensitive ectodermal cell exposed to extrinsic accidents, and a muscular, highly contractile cell (or a muscular process of the same cell) buried at some distance from the surface of the body, and thus less susceptible to external influences. If in Hydra, we imagine the junction of the ectodermic muscular process with the body of its cell to be drawn out into a thin thread (as is said to be the case in some other Hydrozoa), we should have just such a primary nerve. Since there would be no need for such a means of communication to be contractile and capable of itself changing in form, but on the other hand an advantage in its remaining immobile, and in its dimensions being reduced as much as possible consistent with the maintenance of irritability, the primary nerve would in the process of development lose the property of contractility in proportion as it became more irritable, i.e. more apt in the propagation of waves of disturbance arising in the ectodermal cell. (Foster, 1877)

In reference to the spinal cord, Foster viewed its constituent nerve cells as forming a "functionally continuous protoplasmic network" (Foster, 1877). Romanes found this notion of physiological continuity useful, and he developed it further in relation to the nature of the excitable tissues of jellyfish by distinguishing between two forms of continuity: "By 'contractional continuity' I shall wish to be understood such a condition of contractile tissue as admits of the uninterrupted passage of contractile waves; while by 'excitational continuity' I shall wish to be understood such a condition of the contractile tissues as admits of one part responding to stimuli applied at another part, *whether or not contractile waves are able to pass along the intervening parts*" (Romanes's italics). Romanes would only go a short distance toward conveying an evolutionary subtext in his definition of the nature of nervous tissue: "And as it is further evident that this distinction has reference to the most fundamental quality wherein the function of nerve is distinguished from that of muscle, viz. the power of setting up responsive contractions at a distance from the seat of irritation, it will be understood that by the term 'excitational continuity' I intend to denote the first indications we can perceive in the animal kingdom of the distinguishing function of nerve-tissue."

French (1970a) stressed that this kind of intellectual musing on Romanes's part was typical of the preoccupation of Foster's Cambridge School of Physiology with the question of the evolutionary relationship of structure and function. It was far from the outlook of the anatomists and it courted potential pitfalls. French quotes Burdon-Sanderson who, despite his deep friendship with Romanes, cynically noted that, should the physiologist fail to find the nerve behind the function, he could always "fall back on that worn-out *Deus ex machina*, protoplasm, as if it afforded a sufficient explanation of everything which cannot be explained otherwise." To Romanes's credit, he tried hard, but not always successfully, to avoid this trap and he managed to stay away from the murky thinking the intractable topic all too often invited.

Let us now go back to the improvised seaside laboratory. If Romanes selected jellyfish for no other reason than that they provided a useful analogy for the heartbeat problem, then he was incredibly fortunate in his choice, for it turned out that jellyfish have an uncanny ability to survive surgical tissue ablation and electrical stimulation. He was quick to gain substantial insights from poking questions at the pliable and cooperative animals. In his *Preliminary Observations*, the object of the Croonian Lecture at the Royal Society of London on 16 December 1875 and published in 1876, one gets a taste of the excitement that Romanes experienced as the outcome of his experiments unfolded before him.

On the basis of simple but cleverly designed cutting experiments, Romanes was able to show in naked-eyed jellyfish – today's Hydromedusae – that *"excision of the extreme margin of a nectocalyx* [swimming bell] *causes immediate, total, and permanent paralysis of the entire organ"* (Romanes, 1876). After careful assessments of repeated trials he was able to deduce the incontrovertible meaning of this result:

> From this experiment, therefore, I conclude that in the margin of all the species of naked-eyed Medusae which I have as yet had the opportunity of examining there is situated an intensely localized system of centres of spontaneity, having at least for one of its functions the origination of impulses to which the contractions of the nectocalyx, under ordinary circumstances, are exclusively due. And this obvious deduction is confirmed (if it can be conceived to require confirmation) by the behaviour of the severed margin. This continues its rhythmical contractions with a vigour and a pertinacity not in the least impaired by its severance from the main organism; so that the contrast between the perfectly motionless swimming-bell and the active contractions of the thread-like portion which has just been removed from its margin is as striking a contrast as it is possible to conceive. (Romanes, 1876)

By analogy with heart rhythmicity, Romanes had localized the central pacemaker of the rhythmic contractions, which was both necessary and sufficient to control the swimming movements in Hydromedusae. But at this point he cautiously refrained from attributing a nervous nature to this centre.

However, when he turned his attention to the covered-eyed jellyfish, currently known as the Scyphomedusae, Romanes was in for a surprise. He found that excising the swimming bell margin either had no effect on the normal swimming movements or caused at best a temporary paralysis from which the jellyfish recovered after a variable time. He also found bewildering differences of responses to treatment between individual specimens within a species or between species. In addition he found in *Aurelia aurita*, the only species in which it was possible to perform the required surgery, that ablating only the rhopalia – a sensory organ at the bell margin that includes a sensory epithelium with an ocellus and a statolith – was sufficient to cause the above-mentioned temporary paralysis. Romanes could not trivialize such a dichotomy of response between Hydromedusae and Scyphomedusae. He offered the following interpretation:

Hence, in comparing the covered-eyed group as a whole with the naked-eyed group as a whole, so far as my observations extend I should say that the former resembles the latter in that its representatives usually have their main supply of locomotor centres situated in their margins, but that it differs from the latter in that its representatives usually have a greater or less supply of their locomotor centres scattered through the general contractile tissue of their swimming-organs. But although the locomotor centres of a covered-eyed Medusa are thus, generally speaking, more diffused than are those of a naked-eyed Medusa, *if we consider the organism as a whole*, the locomotor centres in the *margin* of a covered-eyed medusa are less diffused than are those in the margin of a naked-eyed Medusa; for, so far as my observations extend, I find that excision of the marginal bodies [rhopalia] alone produces a greater comparative effect in the covered-eyed genera of Medusae than it does in the genus *Sarsia*. (Romanes, 1876)

So the central pacemakers differed between the two groups of jellyfish in their organization and location at the bell margin, but what happened downstream from these centres? Romanes sought answers by means of three experimental approaches: (1) mechanical and electrical stimulation, (2) elaborate cuts through the bell and the "locomotor centres," and (3) effects of drugs known to have specific actions on nerve and muscle cells of higher animals. With regard to mechanical stimulation, he found that "every Medusa, when its centres of spontaneity have been removed, responds to a single stimulation [poking with a forceps] by once performing that action which it would have performed in response to that stimulation had its centres of spontaneity still been intact." The results with electrical stimulation, performed thanks to equipment borrowed from his friend Burdon-Sanderson, confirmed that the entire bell of the jellyfish was excitable, not just the locomotor centres, and that this excitability could be revived by local stimulations in animals paralysed by the removal of the locomotor centres.

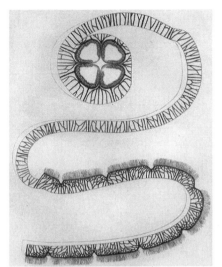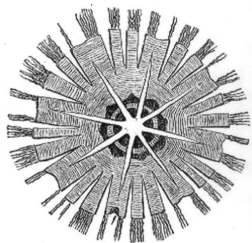

5.2 · Two representative examples of cuts made by Romanes through the jellyfish swimming bell to test the ability of the excitation wave to still travel across the muscle sheet. (*Left*) A specimen of *Aurelia aurita*, with manubrium and all but one rhopalia removed, submitted to spiral incisions. (*Right*) The swimming bell, with all but one rhopalia removed, submitted to all fashions of radial incisions. From Romanes (1876).

What was even more remarkable was the effect of making numerous cuts in different orientations through the tissues of the swimming bell and then stimulating at one point to see if and how far the resulting contractile wave (or excitation wave) would travel. Romanes went far to test the "amount of section which the contractile tissues of Medusae will endure without suffering loss of their physiological continuity." Two eloquent examples of "extreme incisions" performed by Romanes are shown in Figure 5.2.

By removing all rhopalia, which are the suspected swimming pace-makers, but one, and using this rhopalia as the start point of a spiral incision through the bell, resulting, as shown in Fig. 5.2 (left), in a strip up to a metre long, Romanes was able to demonstrate that in many cases the excitation of the rhopalia managed to travel across the entire contorted strip. If he chose instead to inflict overlapping radial cuts to the bell with again only one rhopalia remaining, as illustrated in Fig. 5.2 (right), a similar result would ensue; that is, the propagation of the excitation wave is often not impeded in the least. Such startling observations should immediately suggest that whatever conducting tissue there is can only be of a diffuse nature; if it was of a discrete cable nature it would have been severed by the surgical interventions and the excitation would have failed to spread. But, while acknowledging the importance of his finding, Romanes wavered as to its interpretation:

> What is the nature of the general contractile substance of Medusae? Is the contractile tissue of the swimming-organ pervaded by a definite system of sensory and motor tracts, so to speak, radiating respectively to and from the marginal centres? Or is the contractile tissue of the swimming-organ of a more primitive nature, the functions of nerve and muscle being more or less blended throughout its substance? Now, for my own part, I deem this question the most interesting one with which the present paper is concerned; for the evolutionist, no less than the physiologist, will recognize its importance as of the highest ...
>
> From the observations already detailed it might well be concluded that the method of inquiry by section has already settled the question before us, seeing that this method has apparently reduced the hypothesis as to the presence of definite lines of discharge to an absurdity. A moment's thought will render obvious how very trying the spiral form of section already described must be to anything resembling a nervous plexus, while a glance at fig. 3 [our Fig. 5.2] would seem to render the supposition as to the presence of any such plexus almost impossible. Nevertheless there is a weighty body of evidence to be adduced on the other side ...
>
> All we have to assume is that there exists *a more or less intimate plexus* of such lines of discharge, the constituent elements of which are endowed with the capacity of vicarious action, and that in some cases the section happens to leave a series of their anastomoses in a continuous state ... Moreover I think that the difficulty of accepting this explanation will diminish if we cease to regard the hypothetical plexus as presenting the high degree of integration characteristic of a properly nervous plexus; but in this preliminary paper I cannot, without undue length, discuss this subject ...

I must nevertheless assert my persuasion that, so far as my observations have yet gone, a legitimate deduction from them appears to be, that in every individual of this species [*Aurelia aurita*] (and so from analogy, as well as observations on other species, probably in all Medusae) these slender lines of differentiated tissue are present, that through their mediation the spontaneous impulses originating in the marginal centres are communicated to the contractile tissue of the swimming-organ, and therefore that these slender lines of differentiated tissue are functionally, if not structurally, nerves. (Romanes, 1876)

Straining to err on the side of caution, Romanes resorted here to something resembling sibylline argument. But he had finally released the word "nerves." Once the word was uttered, he looked for "additional facts tending to prove the identity of the specialized marginal tissue of Medusae with nervous tissue in general." The presence of sensory structures such as "eye-specks" (ocelli), whose role in attracting jellyfish to light sources (phototaxis) he experimentally demonstrated, was one such fact. The other was the "effects of various poisons on the locomotor system of Medusae." By showing the anaesthetizing effects of chloroform, ether, morphine, strychnine, and curare on swimming, Romanes had gone a long way toward deducing that the jellyfish motor system behaved as expected of a system coordinated by a nervous system.

But a decisive experiment bolstered his case even further. He devised an experimental setup by which only part of the swimming bell (in the middle) was exposed to curare; when he stimulated the preparation on one of the intact sides, the elicited contractile wave was interrupted in the curare-exposed segment, only to re-emerge on the opposite, intact side of the preparation. This behaviour tended to confirm that whatever carried the excitation wave (nerve plexus?) was physically and functionally separate from the contractile elements. Romanes seemed quite excited about the significance of this observation:

I think it would be difficult to overrate the importance of these results: to my mind they are perhaps the most interesting which are contained in this paper. They not only prove that curare poison is consistent in manifesting its remarkable property when applied to these the lowest forms of life that present the beginnings of a nervous system; but they prove what is far more important, that in animals which, as we have seen from other evidence, present us with the first indications of a nervous system, the latter appears to have already undergone a differentiation of its functions, such that it is capable not only of influencing contiguous contractile parts, but also of being influenced by distant excitable parts. (Romanes, 1876)

At the end of his first two summers spent toiling on these experiments, Romanes took a step back and pondered his next move. Throughout his investigations in Dunskaith, he had kept in touch with two of his friends at University College in London. One of them was Burdon-Sanderson, who had volunteered moral encouragement along with suggestions for experiments. Another was Edward Schäfer, who will soon play an important part in our story. Romanes became acquainted with Schäfer around 1874 and struck up a friendship just as the latter was launching his academic career at University College, London. Schäfer was known as "a kind, friendly and compassionate man with a good sense of humour" (Sparrow and Finger, 2001), a personality type that would appeal to Romanes's own, and the addition of Schäfer's first-rate intellect would clinch a meeting of minds and souls.

While his investigations were going on, Romanes sent progress reports to Schäfer and repeatedly asked him for advice on how to interpret this or that observation: "I should like to have your opinion about the meaning of the following facts" or "I wished you would say what you think about this peculiarity in relation to a subject that I have been working up" (letters of Romanes to Schäfer in the spring and summer of 1875: Romanes and Romanes, 1896). He also wanted to enlist Schäfer's help with the histology, as he was not making any progress in that regard and Schäfer was already an excellent histologist as well as physiologically inclined: "I have had no time to do anything at the histology as yet. Would it be worth while for me to send you various species in a little sea water? They would arrive in a tolerably fresh condition, but would require to be examined at once. I might try sending some in spirit and others in chromic acid."

In another letter to Schäfer, Romanes was tinkering with embryonic interpretations of his findings in relation to Agassiz's anatomical observations:

> I know you do not like theory, so I shall return to fact. There can be no doubt whatever that the seat of spontaneity is as much localised in the margin as the sensibility to stimulus is diffused throughout the bell. There *must*, therefore, be some structural difference in the tissue here to correspond to this great functional difference. Agassiz is very positive in describing a chain of cells running round the inner part of the marginal canal. Now although I sometimes see a thin cord-like appearance here, I should not dare to say it was nervous. Gold certainly stains it, but it also stains many other parts of the tissue, and until I can see *cells* here I cannot be sure about a *visible* nervous cord. The cord I do see may be the wall of the marginal canal. I intend to persevere, however, trying your suggestions, also osmic acid. (Romanes and Romanes, 1896)

Schäfer seemed adamant that the histology should be given priority to help Romanes get a comprehensive picture of the jellyfish locomotor system before wrapping up the writing and submitting the paper for publication, but Romanes fended off that advice. It finally came to a head in a letter to Schäfer dated September 1875, which says as much about Romanes's hierarchy of biological disciplines and the publishing practices of the day as it does about the pageantry of friendship:

> I gave my careful consideration to all you said about publishing, and at one time nearly decided to wait another year. But eventually I sent in the paper [to the Royal Society]. It seems to me that the histology can very well wait for future treatment – that its absence is not sufficient justification for withholding the results I have already observed. These results, after all, are the most important; for they prove that some structural modification there *must* be; whether or not this modification is visible is of subordinate interest. Besides, I do not, of course, intend to abandon the microscopical part of the subject altogether. In my view, inquiry

into function in this case must certainly always precede inquiry into structure; for although, when all the work shall have been collected into one monograph, the histology must occupy the first place in order of presentation, very little way could have been made by following this order of investigation.

I also had to reflect, that if I postponed publication, it would be impossible to expect the R.S. [Royal Society] to publish the results *in extenso*, – *i.e.*, I should have to bring out the work through some other medium.

And in addition to all this, there came a letter from Foster preaching high morality about it being the duty of all scientific workers to give their results to others as soon as possible.

As I said before, I thank you very much for the consideration and advice you have given, but I know that you would not like me to feel that the expression of your opinion in a matter with which you are not so fully acquainted as myself should lay me under any obligation to be led by it, after mature consideration seemed to show that the best course for me to follow was the one which I took. (Romanes and Romanes, 1896)

The receipt of the revised manuscript was acknowledged by Thomas Huxley on behalf of the Royal Society on 1 November 1975. According to Romanes's correspondence with Darwin, there were delays in the publication of his first jellyfish paper that obviously bothered him; he had not even received the proofs by early June 1876, so that before he saw the paper in print he had already started a new season of investigations.

His objectives for the summer of 1876 were to consolidate the aspects he had covered in the first paper with more detailed observations, to open up other aspects he had barely or not covered yet, and to react to a preliminary paper by a German competitor, Theodore Eimer (of whom more later), which covered similar ground. He looked at the effects of temperature and gases (oxygen, CO_2, and others) as well as light and electrical stimulation. Using the latter method he found that on repeated stimulation of fixed strength the contractile waves increased steadily in amplitude (staircase effect) and were initiated with latent periods – the time between the arrival of the stimulus and the onset of the response – of steadily shorter duration. He exclaimed: "how surprising is the effect of a series of stimuli, first in *arousing* the tissue, as it were, to increased *activity*, and second in developing a state of *expectancy*" (Romanes, 1877). This phenomenon, known since the early 1870s to investigators such as Henry P. Bowditch, a Harvard physiologist and student of Carl Ludwig in Leipzig, and Edward C. Stirling, a student of Foster at Cambridge, was later to be called *facilitation*, a process by which synaptic transmission between nerve cells or between nerve and muscle cells is accelerated and strengthened. Facilitation is based on the preparation's still "remembering" what happened in the previous stimulus, so that overly delaying the arrival of the next stimulus will erase that "memory," a feature that Romanes also noticed: "it is observable that the tissue has, as it were, completely forgotten the occurrence of the previous series [of stimuli], so that the next staircase has to begin anew from the first step."

Another original finding in Romanes's second paper pertains to reflex action. If he poked the tip of the polypite (manubrium, or mouth organ) with a forceps, the polypite of the hydrozoan *Sarsia* retracts and the swimming bell responds with an escape response. In another species, *Tiaropsis indicans*, Romanes observed a reflex of remarkable complexity. If any point of the swimming bell is poked, the polypite answers with "an exceedingly rapid crouching movement" while the poked part of the bell bends inwardly; this is followed by the polypite being deflected toward, and moving rapidly "with unerring precision, to meet the in-bent portion of the nectocalyx." The analogy with nerve-driven reflexes of higher animals moved Romanes to reflect on the meaning of such a display:

> We have here, then, a curious fact, and one which it will be well to bear in mind during our subsequent endeavours to frame some sort of a conception regarding the nature of these primitive nervous systems. The localizing function which is so very efficiently performed by the polypite of this Medusa, and which, if anything resembling it occurred in the higher animals, would certainly have definite ganglionic centres for its structural correlative, is here shared equally by every part of the exceedingly tenuous contractile tissue that forms the outer surface of the organ. I am not aware that such a diffusion of ganglionic function has as yet been actually proved to occur in the animal kingdom; but I can

scarcely doubt that future investigation will show such a state of things to be of common occurrence among the lower members of that kingdom. (Romanes, 1877)

As we shall later see, his prophecy proved accurate. But for now, what strikes one is Romanes's new readiness to invoke the "N" word. So far, his professional caution had deterred him from interpreting experimental results too narrowly as explainable by the activity of nerves. In this second paper he now comes forward with more explicit references to nerves and ganglions, even speaking of "diffusion of ganglionic function." What brought this change of heart on the question of nerve elements? The investigations reported in the second paper seemed to have tipped the scales. He was well aware of what he needed to demonstrate in order to revise his thinking, as this passage suggests:

For if it can be proved that the contractile tissues of *Aurelia* are pervaded by tissue-tracts which display the essentially nervous function of establishing what I have termed excitational continuity between different parts, then I think we may be more prepared to believe that the passage of contractile waves depends on the presence of tissue-tracks presenting a nervous character. For the evidence being already in favour of the hypothesis that the passage of contractile waves depends on the presence of differentiated elements whose function is presumable nervous, such evidence would be further strengthened if it could be shown that in the very same tissue there occur other differentiated elements whose function is demonstrably nervous. Now that such elements as the last mentioned do occur in all parts of the excitable tissues of *Aurelia aurita* is a fact concerning which there can be no question. (Romanes, 1877)

The only remaining obstacle to Romanes's acceptance of the nerve plexus theory had been the astounding extent of incisions necessary to block the passage of the contractile waves. Like many of his contemporaries, Romanes could not shake off the mindset that the size of ganglions and nerves must have a lower limit set by the higher invertebrates and vertebrates they were familiar with. With that size scale in mind it strained credibility that the narrow channels of passage left by extreme surgical incisions could have spared some of these nerves. But Romanes was coerced to accept that nervous elements were involved, even in these impossibly narrow passages created by incisions, by undeniable evidence coming from a friendly source.

Either Romanes never followed through with his offer to send Schäfer fixed jellyfish material for histological work, or he did but Schäfer could not salvage anything from it, for no document has surfaced to settle that point. But what is certain is that, perhaps in desperation as the clock ticked and there was no histology to show in support of the physiological work, Romanes tried another tack by inviting his friend Schäfer to join him at Dunskaith and work from there on fresh material. In a letter to Darwin dated 18 August 1876 Romanes alludes to this in a tongue-in-cheek manner:

> In "Nature" [Romanes, 1874] I did not express my doubts [about the nervous character of conducting tissues], but it was because I feared there may yet turn out to be a skeleton in the cupboard that I kept all these more or less fishy deductions out of the R.S. [Royal Society] papers. Further work may perhaps make the matter more certain one way or another. Possibly the microscope may show something, and so I have asked Schäfer to come down, who, as I know from experience, is what spiritualists call a "sensitive" – I mean he can see ghosts of things where other people can't. But still, if he can make out anything in the jelly of *Aurelia*, I shall confess it to be the best case of clairvoyance I ever knew. (Romanes and Romanes, 1896)

His skepticism was not warranted. Schäfer's "clairvoyance" turned out to be down-to-earth scientific flair, with skills and hard work to boot. Romanes had dabbled in some histology treatments for his second paper. The paper was read at the Royal Society on 11 January 1877, but the publication in the *Philosophical Transactions of the Society* – in a situation reminiscent of the first paper – was delayed such that Romanes was able to include in September 1877 a note informing his readers that he had decided to remove his own histological findings from the paper so that Schäfer's could be published separately. Romanes finally had to admit that the "ghost" was in reality a *bona fide* nervous system. The idea of an undifferentiated "tissue-track" along the lines of the neuromuscular cells of Kleinenberg was now given up. Foster's group in Cambridge had been holding their breath that Romanes's findings would substantiate their view of the non-nervous nature of the heart pacemaker, but jellyfish were no help. As H. Newell Martin, an ex-student of Foster, put it in a letter of 24 March 1878 to Schäfer, "I feel an unscientific tendency to grieve over the loss of conductivity along special tracts not differentiated into nerve fibres; it was so nice!" (cited by French, 1970a).

Theodor Eimer: Caught in the Finishing Lines of the Publication Race

While Romanes was happily pursuing the first year of his investigations, he was unaware that Theodor Eimer, a German zoologist, had launched his own research on jellyfish. Furthermore, Eimer's research eerily followed the line of Romanes's own; that is, he was attempting a comprehensive investigation of the jellyfish nervous system by way of physiological and histological approaches. The two researchers were on a collision course.

Theodor Gustav Heinrich Eimer (1843–1898) was born near Zurich, Switzerland. His father was a political exile from Germany who practised medicine, and his mother was a Swiss citizen (Churchill, 2008). In 1862 he enrolled in medicine at the University of Tübingen, where he fell under the influence of Franz Leydig, a distinguished histologist famous for his *Lehrbuch der Histologie des Menschen und der Tiere* and for the discovery of the interstitial (Leydig) cells of the testes. Eimer spent time also in Freiburg and Heidelberg before earning his medical degree in Berlin in 1867. In Berlin the famed pathologist Rudolf Virchow, instead of pushing his student toward a medical career, encouraged him to study zoology and comparative anatomy and, under the supervision of Albrecht von Kölliker at the University of Würzburg, Eimer was granted a doctorate on the basis of his histological and physiological research on the absorption of fat by the intestine. Eimer came to investigate jellyfish in a roundabout way, as Frederick Churchill explains:

On 18 July 1870 he married Anna Lutteroth, the daughter of a Hamburg banker, and the following day was habilitated in zoology and comparative anatomy. Immediately thereafter Eimer volunteered for military service as a field surgeon [Franco-Prussian war]. He saw action at the siege of Strasbourg and was joined by his wife, who served as an army nurse. After being decorated for service, he was forced to retire from active duty because of illness. Early in 1871 [for his convalescence] the Eimers journeyed to Capri, where Theodor became familiar with marine organisms. He returned to the island in 1872, 1876, 1877 and 1879. The coelenterates of the Bay of Naples and the lizards of Capri formed the subjects of Eimer's first book-length monographs in zoology. (Churchill, 2008)

Eimer thus developed an interest in jellyfish as a serendipitous result of his illness. In the summer of 1872 he worked on the Capri fauna and the following year he reported his findings, in which a description of the anatomy of comb-jellies – "ribbed jellyfish" in German language descriptions – was given (Eimer, 1873) (see chapter 7). That same year, at Baltic Sea locations, he undertook his first surgical experiments (5 September 1873) on the scyphomedusae – which he also calls "toponeurous Medusae" – *Aurelia aurita* and *Cyanea capillata*. The results of these preliminary experiments, and of more conclusive experiments in September 1874, were published in December 1874 (Eimer, 1874). In this paper Eimer recorded observations remarkably similar to those of Romanes. This is how he describes them, in a later publication and in an edition in English:

The Medusa *Aurelia aurita*, when I had removed from it all the primary-nerve-centres [marginal bodies], became completely motionless. But after the animal thus mutilated had lain several days in clean sea-water, it gradually commenced to exhibit movements again: one day, movements at first trembling and irregular appeared in the umbrella. These movements evidently started from a definite spot which might be at any part of the umbrella, and extended thence over the whole, just as in the uninjured animal they proceed from the nerve-centres in the margin ... Thus a new nerve-centre had been formed in place of the old from the nerve-cells scattered over the body-surface, and had taken upon itself the movement and direction of the animal ...

The movements of Medusa are, as I have shown, involuntary, but they can be retarded or hastened, diminished or intensified voluntarily. As involuntary movements they are respiratory, as voluntary they are the means of locomotion. (Eimer, 1890)

This citation is instructive, in that Eimer was bolder than Romanes in attributing the control of swimming to nerve cells. This may be due in part to Haeckel's intellectual dominance over the younger German zoologists of the era – and Haeckel had after all described nerve tracts at the margin of the umbrella only nine years before. But Eimer's personality may also have coloured his interpretations: notwithstanding his skills as an observer, he was quick to speculate. He did not entertain Romanes's skepticism regarding Agassiz's descriptions of the jellyfish nervous system. While he was performing his program of sectioning experiments to test, in Romanes's words, "the amount of section which the contractile tissues of Medusae will endure without suffering loss of their physiological continuity," Eimer asked himself if the upper nerve ring described by Agassiz could instigate and mediate the passage of contractile waves. To test this he "made two radial incisions proceeding from the centre towards the circumference of the disk. He found, of course, that these sections might be carried to within quite a short distance of the margin before the portion of tissue which was included between them separated from the rest of the umbrella" (Romanes, 1876). The test, in Eimer's estimate, validated the hypothesis.

Romanes was only made aware of Eimer's 1874 paper in February 1876 by the Danish naturalist Christian F. Lütken (1827–1901), an invertebrate specialist. Lütken had read the short note on Romanes's "fundamental" jellyfish experiment published in the journal *Nature* on 12 November 1874. That fundamental experiment, as Romanes himself called it in his *Preliminary Observations*, was the paralysing effect of removing the marginal bodies of the swimming bell. Obviously, Lütken, living in continental Europe and next door to Germany, had easier access to what Romanes considered obscure scientific journals. Reading the two papers, Lütken could not help being struck by the close resemblance, so he decided to alert Romanes to it. Romanes was undoubtedly shocked by this revelation, although no document has surfaced to reveal his feelings on the subject. The manuscript of his *Preliminary Observations* had already gone to press, so he prepared a

5.3 · Theodor Eimer in Tübingen. Courtesy
Universität Tübingen (*Professorengalerie*).

postscript addressing the parallel findings of Eimer. The postscript of
several pages was received by the Royal Society on 24 March 1876, in
time to be included at the end of the *Preliminary Observations* paper.

Romanes dealt with the sensitive matter straightforwardly. He acknowledged that the majority of Eimer's observations agreed with
his own; the major bone of contention was the extent of tissue in the
vicinity of the marginal bodies that has to be removed before the bell
becomes paralysed – in other words, the boundary of the pacemaker.
Eimer had concluded that "the seats of spontaneity are not the litho-
cyts alone, but the entire crescent-shaped interruptions of the margin
in which the lithocysts are lodged." Romanes, on the other hand, stead-
fastly held to his original deduction that the locomotor centre resides
in the lithocysts. He even went further, specifying that it sufficed to
remove "the little sac of crystals composing the central part of the litho-

cysts" (Romanes, 1876) for the contractile zone to cease activity. His meaning is not clear here, but it suggests that he viewed the balance sensory organ, the statolith, as the necessary and sufficient locomotor centre – a strange statement to make, considering the caveats of poor surgical precision and lack of knowledge on the nature of the rhopalia at that time.

But in the end the whole exercise came down to locking horns over priority. Romanes tried to make light of the matter – "as to the *mere matter of priority*" – but his treatment of it suggests that he cared very much about recognition for his work, as is normally expected of scientists past or present. Romanes fired the first shot by stressing that, "as Dr. Eimer's work was done in September of 1874, I have a right to claim precedence, both as to observation and publication of what I have termed the fundamental experiment." Romanes made an honest mistake in thinking that Eimer's own fundamental experiment occurred in 1874. Eimer himself had given a muddled account of the chronology of his experiments in his 1874 paper. However, perhaps in reaction to Romanes's analysis of his work, Eimer became more specific both in his full monograph (Eimer, 1878) and in later work (Eimer, 1890), stating clearly that the jellyfish cutting experiments were started on 5 September 1973 and included the fundamental experiment as Romanes understood it. As Romanes's experiments began early in the summer of 1874, priority should have gone to Eimer were it not for his muddled chronology and post-hoc reshuffling of dates, which raised doubts and kept the issue inconclusive. But as to the matter of priority of publication, there is no doubt that Romanes won by the slimmest of margins – one month!

By the time Eimer was acquainted with Romanes's *Preliminary Observations*, his own jellyfish research was still a work in progress. In 1876 he had done much of the experimental work and was slowly advancing with the histology of the nervous system. It is clear from the tone of his remarks in the 1878 monograph that Eimer resented being relegated to an "also-ran" status due to his circumstances. In 1874 he was appointed inspector of the grand duke's zoological collections in Darmstadt and associate professor of zoology at the Technische Hochschule there. The following year he succeeded Leydig to the chair of zoology and comparative anatomy at the University of Tübingen (Churchill, 2008). This series of moves and the heavy teaching and administrative duties that his jobs entailed – which Romanes never had to face – must have seriously impeded the progress of his research. Eimer also complained that residing and working in a landlocked place made it difficult to organize time and logistics for research at faraway coasts – the Eastern North Sea or the Baltic – so that he had only a few days to capture animals in reasonable shape for experiments, pending weather conditions.

Eimer's predicament was aggravated in 1877 when he read the Hertwigs' preliminary paper on the histology of the jellyfish nervous system. He no doubt felt hemmed in on all sides, and here was another opportunity for him to lose priority of publication! We are accustomed to view the publish-or-perish conundrum as a product of twentieth-century science practice, but there is clear evidence that it took the form of a priority race in the second half of the nineteenth century, even though the scientific disciplines were far from being crowded. Prominent biologists like August Weismann, to whom Eimer dedicated his jellyfish monograph, bitterly commented on the situation: "I hate this breathless racing very much, but it is also aggravating to sign away to others something of the little bit that one can produce in a lifetime" (Letter to J.V. Carus, 9 April 1890, cited by Nyhart [1995]).

From a letter to his colleague Ernst Ehlers, dated 14 October 1877, it is clear that Eimer was eager to vent his spleen in the same tone over the Hertwigs. Lynn Nyhart (1995) provides the gist of the letter: "Theodor Eimer wrote in 1877 that certain preliminary results published by the Hertwig brothers duplicated his own earlier work, which had been awaiting publication until he finished with the "endless physiological experiments" that would round out the morphological results. "I wanted to complete the whole peacefully and fully without preliminary notices – now here come the predators (*Jäger*)."

It was, of course, unfair to characterize the Hertwigs as predators, given that they knew nothing of Eimer's parallel investigations. But Eimer could be excused for using the brothers as lightning rods for his anger and frustration at being a pawn of fate. He finally satisfied himself that the "whole" was attained, and his monograph was published the year after his letter of discontent. The monograph gives a lucid account of Eimer's physiological experiments and, while they confirm many of Romanes's findings, they come short of reaching the level of investigative cleverness, thoroughness, and insightfulness that marked Romanes's contributions. Eimer's illustrations of his cutting experiments, unlike those of Romanes, were of poor quality and too cartoonish to be informative. In contrast, his morphological work was of high quality and went even further than that of the Hertwig brothers in characterizing the nervous system of jellyfish, particularly at the cellular level. This comparison simply reflects the suspicion that Eimer, as suggested by his letter to Ehlers, used the physiological work as support for the neuroanatomy and not the other way around, which was Romanes's perspective.

There is another issue that separated Eimer's and Romanes's approaches. Romanes entered into discussions of jellyfish neuroanatomy only after he had determined for himself clear and rigorous criteria by which to identify nerve cells. Eimer acknowledged that in such animals with "laminar central nervous systems" one could expect difficulties in distinguishing nerve cells from other cell types. But rather than be impelled like Romanes to face these challenges by providing a somewhat scientific definition of nerve cells, Eimer contents himself with sophistry of the kind exemplified here: "The nerve-cells in the Ctenophora and Scyphomedusae are so far from being differentiated and recognisable, that I have been accused of mistaking connective-tissue-cells for nerve-cells. It is, however, obviously, on my view of the matter, necessary that nerve-cells should at the commencement of their evolution be similar to other cells. It is a known fact that even in higher animals the nerve-cells in the embryonic condition cannot be distinguished from other embryonic cells" (Eimer, 1890).

When Eimer introduces more rigour to his discourse, it appears to be a retrospective concession to what Romanes had expressed before him:

Only function could impress upon nervous, as on other cells, a definite morphological character. Thus I sought at first in vain for nerve-cells or brains in the Scyphomedusae and could only discover the spots at which the latter exist by the section-experiments above mentioned. Such experiments afterwards completely confirmed my description of the nervous system of Beroë. In both cases, in Beroë as in Scyphomedusae, the presence of a number of nerve cells, or of brains, could be recognised by the fact that the parts in question, when separated from the rest of the animal, alone, or at least in a pre-eminent degree, exhibited life (movement). (Eimer, 1890)

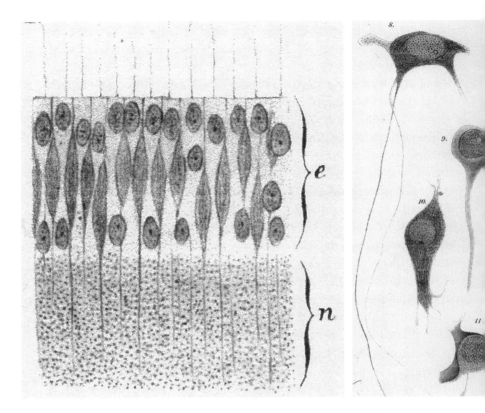

5.4 · (*Left*) Ectodermal epithelium (e) and nerve fibre layer (n) of *Aurelia aurita*. Note the epithelial sensory cell (s). (*Right*) Enlarged view of representative "ganglion cells" lying underneath the epithelium of the bell margin and nerve ring in *Carmarina hastata*. Note the cytological details of the cells. From Eimer (1878), Plates IV and VIII. Courtesy Mohr Siebeck GmbH & Co., Tübingen.

In his anatomical descriptions Eimer paid special attention to the sensory cells of the ectoderm epithelium in *Aurelia aurita*, which he characterized as more or less cylindrical, flagellated cells situated in the cellular layer, over the fibre layer in the inner part of the epithelium (Fig. 5.4, left). He shows these cells as presenting a flagellum (which is actually a cilium projecting to the outside) and a process, directed inward and bifurcating as it enters the nerve fibre layer. To these he assigns a tactile function. The majority of Eimer's neuroanatomical work, however, was done on a different species, *Carmarina hastata,* a hydrozoan jellyfish. In contrast, as stated earlier, all his physiological work was done on scyphozoan jellyfish (*Aurelia* and *Cyanea*). To Eimer, hydrozoans seemed more amenable to neuroanatomy than scyphozoans because of their relative simplicity and tissue tractability. But this approach had potential pitfalls. Romanes, upon reading the Hertwigs' 1878 monograph, remarked in his concluding observations (Romanes, 1880) that "there is so great and fundamental a difference between the nervous system of the naked- and of the covered-eyed Medusae, that a simultaneous description of the nervous system in both groups is not by these authors considered practicable."

So it was perilous on Eimer's part to have attempted to find confirmation of his neurohistological results obtained from hydrozoans in the experimental results obtained from scyphozoans. Nevertheless, the cytological quality of the neurohistology done with *Carmarina hastata* is excellent and superior to any produced by the Hertwigs. The method Eimer used so successfully consisted in immersing the jellyfish tissues in potassium dichromate mixed in seawater, followed by a thorough rinse in distilled water and staining with carmine. He then used glycerol over several days to clear the preparation so as to allow the best visualization of nerve cells. An example of the remarkable cytological detail Eimer could achieve is shown in his rendering of the "ganglion" cells present at the margin of the swimming bell (Fig. 5.4, right). Here we see the details of the large nucleus of the cell bodies (including the nucleolus) and of the peri-nuclear cytoplasm. The latter clearly shows the granular content later known as Nissl bodies. It also allows us to appreciate the diversity of these nerve cells on the basis of the organization of their processes: unipolar, pseudounipolar, bipolar, and multipolar.

In the marginal bodies of the hydromedusan *Carmarina hastata*, Eimer describes the statocyst sitting over the ring canal (Fig. 5.5, left). And just underneath the ring canal he located one of the nerve rings. In another drawing (Fig. 5.5, right) he depicts what he calls the radial ganglion and the pin-like ganglion (*Spangenganglion*), which are the aggregations of ganglion cells connected with the lower and upper nerve rings, respectively. He located the nerves that connect the statocyst with the lower nerve ring.

Eimer also describes the intimate association between nerve cells and the strands of muscle fibres in the umbrella and velum of the swimming

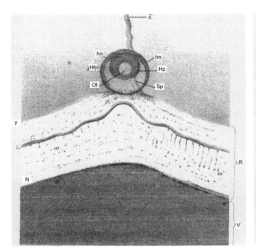 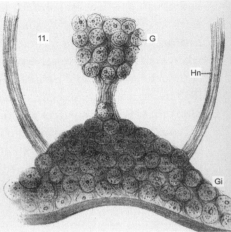

5.5 · (*Left*) Marginal body in the subumbrella of *Carmarina hastata* in which is found the marginal ring (R), and across the mesoglea and ring canal is found the nerve ring (N). Hbl, statocyst; hn, statocyst nerve; Hz, statocyst cells; n, radial nerve; Ot, otholith, Sp, radial ganglion; V, velum. (*Right*) Radial ganglion (Gi) from which the statocyst nerves (Hn) arise, and to which a pin-like ganglion (*spangenganglion*, G) is connected. From Eimer (1878), Plate VII, Fig. 2, and Plate IX, Fig. 11.

bell (Fig. 5.6, above). In this drawing he clearly shows the nerve cell processes reaching into the muscle fibre bundles. In another drawing he also shows that ganglion cells feed their processes into the nerve ring of which they are a constitutive part (Fig. 5.6, below).

Eimer did not hold back from playing out the significance of his observations. Foremost in his mind was the question of the origin of the nervous system and what the jellyfish findings could contribute to resolving this question. Eimer's taste for speculation pushed him to elaborate far beyond the point to which Romanes allowed himself to go, while his interpretations converged more closely with those of the Hertwigs. First, he emphasized that primitive larvae or animals retaining an embryonic frame as adults and lacking a nervous system are able to sense their environment and respond with some kind of motor activity. He asked whether the ectoderm epithelium was the seat of these pre-nervous "sensory-motor" goings-on. He drew an analogy with *Hydra*, which he characterized as "simply a gastrula-sac consisting

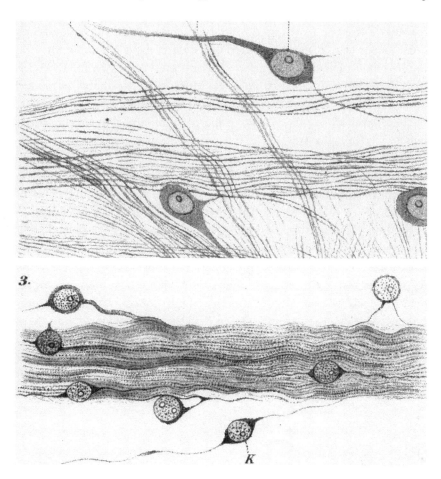

5.6 · (*Above*) Muscle and nerve elements next to the radial muscle sheet in the upper part of the velum of *Carmarina hastata*. (*Below*) One of the nerve rings of *C. hastata* with its associated ganglion cells. From Eimer (1878), Plate XI, Figs. 1 and 3.

of two layers, the neuromuscular-cells (the ectoderm) and digestive cells (the endoderm), and provided with prehensile arms." This overly simplistic characterization relied on Eimer's view that Kleinenberg's neuromuscular cell could be a precursor of the nerve cell. From his assumption "that ectodermal cells in the lower Metazoa are the seat of volition," he went on to assert that "a morphologically recognisable nervous system consisting of separate nerve-cells and nerve-fibres must, according to these considerations, in its most primitive form lie immediately beneath the epidermis, connected on the one hand with the latter, on the other with muscles. Moreover, it probably extended at first as a layer all over the body" (Eimer, 1878, 1890).

That "layer all over the body" Eimer saw in jellyfish, as represented in Fig. 5.5 (left) above. This layer, in his view, was the first step in the original construction of a nervous system. The second step, the clumping together of nerve-cells leading up to a brain rudiment, he saw played out in jellyfish, where it was seeded at the bell margin and in the vicinity of sense organs. Thus "brains connected by a ring-nerve" have their cells and fibres "demonstrably formed of ectoderm cells, which are connected with the ectoderm, but which have come to lie beneath it." He concludes with a general view of this evolution: "Thus we have in the Medusae a laminar central nervous system extending over the body, whose cells are commencing to concentrate at spots particularly suitable for communication with the external world, and there to form definite brains. In the higher Metazoa, in Vermes, Mollusca, Arthropoda and Vertebrata, these brains, or ganglia, are completely formed, but their embryonic condition still indicates the epiblast [ectoderm] as their original place of origin" (Eimer, 1890).

The Legacy of the Physiologists

One cannot overemphasize the importance that Romanes's contemporary colleagues attached to the experimental work he was conducting. England, as Geison (1978) persuasively demonstrates, had lagged behind Germany and France in developing strong academic programs in experimental physiology. The first seed to germinate in England, was the founding of the Cambridge School of Physiology by Michael Foster, and that impulse later spread to other British universities. That Romanes happened to yearn to do biological research at Cambridge shortly after Foster had established his laboratory there proved to be one of those remarkable historical coincidences that become fertile ground for scientific achievements. Another fertile coincidence was, of course, the rising awareness of the analogy of the jellyfish swimming beat with the question of the heart beat, which engrossed Foster and dominated his research program entirely. Thus, Foster and his disciples could not help but regard Romanes's investigations as pivotal for the interpretation of their own experimental findings on heart function. But as much as Romanes's disclosure of his results was awaited by the Cambridge physiologists, other investigators were no less interested in their paramount importance to zoology. Both Darwin and Thomas Huxley not only saw the importance of the jellyfish investigations for evolutionary theory but they also actively encouraged Romanes in his pursuit – Darwin as a father figure to his protégé and with vested interest in evolutionary thought, and Huxley as corresponding editor for the *Transactions of the Royal Society* and an expert on coelenterates himself.

Eimer, in contrast, was denied the nurturing environment that Romanes enjoyed. Although physiology as a biological discipline had flourished earlier in Germany than in Britain, its practising stars such as Carl Ludwig and Emil Du Bois-Reymont pursued interests remote from Eimer's, and Einer never connected academically with them or their discipline. Ludwig's Physiological Institute in Leipzig was devoted principally to cardio-vascular research of medical relevance, whereas Du Bois-Reymont's physiology Chair in Berlin developed the field of electrophysiology. So Eimer was already isolated from mainstream physiology when he began his experimental investigations of jellyfish. His sense of isolation, as alluded to earlier, extended also to his difficulties in obtaining his jellyfish specimens and was aggravated by the heavy demands of his academic position on his time. He lacked a network of colleagues with whom to share his excitement and preoccupations, so that he went from a sense of isolation to one of alienation. To make matters worse, his resentment grew as he became engaged in a publishing race with both Romanes and the Hertwig brothers. As a result of these accumulated frustrations, Eimer was unable to get the recognition he had anticipated for his jellyfish work, and the impact of his research became subordinated to that of Romanes.

The impact of Romanes's jellyfish investigations was far-reaching. The insights on the organization and functioning of the jellyfish nervous system that he gained from his shrewdly conducted experiments earned him the respect and admiration of his contemporaries. But more important, the validity of his results stood the test of time. No one has expressed this better than the future jellyfish electrophysiologist Leonard Passano:

Even though they were carried out 90 years ago there is still
a marvelous freshness about the discoveries of G.J. Romanes
concerning what we would now call the behavioral physiology
of coelenterates. When periodically rereading his three classic
papers in the Philosophical Transactions (187, 1877, 1880) or his
popular book (1885), culled from these papers, I am reminded
again of his achievement; indeed a disquieting thought (which
other investigators may have experienced also) is that Romanes
has already observed and recorded all of the facts of the matter,
and that as our knowledge of these animals continues to increase
it only sustains his original insights. The more I learn of jellyfish,
the more I understand what Romanes has to tell us of them.
(Passano, 1965)

The year that Passano's tribute was published also saw the publication of Bullock and Horridge's comprehensive treatise *Structure and
Function in the Nervous Systems of Invertebrates* (Horridge and Bullock,
1965), which was critically acclaimed by the academic community. The
co-authors had previously published on jellyfish, and in their chapter
on coelenterate nervous systems they conducted an exhaustive and authoritative review of the literature. It is clear from their remarks that
they were profoundly influenced by what they called the "classical
demonstrations" of Romanes and Eimer. Indeed, they regarded these
demonstrations as the solid foundation on which they and others built
their research programs. A relatively recent paper by Richard Satterlie
summing up our current knowledge of the control of swimming by
the jellyfish nervous system gives further resonance to the view that
Romanes unwittingly anticipated developments in the field:

Romanes (1885) was aware that the nerve net is made up of individual discrete neurons, and that these neurons share some type
of physiological connectivity throughout the nerve net. From his
writing, it is evident that similar physiological connectivity had
been "discovered" within the spinal cord of vertebrates, even
though the concept of the chemical synapse was not yet formulated. In Romanes' words, "... there can be scarcely any doubt
that some influence is communicated from a stimulated fibre *a*

to the adjacent fibre *b* at the point where these fibres come in close apposition." Then, "… much more remarkable does this fact become when we find that no two of these constituent fibres are histologically continuous with one another." Romanes recognized that "no anatomical continuity exists, but … physiological continuity is maintained by some process of physiological induction …," although he revealed that it was "premature to speculate" on the mechanism of physiological induction. Once again, Romanes' cutting experiments demonstrated the nonpolarized, diffuse nature of conduction through the subumbrellar nerve net (Romanes 1885) … Romanes (1885) found that neuromuscular transmission exhibits frequency-dependent facilitation, a fact that was elaborated by Bullock (1943), Pantin and Vianna Dias (1952), and Horridge (1956). (Satterlie, 2002)

Satterlie concludes his paper by inviting the reader to "simply envision what Romanes would do if he had had the tools of the twenty-first century at his fingertips." It would be hard to formulate a better tribute.

When Romanes published a popular book on his jellyfish investigations in 1885, his interest in pursuing research on jellyfish had already waned. However, his personal reflections on the jellyfish findings as well as his use of Darwin as a sounding board for his inquiries launched him on a scholarly journey that resulted in the publication of *Animal Intelligence* (1882) and *Mental Evolution in Animals* (1883), books that pioneered the field of comparative psychology. Soon afterward, he added a physiological dimension to Darwin's theory of natural selection by publishing a book called *Physiological Selection: An Additional Suggestion to the Origin of Species* (1886). He also wrote on Darwin and his legacy as well as on religion. His prodigious intellectual output was prematurely cut short by his death due to a brain tumour in 1894.

Eimer's intellectual output, in contrast, took a radically different turn after his jellyfish investigations parallel and independent to those of Romanes. The jellyfish work had already spurred philosophical musings that led Eimer to espouse Lamarck's theory of inherited characteristics and the biogenetic law articulated by the *Naturphilosoph* Lorenz Oken (1779–1851), according to which an organism is the result of combined environmental influences on the individual from conception.

This was a clear departure from the contemporary Darwinian outlook. Eimer later became a champion of the far-fetched theory of orthogenesis (directed evolution), as Churchill explains:

> Eimer extended his observations [on a subspecies of wall lizard in Capri] to the markings of birds, mammals, and especially butterflies. He consolidated his findings in *Die Entstehung der Arten* (1888–1901). In a highly speculative way, he argued that the formation of the organism was determined by the operation of four laws of growth: (1) that a directed evolutionary process is preceded by changes in the ontogeny of the individual; (2) that new characters first appear in the mature males and may be transmitted to the rest of the species by heredity – this is what Eimer called the law of male preponderance; (3) that these new characters usually appear at the posterior end or on the inferior side of the male; and (4) that varieties are simply sequential stages in the development of a species. (Churchill, 2008)

These unorthodox views earned Eimer the disapproval of many colleagues, including the leading evolutionist at the time, August Weismann, to whom Eimer had dedicated his jellyfish monograph. He spent his remaining years locked in often acerbic and bitter arguments with his opponents, and like Romanes died prematurely (at fifty-five) in 1898, from complications after surgery for a severe intestinal disorder.

CHAPTER SIX

Edward Schäfer, Jellyfish, and the Neuron Doctrine

> In the further course of evolution a certain number of these specialized cells sank below the general surface ... they became nerve cells.
>
> Edward A. Schäfer, 1912

The friendship struck between George Romanes and Edward Schäfer during Romanes's jellyfish project was pivotal. Not only did Schäfer help his friend provide neurohistological support for his experimental investigation of jellyfish swimming control but he was also instrumental in planting the first seed that led to what became known as the neuron doctrine. The neuron doctrine has been heralded as "the fundamental organising principle of the nervous system and the principle upon which all manifestations of nervous function depend" (Jones, 1994). Its historical importance is such that Schäfer's contribution deserves to be examined at length. But before being able to appreciate the role of Schäfer and other invertebrate neuroanatomists, we must first look at the historical background of the doctrine.

The Earliest Inklings of the Neuron Doctrine

What the early neuroanatomists saw through the microscope by way of nerve cells was often fuzzy, much constrained by the availability of appropriate histological processing methods and the optical quality of microscopes at the time. Not until the 1860s were lenses perfected enough, thanks to the collaborative efforts of Ernst Abbe, the optical science theoretician, and Carl Zeiss, the microscope maker, to discern cellular material clearly. Until then, understanding of the cellular organization of the nervous system was too limited to entertain any insightful thought about the connectivity of nerve cell processes (branching axons and dendrites) with each other. However, even without incontrovertible evidence at their fingertips, many were inclined, if only to show their adhesion to cell theory, to believe that nerve cell processes overlap, rather than anastomose, with each other.

Otto Friedrich Karl Deiters (1834–1863), a young German investigator lost to typhus in his prime at the age of twenty-nine, was the first to focus on the question of nerve cell connectivity in his posthumous monograph on the brain of mammals, including man (Deiters, 1865). Taking advantage of an improved method introduced by his mentor in Bonn, Max Schultze – which consisted in chemically fixing brain cells in osmium tetroxide, macerating and microdissecting them, and then staining them in carmine – Deiters was able to discriminate between the axon, as issuing from the soma (nerve cell body), and the dendrites, which looked like small extensions of the soma (protoplasmic processes). He also clearly stated that these processes never merged with processes from neighbouring nerve cells, thereby boosting the view that each nerve cell is a separate entity, as cell theory requires. But he was unable to resolve the discontinuity between dendrites and axon branches from another nerve cell abutting on these dendrites, so he viewed the complex as a continuous axon and in so doing gave his readers the impression that anastomoses occurred as predicted by the continuous network (or reticular) theory (Jones, 1994). In spite of Deiters's keen desire to settle the issue of how nerve cells physically transacted with each other, the ambiguities of his descriptions left the issue unresolved.

From this period forward, the network theory picked up momentum and found new adherents. The latter were influenced by the published observations of Joseph von Gerlach (1820–1896) of the University of Erlangen, on the spinal cord and cerebral cortex (Gerlach, 1872). Scholars largely agree that Gerlach interpreted his observations of nerve cells in these structures to mean that "the whole nervous system represents a protoplasmic *continuum* – a veritable *rete mirabile* [wonderful net]" (Shepherd, 1991). Gerlach found comfort in thinking that such a continuum would be an ideal medium to propagate the nerve impulse – the "action current" previously discovered by his compatriot

Du Bois-Reymont – without resistance. Another influential figure in this regard was Franz Leydig (1821–1908), mentioned earlier as a mentor of Theodor Eimer. Leydig's comparative work on a variety of invertebrates led him to view the neuropile at the core of their ganglia as a mass of nerve fibres in continuity with each other. In his highly popular histology textbook (Leydig, 1857) he referred to the neuropile as a molecular mass or dotted mass, and asserted that "the essential substance of the nerve centres consists of the molecular mass, in which smaller and larger cellular elements are embedded" (translation in Shepherd, 1991). His use of the words "mass" and "embedded" conveys the notion of an entanglement akin to a merging of elements.

Now if a luminary of Leydig's stature sided with the network camp, it is not surprising that other students of invertebrate nervous systems followed suit. Although Romanes, the Hertwig brothers, and Eimer did not hold strong views on the matter, it is implicit from their texts that they went along with those who supported the network theory. After all, Michael Foster, Romanes's mentor at Cambridge, was rather explicit in his textbook of physiology when he mentioned fusion between nerve cells, and between nerve and muscle cells, citing Eimer, the Hertwigs, and Romanes. The Hertwigs also noted in their sea anemone monograph that at least some of the ganglion cells they described were anastomosed. Thus, when Schäfer entered the fray, continuity between nerve cells seems to have been consensually accepted as dogma among coelenterate and other neuroanatomists. What started as a favour to his friend Romanes turned into a paradigm-changing discovery.

Schäfer Lends a Hand to Romanes

Edward Albert Schäfer (1850–1935) was born near London to a father who, although born in Hamburg, had become a naturalized Englishman before Edward's birth (Sparrow and Finger, 2001). He was a gifted student and after he enrolled at University College in 1868 his gifts were soon detected by Professor William Sharpey, who "greatly admired Schäfer's intellect and skill, and treated him more like a valued colleague than a student, despite their differences in age and experience" (Sparrow and Finger, 2001). Consequently, Schäfer became the first recipient of the newly established Sharpey Scholarship in 1871, and the twenty-one-year-old *Wunderkind* immediately began his

6.1 · Edward Albert Schäfer in 1923. Courtesy
National Library of Medicine.

research. Between 1872 and 1876 he published articles on anatomical/
histological and physiological topics of an eclectic nature. After graduating from the University College Hospital Medical School in 1874,
Schäfer was appointed assistant professor of physiology and worked
under John Burdon-Sanderson.

By the time Schäfer responded to Romanes's call for help and joined
his friend at Dunskaith in August 1877, he was a busy man. On the
domestic front he was engaged to marry Maud Dixey the following
year (Hill, 1935). Professionally, in addition to his academic duties and
research schedules at University College, he had just published his first
book, *A Course of Practical Histology* (Schäfer, 1877). If there had been

any need to bolster Schäfer's histological skills ahead of his work on jellyfish, the publication demands of this timely book would have been just the ticket. Although his future record represents him as one of the most distinguished physiologists of his period, Schäfer believed that histology was a necessary partner of physiology. As Leonard Hill (1935) notes: "Schafer held strongly to the view that histology should be part of physiological teaching and deplored the modern tendency to relegate it to the anatomist. Up to Sharpey's time Anatomy had been considered to be the foundation of medicine and surgery; the young Schafer pointed out that it was only justifiable if the object of medical education were to provide physicians not for the living, but for the dead."

Well, Schäfer put his belief to work quite early in his career (at twenty-seven) by supplementing Romanes's experimental results with his histological fireworks. At the time of his stay in Dunskaith, the scyphozoan *Aurelia aurita* was the best available jellyfish to catch, and since many of Romanes's experiments involved this species, it seemed a natural decision to give his attention to it. Besides, he wanted the investigation to progress quickly so that he could soon return to his hectic professional life in London. In approaching the task he first tried to determine which methods would yield the best results for microscopic visualization. He narrowed his choice to gold chloride and osmic acid, and in his experience with *Aurelia* the two chemicals turned out to be complementary:

The reagents which are ordinarily employed for the demonstration of nervous tissues, and especially the chloride of gold, bring the structures in question to view in the most striking manner possible. As in the higher animals, the nerve-fibres and the substance of the ganglion-cells become stained of a deep violet colour by this reagent, so that the fibres may be followed with ease over large tracts of the surface ... Osmic acid preparations lack this distinctiveness of colouration, so that the nerves are scarcely better exhibited in situ than in the fresh preparation. But whereas after treatment by the chloride of gold method the fibres appear markedly smaller than in the fresh tissue, and seem to have become somewhat shrunken, in the preparations made with osmic acid they preserve their original size and for the most part their pristine appearance. (Schäfer, 1878)

With the aid of these highly suitable techniques, Schäfer was able to readily survey the nervous system of *Aurelia*, distinguishing three components of its nervous system: the marginal bodies already identified by Romanes, the "nerve-epithelium" in the vicinity of the marginal bodies, and the subumbrellar plexus covering the entire undersurface of the muscle sheet. The latter was his prized discovery, a nerve plexus that had escaped the notice of his predecessors, including the Hertwig brothers. He felt confident in recognizing its nervous nature because the nerve fibres of the plexus looked similar to those of the sympathetic nervous system of vertebrates, including the swellings along the fibre that are now known as nerve varicosities. Even to a cautious investigator like Schäfer: "These appearances are so obvious as to allow of no question that we have before us undoubted nerve-fibres, and bipolar ganglion-cells" (Fig. 6.2, left). He described the orientation of the nerve fibres as overall radial along the main axis of the swimming bell. The nerve cell bodies had a "roundly fusiform" shape and possessed a spherical or ovoid nucleus. The end of the nerve fibre that projects to the muscle sheet expands into what Schäfer likened to "a primitive form of the motorial end plate." He noticed inclusions in these end plates that he erroneously associated with nuclei. The other end of the nerve fibre is sometimes bifid. He also saw similarly stained nerve cells with shorter fibres or altogether devoid of fibres, which he correctly characterized as being "in process of development into nerve-cells and nerve-fibres." These are known today as interstitial stem cells.

Schäfer described the lithocysts of *Aurelia* as "eight thumb-shaped projections at the circumference of the umbrella, each being situated in one of the marginal bodies and having a horizontal direction with a slight upward curvature." The projections appear as specialized ectodermal infoldings in which the gastrodermal canal becomes inserted (Fig. 6.2, right). The ectoderm and endoderm components of the lithocyst constitute what Schäfer called the nerve-epithelium. In the ectodermal part of the lithocyst, he was able to notice peculiar sensory cells, with a basal nerve cell body and a distal pigmented enlargement extending further distally into a "very long filament of exquisite fineness," probably a sensory cilium. These Schäfer suspected were light-sensitive cells. Teased from the endodermal part (gastroderm) of the lithocyst are otolith cells described previously by the Hertwigs and others.

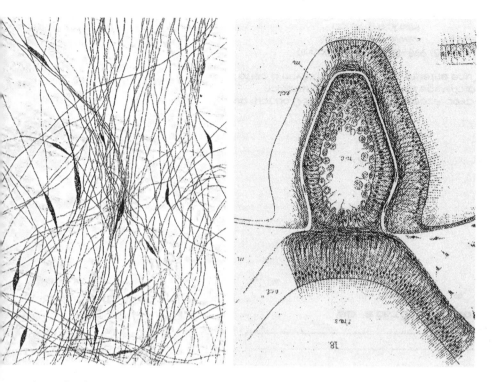

6.2 · (*Left*) A subumbrellar nerve plexus situated between two "nutritive tubes." "At certain parts the fibres come together to form wonderfully intricate interlacements" (Schäfer, 1879). (*Right*) Vertical section through the base of the lithocyst, showing the ectodermal covering of the lithocyst, the ectodermal covering of the upper surface of the umbrella and the nutritive canal or gastroderm slipping into the interior of the lithocyst. From Schäfer (1878). Plate 50, Fig. 7, and Plate 51, Fig. 10.

While Schäfer did not hold back from expressing his awe at the evolutionary implications of his findings, especially regarding the subumbrellar plexus, in the end he offered a far more tepid and mundane pronouncement than Romanes, Eimer, and the Hertwigs ever did. Perhaps he felt that his colleagues had said it all and he had little to add. He simply remarked: "It is interesting to observe, even so low down in the metazoic scale as the Medusae, that the textures, which in the higher animals are generally looked upon as the most highly differentiated, should have already attained a degree of structural complexity and of functional activity in many respects scarcely inferior to the nervous and muscular tissues of Vertebrates." He was more loquacious when reflecting on the possible function of the subumbrellar plexus. Its role, he proposed, was twofold. At a basic level, each nerve fibre served to connect a muscular field at the bell margin with another muscular field nearer the top (centre) of the bell. At a holistic (organismic) level, the profuse interlacement of the nerve fibres he observed would serve to ensure that nerve impulse transmission through this mass of fibres brings about a general coordination of the contractions and thereby a harmonious swimming response.

Schäfer and the Discontinuity of Nerve Fibres

Schäfer was taken aback when he actually saw under the microscope that the nerve fibres of *Aurelia* were not anatomically continuous with each other. Under the influence of the prevailing reticularist theory and of his friend Romanes, whose investigations supported a physiological continuity, he was prepared to see a corresponding anatomical continuity. He was seized enough by the importance of the observation to dedicate six drawings to the portrayal of a long nerve fibre as he followed it and neighbouring nerve fibres under the microscope, unable to witness any "protoplasmic" fusion between them (Fig. 6.3). The text of his accompanying description could hardly be more unequivocal and methodical:

> If we trace out the course of the individual nerve-fibres more
> closely [Fig. 6.3], we are struck with certain remarkable facts. In
> the first place, each fibre is entirely distinct from, and nowhere
> structurally continuous with, any other fibre. Secondly, each
> fibre is provided with a bipolar nerve-cell [Fig. 6.3, top row, left],
> which is interpolated in or near the centre of the fibre, each end
> of the fibre representing the prolongation of one of the poles
> of the nerve-cell. Thirdly, each nerve-fibre is of limited length
> (seldom exceeding 4 millims. from end to end), and in most cases
> tapers at either extremity to a gradual termination. Lastly, it may
> be mentioned that the fibres are rarely branched; and where they
> are so [Fig. 6.3, top row, middle] the branches do not join with
> other nerve-fibres, but after a longer or shorter course end in a
> tapering extremity like the unbranched fibres. (Schäfer, 1878)

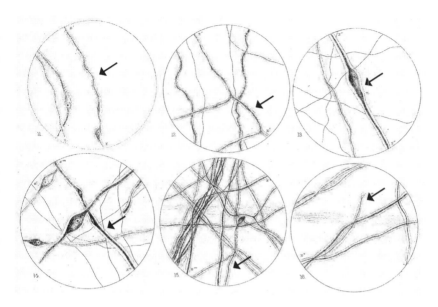

6.3 · Tracing the path of a nerve fibre (arrows) from the subumbrellar plexus of *Aurelia* in six successive visual fields of the microscope. Other nerve fibres come in the way and cross over each other. From Schäfer (1878), Plate 51, Figs. 11–16.

The observation was so unexpected and potentially so controversial as to raise fear in Schäfer that Romanes would not take it lying down. Possibly in an attempt to avoid a painful showdown with his friend, Schäfer strained to assess the significance of his observation in the best light possible:

> It seemed at first sight almost incredible that with such a prodigious number of nerve-fibres, exhibiting so close an interlacement, there should be no actual junctions of the intercrossing nerves. And it was especially difficult of credence because some of the experiments of Mr. Romanes, performed with the view of testing the amount of section which the tissue could endure without loss of nervous (or excitational) continuity, seemed to point to the existence of a structurally continuous network of nervefibres. Nevertheless, there can be no doubt that the fibres do not come into anatomical continuity. On the other hand, it can

readily be seen that each nerve-fibre comes at one or more points of its course into very close relations with other nerve-fibres. Two fibres, for example, may sometimes be observed to bend towards each other out of their previous course, in order to run closely side by side for a greater or less distance, and in such cases one fibre may hook round the other ... or they may be two or three times intertwined. At other places a number of fibres come together from different parts and join in a very close entanglement, the fibres in which run for the most part parallel ... and it is only with difficulty that the individual fibres can be followed. So that although there is no actual anatomical continuity, abundant opportunity is afforded for inductive action, whether electrical or of some other kind. That physiological continuity is thus maintained it seems as yet premature to conjecture. (Schäfer, 1978)

This passage reveals much about the possibilities that Schäfer allowed himself to muse upon, in the context of the science of his time. His use of the word "junctions" is ambiguous. At a junction two elements can meet and join, but not necessarily fuse with each other. While the word obviously carried the meaning of fusion in his mind, he does not explicitly raise the alternative of a physical contact between two "intercrossing nerves." And yet his drawings show numerous nerve fibres crossing over each other. These crossing points are now known to be en-passant synaptic junctions over which transmission of impulses from one nerve cell to another takes place, usually by chemical means. Later in the passage he allows for close contacts between nerve fibres to become transmission zones for inductive action, but only for fibres running parallel to each other. It is noteworthy that Schäfer does not confine inductive action from one fibre to another to some electrical conduit, suggesting "some other kind." Nowadays, of course, the other kind is known to be the dominant one: chemical transmission.

On hearing of his friend's histological results, Romanes indeed did not take it lying down. We owe our understanding of much of what transpired from Schäfer's travails and the publication of the *Aurelia* work to a well-researched article by Richard French, the major source of what follows (French, 1970b). In a letter to Schäfer dated 19 September 1877, shortly after Schäfer had completed the research and six weeks before he was to submit it to the *Philosophical Transactions,* Romanes drew the line in the sand: "I am much interested in what you say about the nerves of Aurelia. The subject is really most important but it will take a great deal of demonstration to convince me that, at any rate a large percentage of the fibres, do not communicate" (cited in French, 1970b, from the Sharpey-Schafer Collection, the Wellcome Institute of the History of Medicine). Even after he let himself be swayed by Schäfer's published paper, Romanes harboured some lingering doubt, as he expressed in his popular 1885 book: "But what the nature of the process may be whereby a disturbance in the excitable protoplasm of [fibre] *a* sets up a sympathetic disturbance in the anatomically separate protoplasm of [fibre] *b, supposing it to be really such* – this is a question concerning which it would as yet be premature to speculate" [italics mine].

It is no exaggeration to say that the publication process of Schäfer's paper was a saga of historical importance, as it touched on the pivotal question of the nature of his contribution to the origin of nervous systems as well as to the emergence of the neuron doctrine. The paper, relatively short in comparison with the near-monograph lengths of other neuroanatomists, was submitted to the *Philosophical Transactions* on 31 October 1877 through the channel of Schäfer's old mentor and Royal Society member, William Sharpey, and read at the Royal Society meeting of 10 January 1878. A week later two referees were assigned to evaluate the scientific quality of the paper and to make a recommendation on the advisability of publication in the journal. The first was E. Ray Lankester (1847–1929), a student of Thomas Huxley who trained in invertebrate zoology and evolutionary biology and became one of the most influential zoologists of his time. He had been appointed professor of zoology at University College in 1874; so, as a colleague of Schäfer in the same institution, he would by today's ethical standards have been considered in conflict of interest. Lankester acted as external referee and specialist in the discipline covered by the submitted paper. The second was Allen Thomson (1809–1884), a specialist of human embryology who had held the Chair of anatomy at the Medical School of the University of Glasgow until his retirement shortly before his assignment to Schäfer's paper. He was picked to referee the paper in his role as member of the Committee of Papers, as tradition required.

Lankester wasted no time rejecting the paper. In his report, submitted on 21 January, he basically argued that the nerve fibres seen by Schäfer could be construed by others to be connective tissue fibres. He attacked head on Schäfer's assertion of the individuality and discontinuity of the nerve fibres of *Aurelia*, thus reflecting a skepticism fed by the pervasive prevalence of network theory adherents: "Against the view that the fibres are nerves is the fact that they do not branch and that they end abruptly at each end without being connected with one another or with any other histological element. Hence if they are nerves they differ *toto coelo* from all other known nerves – and Mr. Schäfer has been obliged to propound a hazardous physiological theory to explain the manner in which they might possibly be conceived to function as nerves" (cited in French, 1970b).

In an ironic twist, the cooperation between Romanes and Schäfer, which had been aimed at finding anatomical support for the interpretation of physiological continuity in *Aurelia*, turned against them in the sense that Lankester saw in Schäfer's paper a repudiation of Romanes's findings. If Romanes was still skeptical of Schäfer's results at submission time, Lankester went beyond skepticism to a naked rejection of Schäfer's claim. Thomson, on the other hand, took a more sanguine view of the paper, both verbally at the meeting of the Committee of Papers convened on 21 February and in his written report, dated 8 March. Despite recommending publication in the *Philosophical Transactions*, however, Thomson expressed some reserve about Schäfer's interpretations:

> I may however remark 1st that the free termination of the fibres in question described by the author as occurring after a course of three of four millimetres, seems to have led him into speculations respecting the transmission of nervous influence between one fibre and another which may be deemed premature in the present state of the inquiry; and 2nd that in the examination of some of Mr. Schäfer's specimens, while some fibres are seen to end suddenly or by free terminations as described by him, there appeared to me to be others in which by fine subdivision the fibres were gradually lost in the substance of the surrounding tissues – a possibility obviously contemplated by the author himself. (Cited in French, 1970b)

In the end the committee, perhaps more persuaded by Lankester's negative report than by Thomson's and influenced by Thomas Huxley, who cast a giant shadow in the committee room, asked Schäfer in a letter dated 8 April to "withdraw the paper pending further research to confirm his conclusions" (French, 1970b). Indeed, Huxley had his own reservations about the conclusions of the paper and had no qualms about expressing them directly to Schäfer in a solicitous letter sent shortly after the committee had conveyed its decision: "I was not one of the referees and must not 'reveal the secrets of the prison house.' But my own private opinion rather inclined to the conclusion that the paper would be all the better for a renewed exploration of the 'nerve fibres' so as to make sure about their terminations. There is not the slightest desire on anybody's part to do other than what is best for you & the paper – nor any doubt about the importance of the observations" (cited by French, 1970b).

While this correspondence was bouncing around, the Hertwigs' monograph on the jellyfish nervous system reached publication. It did not take long for the monograph to make its rounds among the likes of Huxley, Lankester, and Romanes, in addition to Schäfer. It seems almost comical how suddenly, on perusing the Hertwigs' handiwork, they became convinced of the validity of the identification of nerve cells in jellyfish when just weeks earlier they had expressed their reservations to their own countryman. Now repentant – and relieved that corroboration was thus secured – they urged publication of Schäfer's paper without requesting modifications to his conclusions. So Schäfer prepared a postscript that he submitted on 16 October 1878. In it he explained that he had been preparing a review of the literature to accompany the paper as an appendix when news of the publication of the Hertwigs' monograph reached him. The Hertwigs' literature review was so comprehensive, he thought, as to make his own efforts superfluous and he abandoned the scheme. "As to the work of the Brothers Hertwig," he wrote, "it is difficult to do justice to the carefulness of the descriptions and illustrations, and the philosophical way in which the subject is treated" (Schäfer, 1878). This was the sentiment expressed also by many of his contemporaries. Finally, the saga ended to everyone's satisfaction, and Schäfer's paper appeared late in 1878, over a year after submission.

As French (1970b) remarked, the substance of the contributions of the Hertwig brothers and Schäfer hardly overlapped, as "the Hertwigs' paper did not, for example, describe anatomically discontinuous nerve fibres." But more important, Schäfer himself emphasized in the postscript of his paper that the Hertwigs only fleetingly covered the neuroanatomy of *Aurelia*, because of scarcity of specimens in the field, and the brothers concentrated on hydromedusae and less on scyphomedusae, to which *Aurelia* belongs. So even though the Hertwig monograph appeared prior to Schäfer's paper, it in no way diminished the originality of the latter.

The Earliest Evidence for the Neuron Doctrine

Although the importance of Schäfer's paper in challenging the network theory of protoplasmic continuity of nerve fibres did not escape Huxley and Lankester, it had no immediate impact on the wider biological science community. Huxley and Lankester were invertebrate zoologists, so one can argue that "Romanes' and Schäfer's work was simply too circumscribed by the restricted orbit of invertebrate physiology to have had any influence beyond that subject" (French, 1970b). Whether Schäfer shared this view has not been documented, but what is apparent from documents unearthed by French at the Wellcome Institute archives in London is that toward the end of his life Schäfer staked his claim to priority in no uncertain terms (1935). On the envelope containing the original of his 1878 paper Schäfer had scribbled: "So far as I know this paper contains the first account of a nervous system being formed of separate nerve units without anatomical continuity. Previously it was universally held that the nervous system was composed of networks of nerve fibres." So we may ask: what happened between the publication of Schäfer's paper in 1878 and Schäfer's priority claim in 1935 to the doctrine of which he appears to have been the precursor?

The neuroscientists who wrote the major historical analyses of the origin and development of the neuron doctrine never mentioned Schäfer's contribution (Shepherd, 1991; Jones, 1994, Guillery, 2005). It seems they never were aware of its existence and, as they were associated with medical schools, they were unlikely to have sifted through the invertebrate neurobiology literature in any depth. This is not to say, however, that they lacked knowledge of the invertebrate literature altogether, as I will show below. But if even these modern scholars failed to notice Schäfer's paper, it should be less surprising that it went unremarked by Schäfer's contemporaries as well. Nevertheless, these

6.4 · (*Left*) Wilhelm His around the time of his neuroembryological discoveries. Courtesy Wellcome Library, London. (*Right*) Auguste Forel around 1920. Courtesy National Library of Medicine.

three modern scholars remain reliable sources about the roots of the doctrine after Schäfer.

The consensus among modern scholars points to the significance of the work of three post-Schäfer neuroanatomists whose research led them to propose that nerve cells constitute separate, discontinuous units: Wilhelm His, Auguste Forel, and Fridtjof Nansen. Remarkably, the key papers of all three were published within months of each other. His (1831–1904), was Swiss-born but pursued his academic career mainly in Leipzig, where he founded the Institute of Anatomy. He is often credited with inventing the microtome around 1861, a tissue-slicing tool that allowed the reliable and precise production of sections thin enough for light to be transmitted through them during micro-scopic observation. It revolutionized the field of histology and led to landmark observations of the kind discussed here. His was quick to use his invention for studying the course of embryonic development at tissue level, thereby founding a new field called histogenesis. In 1887 His turned his attention to the development of the nervous system and nine years after the publication of Schäfer's paper, he published an article (His, 1887) in which he documented for the first time the growth of axonal processes from nerve cell precursors he called "neuroblasts," a term that remains in use today along with "dendrites," "neurite," and "neuropile," other terms coined by him.

In this article His uses embryological observations to surmise that nerve cells are discontinuous. In the following excerpts translated from German (Clarke and O'Malley, 1968; Louis, 2001), he describes the growth of the axon from the motor neuron cell body:

> Each nerve fibre originating from these cells emerges in the vicin-
> ity of a single nuclear pole with a conically enlarged base and
> shows distinct fibrillary striping at the origin. Besides the single
> axis cylinder [axon], other processes are barely present. The cell
> body only surpasses the nucleus by a little, and wherever it can
> be seen uncovered, it seems to be trimmed with blunt spikes
> [roots of dendrites?]. Therefore, the formation of branched end-
> ings happens much later in time than the formation of the axis
> cylinder. (His, 1887)

This type of observations set in motion a thought process that led to the deduction that nerve cells are separate units. At first His offers only a tentative explanation narrowly based on his own observations:

> If it is true that nerve fibres commonly develop as prolongations
> from ganglionic cells and then gradually advance into their
> central or peripheral target territories, the logical consequence is
> that all connections (other than the link to the parent cell), if they
> exist, must have developed secondarily.

But he then expands his argument by bringing in observations from others:

> We know ... that in an overwhelming large number of end
> structures [i.e., pacinian corpuscles and motor end-plates], the
> nerve fibres as such run outward freely ... and we shall ultimately
> have to accommodate to the idea that transmission of a stimulus
> without direct continuity is equally possible in these organs ...

I believe that one arrives at a simpler understanding of the central nervous organs when one gives up the idea that nerve fibres, in order to affect a part, must necessarily be in continuity with it. As to the course of a single fibre, the law of isolated conduction is incontestable.

And finally, His ends the article by penning a rather bold and assertive conclusion:

As a fixed principle I hereby sustain the following: that each nerve fibre originates as a prolongation from a single cell. The latter is its genetic, nutritive, and functional centre; all other connections of the fibre are either indirect or have formed secondarily.

Almost simultaneously, another Swiss neurologist came independently to the same conclusion. Auguste Forel (1848–1931) was born in a French-speaking canton of Switzerland (Vaud) but spent his entire career in Zurich where he became professor of psychiatry and director of the university-affiliated Bughölzli Psychiatric Hospital. He also became known as an ant specialist and social reformer on the side. His involvement in the development of the neuron doctrine has been documented in several papers (Akert, 1993; Parent, 2003). Forel hit on the notion of discontinuous nerve fibres via an entirely different approach than that used by His. André Parent (2003) explains:

Despite a very busy clinical schedule, Forel took the time to set up a research laboratory at the Bughölzli asylum and initiated a series of experimental studies on the origin of some cranial nerves in rodents. By using Gudden's [Forel's teacher] retrograde cell degeneration method, Forel discovered that the section of motor cranial nerves peripherally caused cell degeneration only in small and specific areas of the brainstem. In contrast to what could have been expected if the nervous system was organized as a reticulum – the dominant view at that time – this sectioning procedure did not affect adjoining neuronal networks. This finding, coupled with observations made on material stained by a method developed earlier by Camilo Golgi, led Forel to realize

that the fibres that he was sectioning belonged to single cells and
that both elements formed the fundamental unit of the nervous system.

Forel himself, in his posthumously published autobiography, vividly
narrated the circumstances of his insight into this matter:

During a holiday which was spent in Fisibach I also worked
out quite a different idea, relating to the anatomy of the brain,
which constituted a complete revolution of our views concerning
the connection of the nerve-elements in the brain. At that time
the problem was still quite obscure. We spoke of anastomoses between the ganglion-cells of the nervous system, without
really knowing how such connections between the elements
of cells which are originally quite independent of one another
could be effected. But recently the Italian anatomist, Professor
Golgi, had invented a new method of colouring, by which he
was able to show that the so-called protoplasmic processes of
the ganglion-cells are blind – that is, their terminations are not
connected to anything. However, this author was so completely
ensnared by the old notion of anastomoses that on showing
that the so-called nerve-processes of the ganglion-cells undergo
ramification, he now assumed a network of anastomoses for these
ramifications, and even drew it. In my laboratory we succeeded
in preparing specimens by Golgi's method, and we saw the blind
terminations of the protoplasmic processes, but not the network
of anastomoses connecting the nerve-processes.
    It was as though scales had fallen from my eyes. I asked myself
the question: "But why do we always look for anastomoses?
Could not the mere intimate contact of the protoplasmic processes of the nerve-cells effect the functional connection of nervous conduction just as well as absolute continuity? I considered
the findings of Gudden's atrophic method, and above all the
fact that total atrophy is always confined to the processes of the
same group of ganglion-cells, and does not extend to the remoter
elements in merely functional connection with them ... The more
I reflected, the clearer it seemed that we had hitherto been sunk
fathoms deep in a preconceived opinion. The longer I considered

it, the more untenable I found the theory of anastomoses. All the data supported the theory of simple contact, so at Fisibach, where I could be quiet, I decided to write a paper on the subject and risk advancing a new theory. I completed the paper, adding the evidence afforded by the experimental atrophies of the motor and sensory nerves (*Nervus facialis* and *trigeminus* – that is, the facial nerves of movement and sensation) and sent it to the *Archiv für Psychiatrie* in Berlin. (Forel, 1937)

In the published paper (Forel, 1887), Forel used the results of his retrograde cell degeneration studies to reiterate what he had mulled over in Fisibach; namely, "that all fibre systems or so-called fibre networks in the nervous system are nothing but the processes of single ganglion cells. This process emerges at the basis of the cell. It then emits nervous fibrils that arborize either close to (type II cells) or at long distances from the cell body by forming ramifications that closely intertwine but never really anastomose with one another." Forel was disappointed that his paper failed to attract any attention. He blamed two unfortunate circumstances. First, he felt that he should have given his theory a name, because "people always like names!" (Forel, 1937). But also he resented the slow pace of publication of the *Archiv für Psychiatrie*, which in his opinion cost him the priority of credit for what he then considered a "new" theory. As a consequence, His's paper had appeared before his own.

Forel wrote that, unknown to him, His had arrived at similar results and had published them in October 1886 and that his own paper had appeared only in January 1887, thus obliging him to concede priority to His (Forel, 1937). It is daunting that a peculiar publication process of the *Königlich Sächsischen Gesellschaft der Wissenschaften* in Leipzig led to this. Articles in this journal were published as soon as they were ready for the press (Michael Huebner, personal communication), so the article by His appeared, as Forel intimated, in October 1886. Once a critical number of these published articles was reached, they were assembled into a volume, in this case as volume 13 under the narrow auspices of the *Mathematisch-Physische Classe*, or volume 22 under the general umbrella of the Society, both in the year 1887. The *Archiv für Psychiatrie*, where Forel chose to publish his paper, had no such flexibility: the paper appeared for the first time along with other papers and only when the critical mass of papers was reached – which seemed to take forever, Forel felt. In fact, according to the archives of Springer-Verlag, the current publisher of *Archiv für Psychiatrie* (under a new name), the journal issue containing Forel's paper appeared in February 1887, not January as Forel asserted. In retrospect, given Schäfer's precedence, His's priority over Forel as to who provided the first evidence to support the neuron doctrine turned out to be of little consequence.

The third post-Schäfer neuroanatomist to make observations discrediting the network theory in the same year (1887) was none other than the polar explorer, oceanographer, and Nobel Peace Prize winner Fridtjof Nansen (Huntford, 1997; Edwards and Huntford, 1998; Whiteley, 2006). Born in Oslo, then known as Christiana, Nansen (1861–1930) grew up practising winter sports, particularly skiing, for which he devised several technical improvements. But he soon found his calling in marine zoology, with a specialization in the study of invertebrates. In 1882, at only twenty-one years of age, he was appointed curator of the Bergens Museum, a private institution devoted to research and educating the community in the natural sciences. He was in charge of its collections and was expected to work on their taxonomy, a task he soon found tedious. Under the influence of distinguished visitors to the museum, such as the German zoologist Willy Kükenthal, Nansen's interest turned to invertebrate neuroanatomy. Knowing that any headway in this field required the best microscope available, with the financial help of his father he purchased a top-of-the-line Zeiss microscope, equipped with an oil immersion lens that ensured great magnification without loss of resolution or unwanted increase in optical distortion.

After studying the nervous system of an obscure group (myzostomes), Nansen turned to ascidians and cyclostomes (Fig. 6.5, below) in order to trace the evolutionary transition of the nervous system from invertebrates to vertebrates. In the English translation of the first paper that resulted from that project (Nansen, 1886), he expressed his dissatisfaction with all the staining methods he tried. Furthermore, probably owing to technical limitations, Nansen could not resolve the issue of continuity versus discontinuity of nerve fibres, one of the stated objectives of the study. He is explicit about his difficulty: "Anastomoses or union between the different ganglion-cells by their processes ... I have been unable to demonstrate with certainty in the groups of animals investigated by me, at any rate as a rule." John S. Edwards and Roland

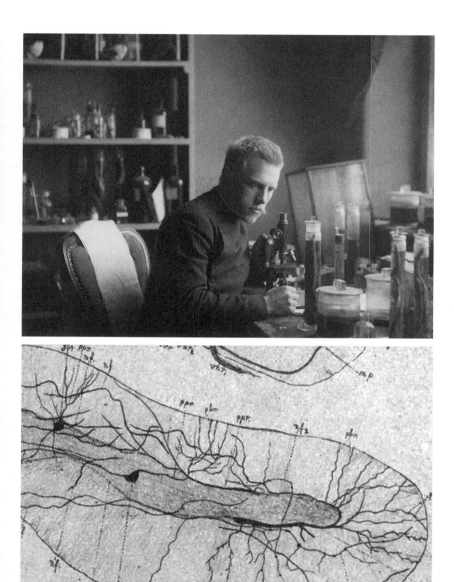

6.5 · (*Above*) Nansen at his Bergen Museum laboratory. From Huntford (1997). By permission of The Picture Collection, University of Bergen Library (photographer Johan v.d. Fehr). (*Below*) A lithograph of a transverse section through the spinal cord of the hagfish, *Myxine glutinosa*, stained by the Golgi method. From Nansen (1887), Plate 10, Fig. 94.

Huntford (1998) took this statement as an endorsement of the discontinuity theory, which led them to submit that Nansen should have had priority of credit over His and Forel. But their claim cannot be substantiated by a sentence that merely records a negative result without validating the discontinuity hypothesis.

Nansen had tried his hand at the Golgi staining method but, unlike Forel, he could not make it work for himself. So he made the bold move to travel to continental Europe in the winter and spring of 1886, visiting laboratories and institutions to break his isolation, but mainly to learn the Golgi technique from the master in Pavia, Italy. In a matter of days, writes Huntford (1997), "Nansen had learnt Golgi's method of staining – the 'reazione nera' – and, in his own words, 'never in my life had I imagined it was possible to prepare such elegant and distinct nerve sections.'" Propelled by his enthusiasm, he travelled from Pavia directly to Naples and managed to gain admission to Anton Dohrn's Stazione Zoologica, where he promptly applied the technique he had mastered on a variety of the marine forms present in abundance in the Bay of Naples.

After his return to Bergen, Nansen completed his anatomical descriptions and presented them as a doctoral thesis submitted to the University of Christiana. A version of the thesis was published in Norwegian, and another, written in English by Nansen himself, was published in December 1887 (Nansen, 1887). Following an extensive and critical literature review, Nansen set out his observations on a number of invertebrates (molluscs, polychaete worms, earthworms, and lobsters) as well as protovertebrates (Amphioxus and ascidians) and a primitive vertebrate (hagfish). Interestingly, he announced that he planned to cover the coelenterates in a subsequent paper, which never materialized. Nansen does not cite Schäfer's paper in his monograph; nor is there any evidence that he was ever aware of its existence, although he quotes both the jellyfish and sea anemone monographs of the Hertwig brothers. And it is doubtful that he had access to the papers by His and Forel either, as they were too recently published and scientific publications can only have trickled their way to Bergen, a European backwater for science in those days.

Nansen described the histological techniques he used in considerable detail. The methods that in his view worked best were Flemming's chromo-aceto-osmic acid fixative followed by staining with Delafield's hematoxylin and eosin, and what he called "the black *chromo-silver method* of Prof. Golgi," which he adapted with good results in the lobster and the hagfish. In this way he discovered that unipolar ganglion cells, sometimes quite large, are the dominant nerve cells of invertebrates. He also discovered that the axons of these cells, which he calls nerve tubes, are ensheathed by "neuroglia," glial support cells of the same ectodermal origin as the nerve cells themselves.

After describing the elements of the nervous system of the examined species in minute detail, Nansen dedicates five pages of his thesis to the continuity/discontinuity issue, calling it the problem of "the combination of the ganglion cells with each other." Right from the start he asserts that "a direct combination between the ganglion cells is, as we have seen, not acceptable." He provides substantive and sound reasoning, and his refutation is far from a dogmatic exercise:

In spite of the most persevering investigations I have not been able to find any direct anastomosis of indubitable nature between the processes of the ganglion cells. Where I thought to have found an anastomosis it always on application of the strongest lenses resolved itself into an optical illusion ...

If a direct combination is the common mode of combination between the cells, as most authors suppose, direct anastomoses between their processes ought, of course, to be quite common. When one has examined so many preparations (stained by the most perfect methods) as I have, without finding one anastomosis of indubitable nature, I think one must be entitled to say, that *direct anastomosis between the processes of the ganglion cells does not exist, as a rule.* What previous writers have supposed to be anastomoses is, in my opinion, probably the neuroglia-reticulation generally extending between the ganglion cells, and the fibres of which are often difficult to distinguish from the processes of the latter.

Another objection against a direct combination, and which does not seem to have been thought of by a great many authors, is the existence of unipolar ganglion cells. How is it possible to explain an existence of unipolar ganglion cells, when we believe in a direct combination by anastomosing processes? (Nansen, 1887)

Indeed, it was the size and unipolarity of the ganglion cells of higher invertebrates (crustaceans and molluscs) that allowed their meaningful contribution to this debate. Their large size was such that the discontinuity of their axon was easier to resolve unequivocally with a reasonably powerful microscope. And it was true, as intimated by Nansen, that the premise on which the reticularists or continuity proponents based their argument was the bipolarity or multipolarity of ganglion cells. Bipolarity and even more so tri- or quadripolarity were considered the best cell designs from which to build a network of anastomosing nerve fibres. Nansen here had more solid grounds to refute the reticular or network theory than either His or Forel. While Nansen provided persuasive, direct demonstration of discontinuity, His could only suggest that discontinuity was supported by the fact that nerve cells are individual and separate units early in their development. But what if the same nerve cells, as they mature, have grown nerve fibres that become entangled with other nerve fibres with which they eventually fuse? His failed to address this conundrum directly, thus weakening his position. Similarly, Forel's experimental hypothesis that cutting a peripheral nerve should result in the widespread degeneration of nerve cells in the brain if the continuity theory held true provided at best indirect, circumstantial evidence in favour of discontinuity. These ambiguities may have contributed to the overshadowing of their work by Ramón y Cajal. Forel, however, viewed the neglect of his own and His's contributions in a different light, as he commented churlishly: "our two papers suffered the fate of the majority of new ideas: they were simply ignored" (Forel, 1937).

The Contributions of Santiago Ramón y Cajal and Wilhelm Waldeyer

It belonged to Santiago Ramón y Cajal (1852–1934) to skillfully assemble a body of observations of such scope as to sway the neurological community at large to the side of the discontinuity theory and lead the way toward the most insightful expression of the neuron doctrine. My goal here is not to enter into the details of Ramón y Cajal's legacy as the universally acknowledged founder of modern neuroscience; numerous scholars have done so more eloquently than I would (Shepherd, 1991; Sotelo, 2003; López-Muñoz, 2006, to cite but three). He was awarded the Nobel Prize in Physiology or Medicine in 1906 for his achievement, jointly with the originator of the technique he used, Camillo Golgi.

With regard to the continuity/discontinuity debate, Ramón y Cajal entered it by the same gate as Forel and Nansen, through the mediation of Golgi's staining method. In 1887, a year after Forel and Nansen started applying it, the Spaniard, then professor of anatomy at the University of Valencia, became acquainted with the method while visiting a colleague in Madrid, Luis Simarro, who was dabbling with it. Awestruck "by the beauty and revelatory power of the method," he immediately began a relentless mission to "unravel the complexity of the brain and to deal with the question of how nerve cells communicate" (Jones, 1994). He understood that Golgi's method was his ticket for coming up with direct evidence that nerve fibres end freely without fusing with other processes. Although he became aware of the work of his predecessors His and Forel (but not Nansen) only later, he reflected on the severe limitations of their contributions by stating: "To settle the question [of contiguity *versus* continuity] definitely, it was necessary to demonstrate clearly, precisely, and indisputably the final ramifications of the central nerve fibres, which no one had seen, and to determine which parts of the cells made the imagined contacts" (Ramón y Cajal, 1937).

At first, things did not go well. Ramón y Cajal had to deal with the known capriciousness of the technique and he did so by tinkering with the strength of the silver nitrate impregnation and the schedules, and by selecting material from small or newly born mammals and from birds, which afforded a clearer view of less complex nerve process branching. From his very first publication in a periodical he created as an outlet for disseminating his work, he produced striking lithographs of the cerebella of birds (chicken, pigeon) in which fine details of the nerve cells (Purkinje, stellate, and basket cells) are rendered, and the course of their processes and their ramifications can be distinctly followed (Ramón y Cajal, 1888) (Fig. 6.6, below). In this seminal paper, followed by others of a similar quality on the retina and other brain centres, Ramón y Cajal right away dedicated a section on "Conexiones de los elementos cerebelosos" in which he unabashedly reported that, however much one follows the fine ends of the processes into the white substance of the cerebellum, no anastomosis is observed. On the

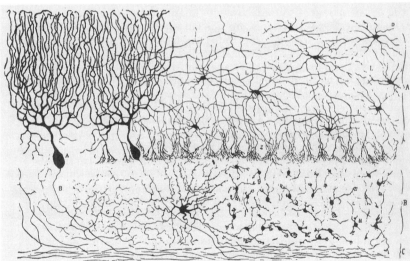

6.6 · (*Above*) Ramón y Cajal in his histology laboratory. (*Below*) The first lithograph of his first publication in May 1888, showing a vertical section through a circumvolution of the cerebellum of the chicken. From Shepherd (1991). By permission of Oxford University Press, USA.

contrary, one clearly sees nerve terminals or contacts (conexiones), later to be called synapses by Charles Sherrington.

No matter how remarkable his anatomical work, in order to gain recognition Ramón y Cajal needed to break the confines of language and his isolation from the main currents of biological and medical research, which at the time was centred in Germany. The opportunity arose in 1889, when he attended the Congress of the German Anatomical Society in Berlin. The pivotal personality in bringing this about was Albrecht von Kölliker (1817–1905), a Swiss citizen who spent most of his academic career in Würzburg, Germany. Kölliker was one of the most influential anatomists in Europe, who decades earlier had written an authoritative textbook of histology that went through several editions and was translated into many languages. Here he recalls ten years later how he crossed paths with Ramón y Cajal:

After visiting Golgi in Pavia in the spring of 1887, and getting acquainted with his methods and preparations, I no longer hesitated to try the method myself ...

These truly and extremely important observations of Golgi's were virtually unknown and unobserved except in Italy; the sole exception being A. Forel who mentions them in the Archiv für Psychiatrie, 1887 [and Nansen, not mentioned]. I therefore take the credit of having made the importance of these contributions known, after first confirming them through my own experiments. Yet, I must permit myself to object immediately to two important statements of Golgi's, namely a) that the protoplasmic continuations of nerve cells are not of a nervous nature and b) that a tangled net of nerves, of axon cylinder continuations and ramified branches consisting of nerve threads of a strictly sensible nature, interlace the whole gray substance.

During further tests a new and vigorous champion soon appeared, D. Santiago Ramón y Cajal, who participated in 1889 at the International Medical Congress in Berlin where he displayed a series of extremely excellent preparations, specifically of the spinal cord. It appeared to me to be an important task to acquaint those knowledgeable in Spanish, and unfamiliar with the German language, with our anatomists, of whom I especially

name His, Flechsig, Waldeyer and Schwalbe. (Kölliker, 1899,
translated in Anderson and Anderson, 1993)

Through Kölliker's account, we learn how His learned of Ramón y
Cajal's work and how the Spaniard was pursuing in new directions what
His had begun. But more important, the account fixes the moment
when Waldeyer focused his attention on the new developments in the
study of nerve cells and firmed up his conceptualization of the neuron
doctrine. In fact, before Ramón y Cajal had the time to fashion his
own thoughts on the matter, which he did for his Croonian Lecture of
1894 (Ramón y Cajal, 1894), Waldeyer had beaten him to it. Wilhelm
Waldeyer (1836–1921), director at the time of the prestigious Anatomical Institute of the University of Berlin and editor of the *Zeitschrift
für Anatomie und Entwicklungsgeschichte*, was celebrated for his mastery
of several disciplines and his ability to digest and synthesize scientific
information (Jones, 1994; Winkelmann, 2007). In the fall of 1891 Waldeyer published a series of six brief articles in the *Deutsche Medizinische
Wochenschrift* (German Medical Weekly), a popular journal that kept
a wide readership abreast of the latest research in a variety of medical
and biological disciplines (Waldeyer, 1891).

Into these articles Waldeyer dropped capsules on "the structure of
cells in the mammalian central nervous system; the terminations of
peripheral nerves; the structure of invertebrate ganglia; the development of the neuroblast; the nature of neuroglial cells ... and the newly
emerging idea of the nerve cell as a component of specific pathways
in the central nervous system, each composed of a series of ascending
and descending linked chains" (Jones, 1994). It seemed as if all these
articles led up to a climax in the final one, published on 10 December
1891, in which the doctrine was born and the term "neuron" was first
introduced in scientific literature. Waldeyer emphasized the importance of the following key paragraph by highlighting it with a different
type font in the original paper:

The nervous system consists of numerous nerve units (neurons)
that are neither anatomically nor genetically connected with
one another. Each nerve unit is composed of three parts: the
nerve cell, the nerve fibre and the fibre arborization (terminal
arborization). The path of physiological conduction can go in the

direction from the cell to the fibre arborization or in the reverse direction. The motor conduction occurs only in direction from the cell to the fibre arborization, the sensory now in the one, now in the other direction. (Waldeyer, 1891)

Waldeyer here not only acted as "midwife" for the birth of the neuron doctrine but also made a statement on dynamic polarization (that is, the direction of impulse propagation), a subject that was later revisited and expanded by Ramón y Cajal (1894), leading to his drafting of the Law of Dynamic Polarization. Despite the exposure given to the neuron doctrine by Waldeyer's exercise of popularization and through the efforts of Ramón y Cajal, reticular theory still held fast in part of the scientific community for decades to come. One reason for this was the discovery of neurofibrils within nerve fibres and in nerve terminals, thanks to the introduction of reduced silver nitrate to the Golgi method, a development in which, ironically, Cajal himself had participated (Jones, 1994). Under the microscope the neurofibrils in terminals seemed to continue into the contiguous neurons, thus giving the reticularists ammunition against discontinuity. It did not help, of course, that supporters of the neuron doctrine could not shore up evidence of means by which nerve conduction is transmitted at contact zones. So the controversy festered until the 1950s when electron microscopy ultimately supplied the necessary power of magnification to expose the gap between membranes of contacting neurons (Palade, 1954; De Robertis and Bennett, 1955), thereby dealing the decisive and fatal blow to the reticular theory.

The Vindication of Edward Schäfer

In one of the great ironic twists in the history of science, it fell to Schäfer, the first to provide evidence of discontinuity among nerve cells, to be the earliest among physiologists and anatomists to broadcast Waldeyer's neuron doctrine to a wider (Anglo-Saxon) community. Since his jellyfish work of 1878, Schäfer's career had progressed considerably, especially after his appointment as Jodrell Professor of Physiology at University College, London, in 1883. His new role signalled a shift of research interest from anatomy/histology to physiology, although he continued to give importance to histology as a complementary support to physiology. Between 1884 and 1888 his laboratory had undertaken a series of studies on brain localization of function which earned him a reputation as a neurologist to be reckoned with (Sparrow and Finger, 2001). As a result, he was invited to submit reviews on the subject to journals and books. Less than two years after the publication of Waldeyer's review, Schäfer responded by discussing the newly spawned neuron doctrine in two publications (Schäfer, 1893a, 1893b).

In the first of these publications (Schäfer, 1893a) he acknowledged Waldeyer's seminal contribution to the promotion of the unity of the nerve cell: "I do not know why one should restrict the term 'nerve-cell' to the body of the cell and thus exclude from that term the nerve-processes. This is not done for any other kind of cell, and it appears to me that the custom that has hitherto prevailed with regard to nerve-cells in this matter is not only inadvisable but even misleading. Waldeyer [1891] has used the term 'neuron' in this way to denote the whole nerve-cell, including all its processes."

In the second publication, a book-length anatomical description of the central nervous system (Schäfer, 1893b), Schäfer relies heavily on Ramón y Cajal for descriptions of the cellular origin of nerve cells during development and for mapping the distribution of neurons in the spinal cord and different parts of the brain. In addition, Schäfer reproduced no fewer than fifteen lithographs from publications by the Spaniard.

In none of these publications did Schäfer mention his own contribution to the neuron doctrine as having been couched in his 1878 jellyfish paper. This is strange and difficult to account for. Was it an oversight or a deliberate omission? The possibility of oversight should be dismissed in view of the explicit statement in the 1878 paper that the nerve cells in the subumbrellar plexus are independent and consequently that some inductive action could account for signal transmission between nerve cells, a finding that so affected Schäfer that he remained conscious of it to the end of his life (as seen above). Then, if it was a deliberate omission, why did he choose to remain quiet on the subject? This question will never be definitively answered for lack of "incriminating" evidence, although it is quite possible from what we know of his character that Schäfer had no ambition for priority. But it is clear that by not acknowledging his contribution publicly, he destined it to be overlooked by his peers, even though it was published in a distinguished and well-read journal.

One of Schäfer's peers, however, who happened to be the most distinguished British neurophysiologist of his day, was privy to Schäfer's jellyfish work from early on, according to a historical analysis presented by French (1970b). Charles S. Sherrington (1857–1952) studied under Michael Foster and Walter Gaskell at Cambridge in the early 1880s, and it is likely that the buzz created by Romanes's and Schäfer's works within those walls, only two or three years earlier, had reverberated in Sherrington's ears. In the seventh edition of Foster's *Textbook of Physiology* (Foster, 1897) Sherrington introduced the word "synapsis" to designate the process of contact at the junction between two neurons (Tansey, 1997). As Schäfer's own physiology textbook was being prepared, he entered into a lengthy correspondence with Sherrington about what to call the junction itself where synapsis occurs. It was then that Sherrington prevailed by proposing the logical offshoot of synapsis, the word "synapse" which is found on page 607 of the book, in the chapter by Schäfer on the nerve cell (Schäfer, 1900). Sherrington himself contributed four chapters to the book, and in the chapter on the spinal cord he refers specifically to Schäfer's 1878 paper as well as to Romanes. This is evidence enough of how well acquainted Sherrington was with the seminal contribution of Schäfer to the advocacy of the nerve cell as an independent entity. It is French's (1970b) contention that Sherrington's early exposure to the jellyfish work of Romanes and Schäfer prepared him to quickly absorb the views of Ramón y Cajal and Waldeyer in the late 1880s and early 1890s and to provide a comparative scope for the masterwork that above everything else earned him the Nobel Prize in Physiology or Medicine (Sherrington, 1906).

Soon after Schäfer acknowledged in print the contributions of Ramón y Cajal and Waldeyer, his research interests turned away from neurology thanks to his discovery that extracts of the medulla of the adrenal gland contracted the wall of arteries and raised blood pressure (Oliver and Schäfer, 1894, 1895). The active agent in the extracts was later identified as adrenaline. This discovery led in short order to Schäfer expounding a theory of internal secretion and to the emergence of a new physiological science, endocrinology (Borell, 1978; Sparrow and Finger, 2001). It also opened the door to the discovery of the autonomic nervous system and thereby to the notion of neuronal communication by chemical transmission. At the turn of the century Schäfer published his *Text-book of Physiology*, which went through several editions and competed for student readership with the corresponding book by Michael Foster. Some have considered Schäfer the equal of Foster in encouraging the development of physiology in Britain and its eventual catch-up with continental Europe (Germany, France, and Italy). It can be said that Schäfer accomplished this by dint of personal investment in research and active recruitment of promising students, whereas Foster did it more by acting as a promoter, creating the Cambridge School. Foster was sucked too early and fully into teaching and administrative duties, leaving him too little time for direct involvement in research. In a letter he warned Schäfer against following his example: "don't do too much lecturing – it *destroys a man* as I know. I have been driven to lecturing from my youth upward. You are not *obliged* to; don't do it. Give all your energy to research" (cited by Geison, 1978). There is every evidence that Schäfer heeded Foster's advice.

Although Schäfer's career as an experimental physiologist and scholar is exemplary, his personal life suffered terrible tragedy. His wife Maud died prematurely in 1896, after eighteen years of marriage, in which they had raised four children. It is believed that this loss was a motive for his removal from University College to the Chair of physiology at Edinburgh University in 1899. The following year, he remarried, choosing Ethel Maude Roberts, the daughter of a fellow scholar (Sparrow and Finger, 2001), as his second wife. As the First World War was approaching, and early in the conflict, Schäfer fell victim to anti-German prejudice in Great Britain. It did not seem to matter that in 1916 his two sons paid the ultimate price for patriotism when they died in combat. To compound his distress, one of his daughters also suffered a traumatic death in the same year, 1916 (Sparrow and Finger, 2001). He was left with one remaining daughter. To avoid any further opprobrium resulting from his German descent, Schäfer decided to anglicize his name as soon as the war ended. He did it by dropping the umlaut from his original name and adding the name of his beloved mentor; from now on he would answer to the last name Sharpey-Schafer. He retired from Edinburgh University in 1933 and died two years later at the age of eighty-five.

In an obituary paying tribute to the man he considered the leader of British physiological research, Sherrington revisited Schäfer's jellyfish paper, remarking that despite later papers by others that revived the "nervous syncytial continuum theory" in coelenterates, Schäfer's contribution was vindicated by the time of his death (Sherrington, 1935). Schäfer's seminal contribution to the neuron doctrine was not his only contribution to suffer lack of recognition; his textbook contributions were similarly under-recognized, according to Sherrington (1935): "Thus it was so with some of the descriptions of tract degenerations in his neurological volume of Quain's *Anatomy*. This unusual course led more than once to discussions on priority with continental workers, who, however, replied that they had not yet seen the textbook; whereat he would remark to his friends, 'But they ought!' with an amused smile."

Perhaps they – the continental workers – should also have seen Schäfer's *Philosophical Transactions* paper in 1878, but having not done so, they had a lower estimation of Schäfer's scholarly professionalism than he deserved. The story does not say if Schäfer really found it amusing, but one is inclined to doubt it.

CHAPTER SEVEN

In the Footsteps of the Giants

It is amidst compact tissues, between epithelial, muscular, glandular and sting cells, that the observer must find nerve cells which have no salient feature but their fine extensions lost in an inextricable mess.

Jules Havet, 1901

Following along the trails blazed by such outstanding investigators as the Hertwigs, George Romanes, Theodor Eimer, and Edward Schäfer, the time period around the turn of the century brought forward a fresh generation of academics who revisited the neuroanatomy of coelenterates, using new methods and introducing a greater variety of species. These new scholars emphasized the comparison of species and classes of coelenterates rather than the search for model organisms to address key evolutionary and functional issues in the manner of their predecessors. The great majority of the new generation, overwhelmingly European, flocked to the popular Naples Zoological Station, already a favourite of their predecessors for its bewildering marine diversity, its well-suiting research logistics, and its convivial atmosphere. None of these new contributors, however, made it a lifelong career to investigate coelenterates, and some of them stopped investigating coelenterate nervous systems right after their doctoral thesis. The long shadow cast by the emergence of the neurosyncytium *versus* neuron controversy enveloped them as much as the rest of the neurobiological community, and in some cases it coloured the evolutionary interpretations of their findings.

Attending to Distant Relatives: Ctenophores

Although coelenterates were meant to include the phyla Cnidaria and Ctenophora from the very first use of the word, we have seen that the pioneering neuroanatomical studies tended to concentrate on cnidarian species that were more familiar to the layman, particularly jellyfish but also sea anemones. Ctenophores were lumped in with the cnidarians – especially the equally pelagic jellyfish – on the basis of a few shared traits. However, what separates them seems more striking than their common characteristics, especially when we consider the implications for the organization of the nervous system. Unlike jellyfish, ctenophores lack stinging cells, but more important, they rely on eight ciliary comb rows radiating over their often gooseberry-shaped bodies to move in the water column. While jellyfish use contractile pulsations of their bell-shaped bodies to propel themselves, ctenophores use the beating of the cilia forming their eight combs in the manner of rowing teams to slide in the water; this is how they came to be known as comb-jellies.

Before ctenophores attracted the attention of neuroanatomists, they had been used as experimental models by embryologists interested in cell patterning during early development (Kowalevsky, 1866). The work of Alexander Kowalevsky (1840–1901) and others spurred some zoologists to give attention to other aspects of these animals, as even their descriptive anatomy was poorly studied. The first to look seriously at the nervous system of ctenophores was Eimer, in a study that describes comprehensively the anatomy of a comb-jelly, including a lengthy description of the nervous system that predates his monograph on jellyfish (Eimer, 1873). Interestingly, the first citation by Theodor Eimer in this, his first monograph, is Kowalevsky's seminal article, a clear indication that Kowalevsky, who did his embryological work in Naples, was the inspiration for Eimer's interest in ctenophores. Eimer collected the ctenophore *Beroë ovatus*, on which all his anatomical and physiological investigations were made, around the island of Capri where he frequently vacationed with his family after illness had initially brought him there to assist his recovery.

Eimer (1873) found that the optimum method for showcasing their nerve fibres was to fix the delicate comb-jellies in a diluted (0.2%) solution of acetic acid, and use specific gold chloride preparations. What he saw under the microscope fit his concept of how the first nervous systems evolved, with the premise borrowed from Nicolaus Kleinenberg, later invalidated, that *Hydra* does not possess a distinctive nervous system:

The question how the Hydra can exercise volition [motor command] as it certainly does, without a separate nervous system is not discussed by Kleinenberg. Yet, when we observe the motions and general behaviour of many free-swimming ciliated larvae, the conviction is forced upon us that although they consist only of epiblast [ectoderm] and hypoblast [endoderm], they pursue definite purposes, that their actions are in some degree directed by a will. It must therefore be assumed, whether nervous cells are somewhere present in Hydra or not, that ectoderm cells in the lower Metazoa are the seat of volition. But a morphologically recognisable nervous system consisting of separate nerve-cells and nerve-fibres must, according to these considerations, in its most primitive form lie immediately beneath the epidermis, connected on the one hand with the latter, on the other with muscles. Moreover, it probably extended at first as a layer all over the body. (Eimer, 1890)

This is precisely what Eimer had recorded in his 1873 monograph on *Beroë*. In it he depicts a two-dimensional sheet spreading all over the body, in which ganglion cells and fibres lie just underneath the ectoderm, with nerve processes crossing over each other in a nerve-net fashion and continuing through the jelly-like mesoglea where connective fibres reside, to terminate on protruding muscle fibre processes (Fig. 7.1, left). Multipolar stellate nerve cells are clearly illustrated, with their processes showing a series of bead-like swellings; hence their appellation "varicose fibres." However, Eimer seemed baffled by the look of these varicosities, wondering if they contained a nucleus, in which case one would end up with a string of anastomosed nerve cells ("subepidermal nervea"). This, of course, would have been consonant with the prevailing view of the 1870s – syncytial nerve net – to which Eimer adhered and which he addressed directly in his monograph. He concluded that "these primitive nerves look like chains of nuclei connected by conducting fibrils, and resemble a series of telegraph stations connected together for the purpose of renewing or strengthening the current at the beginning of each stage" (Eimer, 1890).

Eimer was searching for anything that smacked of a budding centralization of the nervous system, an aspiring brain of sorts. To him the greater density of nerve cells at the aboral end of the body, where sen-

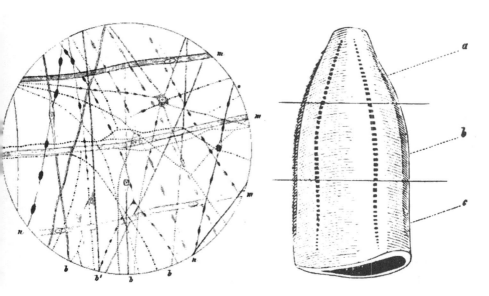

7.1 · (*Left*) View of layer between mesoglea and muscle sheet near the aboral pole of the ctenophore *Beroë ovatus*. N, nerve; m, muscle; b, connective fibres. From Eimer (1873). (*Right*) The cutting experiments conducted by Eimer (1880) on living *Beroë ovatus*. See explanations in text. Courtesy Springer-Verlag.

sory organs are located (statocysts), represented an incipient process of centralization. Likewise, he observed a condensation of nerve fibres forming bundles alongside the ciliary furrow associated with each of the eight comb-rows. He suggested that these bundles might be involved in the control of comb-plate ciliary activity. Another condensation of nerve fibres was visible around the lip of the mouth (perioral nerve tract), which appeared to be closely associated with the circular muscle acting as a sphincter there. With the hindsight afforded by his later observations on jellyfish, Eimer concluded that the nervous system of comb-jellies had reached a less advanced stage of development than that of jellyfish.

Eimer wished to support the neuroanatomy of *Beroë* with physiological experiments as much as he did with jellyfish, but it took him seven more years to see the results in print (Eimer 1880), likely on account of his lack of opportunity to organize fieldwork. In the fall of 1877 he managed to secure a table at the Zoological Station of Naples for that purpose. There he conducted a series of cutting experiments on living *Beroë* specimens, which yielded results largely similar to those obtained for jellyfish:

> *Experiment A.* – I cut five Beroës into three parts transversely so as to form from each three parts of equal height, of which the upper, *a*, contained the aboral pole, with the largest aggregation of ganglion-cells; the lowest, *c*, contained the mouth; while the third, *b*, was the middle portion of the body [Fig. 7.1, right].
>
> After the division, the movement of the swimming-plates [comb-rows] ceased completely in all the pieces, but recommenced after a short time in all those pieces which contained an aboral sense-organ. Shortly afterwards, movement began again in those portions distinguished as *b* and *c*, but in these ceased again after a time, while in the *a* pieces it continued ...
>
> *Experiment B.* – I cut through one of the rows of plates of a Beroë and the subjacent tissue, with a pair of scissors, at a point about 2 cm. below the aboral pole, at the level of *x*. The movements of the plates ceased for a moment in the whole animal. Then the movement recommenced first in the uninjured rows of plates, next in the upper portion of the divided row, *a*, and lastly in the lower portion of the same, *b*. After it was re-established everywhere, it appeared that the motion in the two portions of the row operated upon, above and below the cut, went on independently. Both in *a* and *b* it was rapid, in *b* it could even attain greater rapidity than in *a* ... The fact that the waves in the two sections were independent was established not only by mere observation, but certainly by the following experiment: when I lightly touched the upper portion with a needle the movement ceased in it for a moment, while in *b* it went on as before – and similarly the movement could be stopped in the lower portion while it continued in the upper. (Eimer, 1880, translated in Eimer, 1890)

In this ability of motor activity to surmount surgical obstacles Eimer saw the workings of the underlying syncytial nerve net he had described in 1873. Such a diffuse net, he argued, ensures a plasticity that is impaired or lost when nerve fibres are condensed into a few discrete

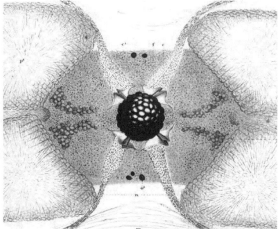

7.2 · (*Left*) Carl Chun in 1899. From Winter (1914). (*Right*) Representation by Chun (1878) of the nervous system of the ctenophore *Eucharis multicornis*, showing from above the sense organ at the aboral pole (statocyst) and the nerves irradiating from it and projecting toward the comb rows. From Chun (1878), Plate 1, Fig. 1.

nerve tracts, as happened in more advanced animals (higher invertebrates and vertebrates).

Unbeknownst to Eimer, German zoologist Carl Chun (1852–1914) had preceded him at the Naples Zoological Station for the express purpose of studying the anatomy of the ctenophore nervous system, in the winter and summer of 1877. Chun was then a student of the distinguished Leipzig zoologist Rudolf Leuckart, considered the founder of parasitology as a biological discipline. Chun later replaced Leuckart as chair of zoology at the University of Leipzig, and led the first German oceanographic expedition to collect deep-sea animals (chiefly fishes and cephalopods) in sub-Antarctic waters from aboard the steamship *Valdivia* in 1898–99.

The results of Chun's investigations into ctenophore nervous and muscle systems were published the year following his stay in Naples (Chun, 1878). In contrast to Eimer, Chun observed only a centralized nervous system, starting with the sense organ (balance organ) on top of the aboral pole of the body. From the balance organ four nerve roots emerge and descend toward the oral end, splitting early to form eight nerve branches (Fig. 7.2, right), each running beneath one of the eight comb rows. The arrangement is such, according to Chun, that sensory feedbacks to the balance organ determine a motor output that reaches the eight comb rows simultaneously, thus ensuring synchronized ciliary beating around the animal, and coordinated swimming. He saw no evidence of the diffuse nerve net (nervea) reported by Eimer. In fact, an astounding twelve of the paper's fifty pages were devoted to a critique of Eimer's interpretations of the nervous system of *Beroë*. In essence, Chun doubted that Eimer's histological methods were adequate to expose the true nervous system of comb-jellies. In his opinion, Eimer's stellate ganglion cells, which were pivotal to his rendering of the nervea, were mere non-nervous ectodermal support cells. For Chun, muscle excitation spread directly from muscle cell to muscle cell, and was not mediated by a nerve net.

But Chun's failure to detect a nerve net was soon corrected by Richard Hertwig, one of Eimer's old nemeses. Richard took a personal interest in ctenophores that, for a change, did not enlist his brother Oskar. Nevertheless, he brought to his investigation the same energy and skills that had served the siblings so well. Richard spent seven weeks collecting ctenophores in Messina in the spring of 1879, and then he removed to the Zoological Station of Naples, where his predecessors had gone. In spite of inhospitable weather he managed to collect four species: *Beroë ovatus*, *Cestus veneris*, *Eucharis multicornis*, and one specimen only of *Cydippe hormiphora*. Later he added other species. While in Naples Richard had to divide his time between this investigation of ctenophores and his concurrent investigation of sea anemones with Oskar. Histological sectioning could therefore only be done after his return to Jena. This work being contemporary with the Hertwigs' other contributions, it is not surprising that the chemical treatment of the material was identical with that used for the jellyfish and sea anemone works (see chapter 4).

As with the other productions of the Hertwigs, Richard's ctenophore study was published promptly after the analysis of the material was completed to the brothers' trademark high standard (R. Hertwig, 1880). The quality of the work and the clarity of the illustrations far surpass those of any competitors. The sensory bodies at the aboral end are brought into sharper salience and described in greater detail than in Chun's paper, showing the statocyst, the four sensory pads, and as-

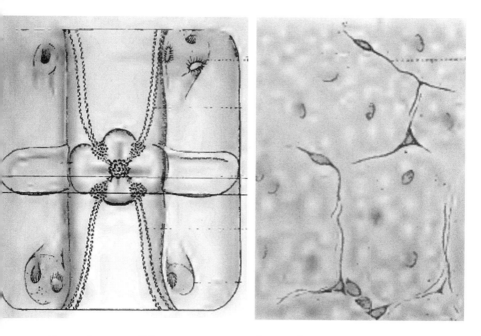

7.3 · Some elements of the nervous system of ctenophores as represented by R. Hertwig (1880). (*Left*) Aboral end of *Callianira bialata* with the sensory bodies of the two polar plates and the beginning of the eight comb-rows. (*Right*) Nerve plexus of *C. bialata* at the base of the ectoderm layer. From R. Hertwig (1880), Plate 19, Fig. 4, and Plate 15, Fig. 11.

sociated ciliated furrows of the comb rows (Fig. 7.3, left). Richard also provides a more accurate description of the nerve nets than Eimer did, showing two distinct nerve nets – subectodermal and "mesodermal" (mesogleal) (Fig. 7.3, right):

> The nervous system of ctenophores consists of ectodermal and mesodermal parts. The former appears in the form of a plexus of ganglion cells, which is located just below the epithelium and is spread uniformly over the whole body surface ... The meshes of this net are delimited only by sparse (2–3) nerve fibres, which are the extensions of multipolar ganglion cells located at the vertices of the mesh and which branch repeatedly over their trajectory ...
>
> I have included as forming the mesodermal part of the nervous system a large number of fine fibres which here and there

accommodate spindle-shaped nuclei and are enwrapped by a
sheath, the neurilemma. They extend freely through the mesoglea
along with the muscle fibres and terminate on either side by
bifurcating before contacting the epithelia. [My translation from
Hertwig, 1880]

In contrast to Eimer, Hertwig saw no evidence of any centralization
trend in the ctenophore nervous system, but he met Eimer partway
in describing some of the nerve fibres, especially in the "mesodermal"
nerve net, as anastomozing. Hertwig saw the mesodermal nerve net
as an innovation of the ctenophores that distinguishes them from the
other coelenterates. This nerve net, he argued, develops from the ectoderm, but invades the mesoderm/mesoglea as it follows its muscle tar-
gets, which have developed from mesoderm as in higher, triploblastic
animals. This and other features of ctenophores Hertwig regarded as
evidence against any direct affiliation between them and other coelenterates, as he could not see how one would have evolved from the
other. He believed that at best they were very distant descendants of a
common ancestor.

Given the track record of the Hertwigs and the general consensus
among their contemporaries that their observational skills were keen
and reliable, it is astounding that a young newcomer to the field found
fault with Richard Hertwig's results as well as with those of Hertwig's
predecessors. Paul Samassa (1868–1941), the son of an industrialist in
the Slovenian part of the Austro-Hungarian Empire (Schödl, 1988),
was a twenty-three-year-old doctoral student of Otto Bütschli in Heidelberg when he went to the Zoological Station in Naples in the winter
of 1890–91 to study the histology of ctenophores. In the publication
that resulted (Samassa, 1892), Samassa flatly denied the existence of a
nervous system in ctenophores, apart from the aboral sensory complex.
The previously described nerve nets and their varicose fibres were, according to him, artifacts caused by the chemical treatments used for
histological processing.

Three years later, however, a doctoral student of Richard Hertwig
in Munich and later a highly respected physiologist, produced a short
article in which he specifically addressed Samassa's contention in order
to refute it (Bethe, 1895). Albrecht Bethe (1872–1954) was born in Stet-
tin, Pomerania (now part of Poland), from a family line of Protestant

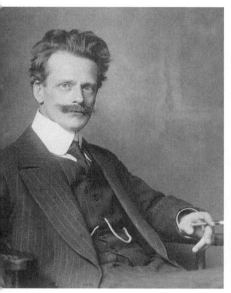

7.4 · (*Left*) Albrecht Bethe in 1914. Courtesy Institut für Stadtgeschichte, Frankfurt am Main (Signature ISG_S7P_1086, photograph HMF C21560). (*Right*) Herbert W. Conn in 1886. From Conn (1948). Courtesy American Society of Microbiobiology.

ministers and schoolteachers, and he was educated at the University of Munich. Bethe states in his article that Hertwig showed him some of his old preparations and that he found in them convincing evidence of the nerve nets his mentor had described fifteen years earlier.

This story is telling in two ways. First, it suggests that Hertwig was rattled and stung by Samassa's challenge to his scientific integrity and reputation. The fact that his brother Oskar was co-editor of the periodical in which Samassa's paper appeared must have dealt him an aggravated blow. And second, it raises the suspicion that Bethe, who may have been used as a proxy to get back at Samarra, cornered himself in a situation of conflict of interest by being seen as trying to defend his doctoral supervisor, a task best left to an independent investigator.

But Bethe decided to dig deeper. He took the opportunity of scheduled summer fieldwork in Helgoland by the North Sea to collect specimens of the genus *Cydippe* and revisit its nervous system. He used methylene blue, a staining method introduced in 1891 by the famous haematologist and immunologist Paul Ehrlich, which picked up nerve cells distinctly even in live tissues. Thanks to this new technique Bethe was able to confirm all of his mentor's earlier findings and dismiss Samassa's claims. He further concluded: "I am now of the opinion that nerve nets, such as occur in the ctenophores, are among the oldest and most primitive forms of the nervous system and that the isolated nerve conduction is a later acquisition. But higher animals, in addition to the isolated nerve conduit used for specific purposes, retained nerve nets whenever very simple and diffuse conductions met their needs." Richard Hertwig's good connections probably helped to hasten publication, as the paper appeared in the fall immediately following Bethe's summer work.

An Assortment of Jellies: Siphonophores

Not unlike ctenophores, siphonophores were considered by nineteenth-century zoologists as sidekicks among coelenterates to the more popular jellyfishes and sea anemones. Although siphonophores belong to the class Hydrozoa, which includes true jellyfishes, anatomists would have considered them more formidable challenges for study than jellyfish, if only because they are colonial animals. They are made up of an assortment of individuals with specialized tasks, which are integrated into a tightly held colonial mass, to the point that the colony is considered to act as a single organism. Siphonophore colonies such as the more familiar Portuguese man-of-war can reach large sizes as they float near the surface of open oceans. The first serious investigator of siphonophores was Ernst Haeckel, who studied the numerous specimens collected during the first great scientific reconnaissance of the oceans, the British Challenger Expedition (1872–76). His publication (Haeckel and Thomson, 1880) helped create an interest in these forms among the new generation of zoologists. Haeckel marvelled at the beauty and complexity of these floating jellies, and treated them as works of art.

But the first to take up the challenge of describing the nervous system of these intriguing forms was Chun, and his two publications (Chun, 1881, 1882) predated any infatuation siphonophores could have stirred up in the wake of Haeckel's masterly work. His only motive, it seems, was to investigate whether the pattern of neural organization already seen in jellyfish, sea anemones, and ctenophores was transferable to siphonophores. Wisely, Chun looked first at the simplest of floating siphonophores, *Velella*, known as the "by-the-wind sailor." He described a plexus of sub-ectodermal nerve fibres connected to "ganglion cells" all over the different parts of the colony, but provided no supporting illustrations (Chun, 1881). He noted no spatial concentration of ganglion cells or centralization of nervous tissue. The next year (Chun, 1882) he expanded his findings in *Velella* and in other species, and this time he noticed a nerve ring associated with the oral disc of gastric polyps, an indication of budding centralization of the nervous system. He also described sub-ectodermal nerve nets similar to the one in *Velella* in the ectoderm of the floaters (air chambers) and in the gastric polyps of *Rhizophysa* and *Physalia* (Portuguese man-of-war). Again, no illustration accompanied the text and the significance of his findings was not discussed beyond the simple conclusion that the organization of the nervous system was similar to that of other coelenterates.

If caution in accepting Chun's results at face value was in order, what with the lack of supporting images, a young American graduate student soon dispelled any reserve. Herbert William Conn (1859–1917; see Fig. 7.4) was born in Fitchburg, Massachusetts, and earned a bachelor's degree in science at Boston University in 1881. While in Boston one of his professors kindled in him an interest for "lower forms of life." He enrolled in the PhD program at Johns Hopkins University in Baltimore and earned his doctoral degree there in 1884. His thesis supervisor was W.K. Brooks, one of the most distinguished and influential zoologists in the United States at the time. His specialty was invertebrate zoology, but his laboratory was not noted for its expertise on the study of the nervous system. Brooks took the young Conn under his wing, and they worked together during several summers of research at the seaside laboratory in Beaufort, South Carolina (Conn, 1948). It was probably during the summer of 1882 that Conn collected specimens of blue buttons, *Porpita*, and processed them for histological examination of their nervous system with the assistance of H.G. Beyer, a surgeon at the nearby Naval Hospital. Their work, which was not part of the doctoral thesis, was published the following year in the publishing organ of Brooks's biological laboratory at Johns Hopkins (Conn and Beyer, 1883). Right off the bat Conn and Beyer signalled approvingly that their results concurred with those of Chun:

Some work which has been done in the Biological laboratory
during the present year, upon *Porpita*, shows that here also is
found a similar system of nerve ganglion cells. The observations
were made without a previous knowledge of Chun's paper, and
are therefore more valuable as confirming his statement as to the
existence of a nervous system among Siphonophora, as well as in
extending our knowledge of the relation and distribution of the
same. (Conn and Beyer, 1883)

Conn and Beyer were successful in observing subectodermal ganglion cells by lightly staining their preparations with osmic acid or by
staining more deeply with haematoxylin. Each of these ganglion cells,
originating in the ectoderm, possesses a small cell body from which
emanate several long processes. They found that tripolar nerve cells
dominated over bipolar and quadripolar cells. The course of the processes over the muscle layer is extensive, each branching into finer fibres
that sink into the muscle mass. Conn and Beyer also noted that often
"the fibres from one cell unite with those of other cells ... thus putting
the different nerve ganglia into communication with each other, and
forming to a certain extent a continuous nerve plexus." In this they adhered, as did many of their colleagues, to the popular syncytial theory.
The processes appear to innervate only muscle cells and no other
ectodermal cell type. As a result, the nerve plexus is particularly well
developed in the most muscular parts of the colony, such as the numerous tentacles and the velum. Contrary to Chun, these authors failed to
detect nerve cells in the feeding polyps. As to the evolutionary or functional significance of these nerve cells, Conn and Beyer offer a rather
tepid endorsement, which seems to renege on their own descriptions:

There is still perhaps some doubt as to whether the structures
here described are really what they have been considered;
whether they may not be some form of connective tissue cor-
puscle without any nervous function. They are, as we have seen,
very few in numbers as compared with any organs which they are
supposed to enervate [sic]; they are connected with no central
system, and simply form a more or less connected plexus of
scattered cells. If they are true nerve elements they are only to be
considered as what may be the beginning of a nervous system. It

can hardly be possible that they play any important function as nervous organs. Always associated as they are with the muscular system, they are to be regarded as muscular [motor?] rather than sensory cells; but the relatively small number of even their fibres, as compared with the number of muscular fibres which each must be supposed to control, certainly indicates that the muscular system cannot be to any great extent dependent upon them for its stimulation. (Conn and Beyer, 1883)

This low density of nerve cells and the lack of centralization led Conn and Beyer to conclude that the nervous system of siphonophores was even more primitive than that of medusae and sea anemones, and less functional. In addition to the nerve plexus, they observed structures over the ectoderm of the velum which form what they considered "a sensory ring extending around the edge of the disc and composed of hundreds of entirely separate organs." A closer look revealed that each of these sensory organs was produced by the invagination of the ectoderm, resulting in a pocket containing large, elongate cells of suspected sensory function. The position of these sensory bodies at the edge of the velum and the exposure of the free ends of the sensory cells to the surrounding water led the authors to believe they had a tactile role. They saw no connection of these organs with the nerve plexus.

Before a year had passed, yet another contribution appeared on the topic. The new contributor, like the Americans, came from outside the German-speaking sphere. Alexei A. Korotneff (1852–1915) was a native of Moscow and graduated from Moscow University in 1876. After a short period in the civil service he returned to the same university, first studying medicine and then zoology, earning his doctorate under Professor Anatol Bogdanov in 1881 (Fokin, 2008). He remained in Moscow for a few more years without a professorship or any other known status – perhaps continuing research under the aegis of Bogdanov. One of these investigations took place in the winter of 1882–83 at the Naples Zoological Station and it involved siphonophores. In the resulting paper, which also covered other histological aspects (Korotneff, 1884), Korotneff expressed astonishment at "the extraordinay wealth of nerves" he found in the specimens of *Physophora*. "You could probably think," he rhapsodized, "that you've met here with a real brain. The ganglion cells and nerve fibres form a dense network that is spread out in the depths of the ectoderm." This statement contrasts sharply with Conn and Beyer's estimation of nerve cell density, and nowhere in their paper did the Americans go so far as to intimate an analogy with a brain-like arrangement. But Korotneff was unaware of Conn and Beyer's paper at the time his own article went to print.

Korotneff observed at the top surface of the swimming bells a dense subectodermal nerve plexus composed mainly of bipolar or tripolar nerve cells in *Physophora*, whereas the corresponding nerve plexus of *Velella* was sparsely distributed and contained many multipolar nerve cells. He noted that the processes of these nerve cells crossed over each other and some even contacted or, he thought, penetrated nerve cell bodies. He erroneously viewed the fibrillar material inside nerve fibres as the axis cylinder (axon) and the granular protoplasm around the fibrils as a myelin sheath (plasma sheath). He also observed a tendency of the processes to bifurcate frequently.

Where Conn and Beyer had seen specialized sensory organs to which they ascribed a tactile function in *Porpita*, Korotneff only found "tactile" sensory cells mixed with gland cells and epitheliomuscular cells, primarily in the ectoderm of the tentacles. He had noticed that touching tentacles with a probe caused a withdrawal of the tentacles from the probe, followed by a flight response by the animal (*Physophora*). From these observations he proposed that touch stimulates the tactile sensory cells, which transmit a signal by way of the tentacle nerve net to the upper ectodermal nerve net of the swimming bells, thus setting in motion the motor responses, first in the tentacle musles (withdrawal), and then in the bell muscles (escape). While Korotneff expanded on the functional implications of his findings, in contrast to his predecessors he was markedly silent on the evolutionary significance.

Eight years later another student of Richard Hertwig took a fresh look at the siphonophore nervous system. Karl Camillo Schneider (1867–1943), the son of a farmer, was born in Germany, near Leipzig (Kühnelt, 1993). He studied at the University of Leipzig and at Munich University, where he earned a doctorate in 1890 under Hertwig's guidance. He then moved to the University of Breslau, where he was granted a research fellowship by the Ministry of Culture of Saxony. While stationed there he travelled to Naples to conduct investigations on siphonophores and other coelenterates over a six-month period. The results of these investigations were published in a sprawling paper notable for its disorganized structure (Schneider, 1892).

Schneider used proven chemical treatments for his histology (osmic/acetic acid, picric acid/carmine), but in his hands these techniques yielded poor results. He criticized Korotneff's paper, which he used frequently as a reference point for comparison, for failing to observe what he himself saw: syncytia of ectodermal cells and of endodermal cells, artifacts that none of his predecessors had noticed. His descriptions of the nerve plexus were more in line with Korotneff's own, except that in several cases he hesitates to identify nerve cells as such, referring to them as "ganglion-like cells." Schneider frequently departs from his predecessors by showing his "ganglion-like cells" squarely in the ectoderm, mingling with the foot-end of the epitheliomuscular cells. The latter are in fact misidentified as "neuromuscular cells," and he attributes their large size to the fusion of nerve and muscle cells. What others would see as a syncytium of several nerve cells Schneider depicts as a single ganglion cell. He miscontrues or overinterprets the existence of sensory cells, on the basis of inadequate histological material. No matter how bad the histology is, Schneider treats his subject narrowly from a histologist's angle. If there was biological significance to his findings, Schneider chose to ignore it.

Haeckel was probably (and understandably) disappointed by the effort of his academic grandson – Schneider was a student of Richard Hertwig, himself a pupil of Haeckel – because he decided to assign a new student of his to a doctoral project narrowly circumscribed to examine afresh the nervous system of siphonophores. The new arrival in Jena was Theodor Schaeppi, a native of Switzerland of whom we know almost nothing save his few scientific publications. In his usual effusive manner Haeckel took the young man under his wing, pulling out all the stops to secure a travel grant for Schaeppi's six-month sojourn in Naples and Messina (winter of 1893–94), where he was sure to collect all the siphonophores he needed. In Italy Schaeppi prepared his specimens by following the histological protocols previously used by Hertwig and Schneider, with few modifications – a shaky portent for someone commisioned to make a fresh start. And yet the paper that emerged from his dissertation (Schaeppi, 1898)) turned out to rise above those of all his predecessors in the quality of its science and illustrative material.

Schaeppi examined physonectid species (*Physophora, Halistemma, Forskalia, Apolemia*), all characterized by a stem around which the upper (nectosome) and lower part (siphosome) of the colony are organized.

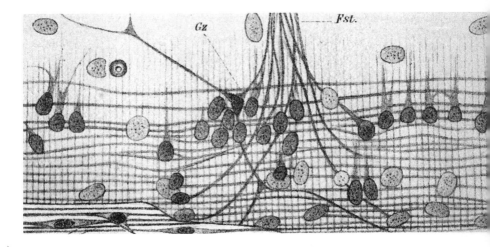

7.5 · Subumbrella of a nectophore of *Forskalia contorta* in which are seen the nerve ring and ganglion cells innervating the sheet of radial and circular muscles as well as the muscle strand, reminiscent of the classical jellyfish organization. From Schaeppi (1898), Plate 26, Fig. 42.

The nectosome contains nectophores, which are really a collection of medusoid swimming bells, and an apical pneumatophore (air bladder) for oriented flottation. The siphosome contains a host of small polyps, including dactylozooids, gastrozooids, and phyllozooids, and reproductive medusoids called gonophores. Concentrating on the stem of the siphosome as his predecessors did, Schaeppi found fault with both Korotneff's and Schneider's findings right from the start. He saw no evidence of the centralized nervous system advocated by these authors in the stem area, and he blamed their confusion of endodermal epitheliomuscular cells for nerve cells for their misguided depictions. He was particularly harsh on Schneider, who, he claimed, had blundered in many of his interpretations of nerve cells.

The osmium treatment allowed Schaeppi to easily visualize the fine fibres of the ganglion cells, but unfortunately also prevented him from clearly discriminating the sensory cells from other ectoderm cells. He obtained clear images of the spread of the nerve net in the ectoderm of nectophores. The nerve net is seen as consisting of mostly tripolar or quadripolar neurons, the fine processes of which often bifurcate and appear to anastomose with each other. Some of these processes come very close to, or make direct contact with nettle cells. This nerve net extends from the underside (subumbrella) of the bell and, through the vellum, to the exumbrella, where it is denser in the lower part of the bell and innervates the radial muscle bundle (Fig. 7.5).

That nectophores are miniature swimming bells is underscored by a neuromuscular organization that appears eerily similar to that of hydromedusae as described earlier in this book. Schaeppi here provides a clear depiction of ganglion cells of the nerve net connecting with the nerve ring at the velum of the nectophores, where a heavy sheet of circular and radial muscles is arrayed (Fig. 7.5). The ganglion cell processes also connect, of course, with the muscle fibres, including to the radial fibre bundles that course from the velum to the apex of the bell. This arrangement, according to Schaeppi, ensures that the exumbrella shortens during the diastole (water filling) of the pulsating cycle via contraction of the radial muscles, and that the nectophore opening is narrowed during systole (water expulsion) thanks to the contraction of the circular muscle. Whereas Schaeppi concurred with his predecessors in the depiction of anastomoses bertween nerve fibres, he appears to stand out from the pack in his clear recognition of a nerve ring similar to that of hydromedusae. He can be faulted, however, for his neglect of sensory elements in the nervous system of these jellies and for his silence on the possible evolutionary significance of his observations, a silence that his mentor Haeckel would have happily broken had he been in charge of the writing.

Does Hydra Possess a Nervous System?

This question arose following the discrediting of Kleinenberg's theory of the neuromuscular cell in *Hydra*, which we discussed in chapter 3. Once the Hertwig brothers and others found it impossible, in the face of their own observations, to support the theory, neurohistologists asked themselves: if the neuromuscular cell does not exist, is there a *bona fide* nervous system waiting to be discovered in Hydra? The simplicity of the freshwater *Hydra* compared with the more complex, marine cnidarian forms gave room for entertaining the thought that perhaps *Hydra*, not unlike sponges, does not require a nervous system to coordinate its motor behaviour. It was just a question of getting on with the research work to settle the matter. Given the fever pitch of the discussion around the issue, it is surprising that serious work on the neurohistology of *Hydra* resumed only eighteen years after Kleinenberg's doomed effort.

But the person who came forward to do the job turned out to be Karl C. Schneider, the very same who, as we have just reported, rivalled Kleinenberg in overreaching his powers of observation at the microscope. However, his research on *Hydra* and marine hydroids (Schneider, 1890) was, by his own admission, suggested by Richard Hertwig – as part of Schneider's doctoral dissertation in Munich – and Hertwig supervised his work every step of the way. This may explain why this paper is of better quality than the paper Schneider produced on the histology of other coelenterates two years later when he was based in Breslau.

Schneider ended up spending the majority of his research time on his cultures of *Hydra* because arrivals of the desired marine hydroids· from Rovigno on the Istrian Peninsula (then part of the Austro-Hungarian Empire) failed to deliver sufficient specimens. In the paper he devoted an inordinate amount of space – by contemporary standards – to describing and analysing the methods used, and focusing on the trials and errors of the apprentice. He tinkered with tissue section thickness, maceration solutions (osmic/acetic acids), carmine staining and duration of glycerol immersion before microscopic examination. Rendering of nerve cells was paramount, but no single technique fit all animal species investigated. Methylene blue, the latest wonder stain for making out nerve cells, failed to deliver on its promises. Both Schneider and Hertwig recognized the critical importance of methodology for settling the thorny issue of the presence or absence of a nervous system in *Hydra*.

The techniques denied Kleinenberg proved extremely helpful to Schneider, who had no difficulty spotting subectodermal ganglion cells in *Hydra*. He found them everywhere, but more abundantly in the mouth region (oral disc) and at the base of the tentacles, with a second zone of relative abundance in the foot (basal disc). The morphology of the ganglion cells corresponded largely with their description in jellyfish by his mentor, Richard Hertwig, and his brother Oskar. Unfortunately, few of his illustrations afford *in situ* views of the nerve cells. When they do, bipolar or, more often, multipolar cells are observed, with branching varicose processes forming a loose plexus. Schneider's attempts to appreciate the position of nerve cells across the ectoderm were often frustrated by obstructing nematocytes (nettle cells), gland cells, and nurse cells. To him ganglion cell processes appeared to contact epitheliomuscular and nettle cells.

Schneider followed the genesis of ganglion cells from what he calls "indifferent cells," which correspond to today's interstitial cells in the ectoderm. He noticed that nematocytes differentiate at the same time as ganglion cells, suggesting that they too are differentiated from interstitial cells. This was an astute observation that has proven valid to this day. Unfortunately, Schneider's astuteness stopped there. He must certainly have surprised his mentor Hertwig by affirming that the ectoderm is devoid of sensory cells. Even he admitted that it seemed a strange affirmation to make, so it is even more astonishing that Hertwig failed to censure his pupil on this matter. Schneider's assertion defied the common sense opinion that the outer ectoderm is the first line of the sensory assessment of the external environment. It was expected that the ectodermal ganglion cells relied on sensory input in order to respond with the proper motor output; without sensory cells, ectodermal ganglion cells would be functionally lame. Instead, Schneider saw what he interpreted as sensory cells only in the endoderm. He mentions that these sensory cells are distributed in the endoderm throughout the animal, with a slightly greater density in the oral disc, but that they are absent in the tentacles. As the tentacles are considered to represent the "feelers" of the animal, this observation seems particularly incongruous. Schneider found endodermal ganglion cells resembling those of the ectoderm, with the difference, of course, that they were connected with the sensory cells.

Despite the shortcomings of Schneider's work, it provided fairly convincing evidence for the presence of nerve cells in *Hydra* that are in no way different in morpholgy and organization from those of other coelenterates. His publication therefore dispelled the notion that *Hydra* lacks a dictinct nervous system and possesses instead a protoneural arrangement in the shape of the mythical neuromuscular cell.

Another investigation of *Hydra* by a young doctoral student came promptly on the heels of Schneider's own. This time an Italian, at last, investigated the nervous system of a coelenterate in the land to which so many foreigners had made the pilgrimage to collect their own coelenterates. Raffaello Zoja was born in Padua in 1869, the privileged son of Giovanni Zoja, the professor of human anatomy at the University of Padua, the very same university where the famous neurohistologist Camillo Golgi worked. The academic pedigree of Raffaello's family extended back to his maternal grandfather, the famed anatomist Bartolomeo Panizza (1785–1867), who discovered the role of the posterior cerebral cortex in vision (now known as the visual cortex) and Panizza's foramen, an opening with a valve connecting the right and left aorta of crocodiles, which is regarded as an important landmark in the evolutionary transition of the circulatory system from reptiles to birds and mammals.

We do not possess a photograph of Raffaello Zoja (1869–1896), but a physical description has survived that tells of a rather handsome young man disadvantaged by fragile health. He is described as "tall, thin, fair, with a rather emaciated, almost ascetic face, ... troubled by a gastrointestinal dyspepsia accompanied by nervous phenomena manifested in rapid cerebral exhaustion which rendered any prolonged mental work impossible" (Mossa, 1898). A rich intellectual environment combined with innate talent led to the precocious development of Raffaelo's academic abilities. He completed his doctoral thesis on the anatomy and physiology of *Hydra* at the age of twenty-two, under the guidance of Leopoldo Maggi, the Chair of comparative anatomy and physiology at Padua and a close colleague of his father. The thesis was published in three instalments in the *Bolletino Scientifico*, a journal co-edited by Maggi and Giovanni Zoja (Zoja, 1890a, 1890b, 1891). The first instalment (1890a) lays down the fundamentals of general morphology for the three species investigated: *Hydra viridis*, *H. grisea*, and *H. vulgaris*. Zoja presents a thorough and balanced review of the literature and constructs a masterly refutation of Kleinenberg's neuromuscular theory – the best so far. In the second instalment (Zoja, 1890b) he specifically investigates effector cells such as the epitheliomuscular cells and nematocysts. In the third (Zoja, 1891) he deals at last with the nervous system.

Zoja identified ganglion cells with long processes in *Hydra* that to him resembled those previously described by the Hertwig brothers in sea anemones. They are localized at the base of the ectoderm and in Zoja's estimate they are distributed evenly throughout the body. Like Schneider, whose work Zoja was unaware of, he found that these nerve cells originate from the interstitial cells. He saw contacts between nerve cell processes and nematocysts, but not with epitheliomuscular cells.

Also like Schneider, he failed to detect specific sensory cells in the ectoderm, but speculated that the cnidocil of nematocysts could act as sensory organs that transmit their signal to the ganglion cells via the connection of nematocysts with ganglion cell processes. His supporting illustrations were of poor quality and, unfortunately, did little to instil confidence in his identification of nerve cells.

If Zoja's histological demonstrations were below par, he distinguished himself from his competitors by providing behavioural observations and physiological experiments of relevance to neural activity. In this he followed the example of Romanes and Eimer, of whom he was well aware. By applying tactile and electrical stimulations and by performing sectioning experiments, Zoja was able to show that the anterior or upper part of the body, especially the mouth and tentacles, respond more swiftly and strongly to stimuli – by contracting the body part away from the source of stimulation – than the posterior or lower part. These results suggested that the kind of nerve centres exposed by Romanes and Eimer in jellyfish exist also to an extent in *Hydra*, even though Zoja's histological observations contradicted this. However, just before Zoja's 1891 paper went to press, the young Italian was made aware of Schneider's paper of the previous year. In a postscript, Zoja takes note of Schneider's observation that ganglion cells are more numerous around the mouth and in the base of tentacles, and this observation reconciles him with his own experimental observations. Although he could not find ganglion cells in the endoderm, Zoja blamed his technical limitations and predicted the cells' existence. He was reassured in his prediction when he read in Schneider's article that endodermal ganglion cells had been observed in *Hydra*.

Zoja also imitated Romanes by applying "poisons" to his preparations. In contrast to Romanes, he found that curare has little or no effect, but the paralytic effects of chloroform and ether concurred with Romanes's results. Zoja went further and looked for evidence of chemical senses and thermal or light sensitivity in *Hydra*. He found evidence of contact chemoreception (gustatory), but for an unexplained reason he believed that the seat of this chemical sense was in the endoderm. He cautiously assumed that the reactions of *Hydra* to temperature are not of a sensory nature, and he found no incontrovertible evidence of light sensitivity. Where his critical reasoning lapsed was in a discussion over the existence of "conscience and will" in *Hydra*. For instance, he

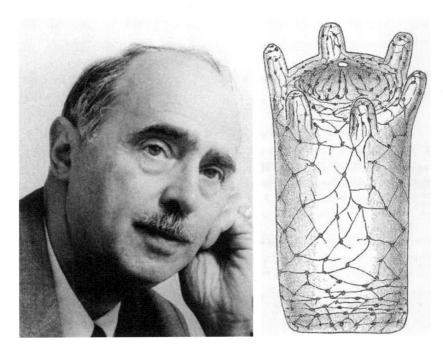

7.6 · (*Left*) Jovan Hadži in his mature years. Courtesy University of Ljubljana (Dr Boris Sket). (*Right*) Representation of a whole-mount preparation of *Hydra viridis* stained with methylene blue. From Hadži (1909), Text-Fig. 2.

regarded observations of *Hydra* pursuing prey with which they have not had physical contact as evidence of a willful search for prey akin to an instinct, a coarse form of psychic life. He came to this conclusion only because he ignored the possibility that *Hydra* is simply attracted by chemical cues released by prey and carried by water to *Hydra*'s feelers. Nevertheless, Zoja's contribution, and particularly the experimental part, helped establish a consensus on the presence in *Hydra* of a nervous system that is organized and functions as in other coelenterates.

Even if the existence of a nervous system in *Hydra* was now no longer in doubt, a frustration lingered that a satisfactory, robust depiction of that nervous system was out of reach. A better histologist using more standardized methods was needed in the field. It was another thirteen years before such an elusive quarry materialized, unexpectedly, in the person of a Serbian. Jovan Hadži (1884–1972) was born to Serbian parents in a town that was part of the Austro-Hungarian Empire but now belongs to Romania. We know very little about his early years, so a great deal must be left to conjecture. He moved to Zagreb, the capital of Croatia, where his career as a zoologist began. He studied at the Zoological Institute, University of Vienna, where, presumably, he earned his doctoral degree. His publication on the nervous system of *Hydra* (Hadži, 1909) probably represents part of his doctoral dissertation, although that is not explicitly stated in the paper.

Hadži's paper very quickly reveals that it stands head and shoulders above all its predecessors for its painstaking scholarship, accuracy of observation, and superb illustrative material. He investigated two species: the "gray hydra," *Hydra fusca*, and the "green hydra," *H. viridis*; green here denotes the presence of symbiotic unicellular algae in the animal. Thanks to his deft use of the methylene blue method and the tiny size of the animals, Hadži was able to provide startling images of the mapped subectodermal nerve net throughout the body (Fig. 7.6, right). This illustration shows in one embracing view that nerve cells are more densely distributed in the oral disc and foot disc than in the body column and tentacles. Nerve cell processes are seen to radiate from the rim of the mouth opening, only to turn into a sparse nerve ring at the periphery of the oral disc. Nerve cell processes show a parallel orientation in the tentacles.

Classical osmic acid maceration and picrocarmine or hematoxylin staining methods allowed Hadži to obtain very convincing slide preparations. He now had no difficulty distinguishing cell types among the sharply contrasted nerve cells. He first made a useful distinction between regular nerve cells (ganglion cells) and sensory nerve cells; that is, thin bipolar cells sending a process to the surface and bifurcating processes inward to muscle feet or to other nerve cells. He found that sensory nerve cells are present only in the ectoderm of the oral disc, the tentacle bases, and the foot disc. Although he reported the presence of nerve cells associated with the endoderm, he could not determine whether they formed an uninterrupted nerve plexus. At any rate, he found that nerve cell processes contacting muscles were generalized in the animal; the same could not be said for nematocytes.

Hadži relied on behavioural and experimental observations to gain insight into the workings of the nervous system, but he failed to mention Zoja's contribution in this regard. These observations led him to conclude that the ectodermal and endodermal muscles must be co-ordinated through connections between the respective nervous systems in the mouth region, but that they acted independently in the column of the body. He suggested – although acknowledging there was insufficient evidence – that the nerve and sensory cell concentration in the foot was a secondary development initiated by the peculiar crawling locomotion of *Hydra*.

Hadži pointed out that, in the end, *Hydra* was no different from the others of its ilk, the marine hydroids, in the development of its nervous system. It is now appropriate to ask how studies of the nervous system of hydroid polyps had progressed up to Hadži's time. These polyps are members of the hydrozoan class that, unlike hydromedusae, are not pelagic and develop only into colonies of polyps. Coincidentally, the first zoologist to pay close attention to the group was, like Hadži, born in a town (Hermannstadt) that is now part of Romania (Sibiu). Carl Friedrich Jickeli (1850–1925) was the son of a wealthy hardware merchant in Hermannstadt. Carl persisted in his desire to pursue a career in natural history despite his parents' opposition. When he was twenty he participated in a scientific expedition to Abyssinia (today's Ethiopia) about which he wrote a book. He collected invertebrates for museums for several years before he felt the need to earn a doctorate, ostensibly to advance his career. He chose Heidelberg, then one of the most esteemed universities in Europe for the natural sciences; and he selected Otto Bütschli for his supervisor, as did several students described in this book before or after him. The results of his dissertation research, which covered all aspects of the anatomy of hydroid polyps, were published first in a short note concentrating on the nervous system (Jickeli, 1882), and a year later in two extensive articles (Jickeli, 1883a, 1883b).

In the first article (Jickeli, 1883a) Jickeli dealt principally with the hydroid *Eudendrium*, giving only passing references to *Hydra*, in which he saw cells resembling the ganglion cells of *Eudendrium*. Relying almost entirely on the classical osmic acid/picocarmine method, Jickeli obtained preparations that emphasized the cytology of ganglion cells – with detailed but irrelevant descriptions of protoplasm and granular content – but he failed to render a general picture of the organization of the nervous system. This lack is also reflected in lithographs that focus on individual ganglion cells from macerated material. He affirms, without illustrative support, that ganglion cell density is greater in a ring of the hypostome (mouth region), where a concentration of gland cells is also found. The most intriguing part of his descriptions relates to what he viewed as close associations between ganglion cells and nematocytes; he saw contacts between these cells that he did not observe between ganglion cells and epitheliomuscular cells. He also noticed that nematocytes in their early stages of development share with ganglion cells the possession of numerous granules in their cytoplasm, but he stopped short of deducing a common developmental origin for the two cell types. He assumed, erroneously, that nematocytes were a specialized form of muscle cell, thereby raising the possibility that they received a motor innervation from free endings of ganglion cell processes.

In his second article (Jickeli, 1883b), which runs to a hundred pages, Jickeli sampled fourteen species of marine hydroid polyps in an attempt to obtain confirmation of his findings in *Eudendrium*. The results confirmed what he had described in *Eudendrium* except for two species, in which nerve cells appeared sparse and showed no connection to either epitheliomuscular cells or nematocytes. In addition, he described ganglion cells that appeared to send fine processs up to the surface of their epithelium. This suggests that he was unable to separate ectodermal sensory cells from ganglion cells because of the inadequate handling of histological methods. The incidence of species characterized by sparsely distributed ganglion cells led Jickeli to suggest that alternative paths of excitation conduction along the body exist in these polyps. As an example he proposed that epitheliomuscular cells may constitute a plexus for that purpose. As it turned out, this was a prescient notion, as we shall discuss later.

Someone who was studying the histology of hydroid polyps and who took notice of Jickeli's short note of 1882 from halfway around the world, decided to respond to it with his own observations. Robert von Lendenfeld (1858–1913) was a personality who loomed larger than life. Born and raised in Graz, Austria, he studied natural history at the local university. He went on to complete his doctorate under Franz E. Schulze, a sponge specialist, again at the University of Graz. But the research topic for his doctoral dissertation was not sponges – although he would become a sponge specialist himself in years to come – but rather a detailed analysis of the flight mechanism of dragonflies. What happened next is best explained by his great-grandson Randolf Menzel:

R. von Lendenfeld left Europe immediately after receiving his doctorate (1881) and headed to Australia with his wife [Anna], whose dowry was important for this adventurous project. As a naturalist and mountaineer, he verified the glaciation of the Snowy Mountain area, determined the highest mountain in Australia, and gave his name to several mountain ridges in the Snowy Mountains. In the New Zealand Alps, several mountains were named by him or carry his name. For example, he was the first person to climb the second highest peak in New Zealand, the Hochstetter Dom. He wrote about his Australian and New Zealand expeditions in his book, *Australian Trip* (1892). This book is a gold mine of lively depictions of Australian flora and fauna, geology, landscape structures, and the way of life of the Australian settlers at the end of the 19th century. (Menzel, 2004)

Lendenfeld spent five years (1881–86) in Australia, where he lectured on zoological topics and, besides his alpine forays, conducted research in marine biology, mainly on sponges and various coelenterates. Among the coelenterates, he had taken time to study the neurohistology of hydroid polyps, especially in species of *Campanularia*. He chose the opportunity of responding to Jickeli's note on the nervous system (Jickeli, 1882) to lay out his own findings and expand on new elements of the hydroid nervous system that had escaped Jickeli's notice.

In the short article that he submitted from Melbourne (Lendenfeld, 1883), he independently confirmed Jickeli's finding of a ring of nerve cell processes associated with a gland cell ring in the oral disc of the feeding polyps, near the base of the tentacles. The large ganglion cells in this condensed area of the nervous system were best visible in gold-impregnated preparations. Lendenfeld emphasized the suspected importance of this nerve ring for controlling a key activity of these sessile hydroid colonies: feeding. But Lendenfeld looked further than his predecessors and discovered, just in the underside of the oral disc and in the endoderm layer, numerous spindle-shaped sensory cells whose endings contact basal ganglion cell processes also forming a ring. Such an endodermal nervous system, he insisted, had no analogue in jellyfish. The finding pushed Lendenfeld to the erroneous conclusion that the brain, which the nerve ring represented at an incipient stage, originated in the endoderm.

The closing contribution to the study of hydroid polyp nervous systems at this time proved disappointing. The young perpetrator of this less than shining work was Max Wolff (1879–1963). German born, he studied medecine and zoology at the University of Jena. There in 1902 he became an assistant to the aging Haeckel, who also acted as supervisor of his doctoral studies. For his dissertation research Wolff chose to investigate the nervous system of coelenterates. He went to the Zoological Station of Rovigno in Italy to sample marine coelenterates for his research. He completed the work and defended his thesis in 1903, but his thesis work was only published the following year (Wolff, 1904), after he had left Jena to become an assistant to the neurologist Oskar Vogt in Berlin.

Wolff's paper leaves the impression that he was in awe of his mentor, but also that Haeckel failed to provide the guidance he needed. The paper tends to ramble and to lack scientific rigour. A glaring blunder is readily spotted in the title, where he lists polyp-bearing Scyphozoa as objects of study; but the paper soon reveals that Wolff meant actinians, which are anthozoans, not scyphozoans. He had it right about the other object of study in the title (hydrozoan polyps), but fell short in their treatment.

The paper starts with an extensive literature review, to which Wolff contributed very little of his own criticism, given that it was written under the aegis of Haeckel, with his biases and prejudices. The long and tedious review and analysis of Kleinenberg's neuromuscular cell theory illustrates the stranglehold this theory had on the minds of two successive generations of neurohistologists. The Hertwig brothers had disposed of the theory on solid ground twenty-six years prior to Wolff's publication, but no neurohistologist after the Hertwigs had missed the opportunity to muse on the theory at varying lengths, and to cast judgment on its validity. There must have been an almost fatal attraction at work here. Wolff, in keeping with the tendencies of his mentor Haeckel, may have wished the theory had it right because it was so beautiful and satisfying. What more satisfying than to conjure up an original cell, the neuromuscular cell, from which nerve and muscle cells would have evolved? But Wolff's blinkers led him to advocate views that flew in the face of the available facts. Contrary to the majority of his colleagues, he ended up not only believing in the existence of the neuromuscular cell, but even worse, convincing himself, with no hint of data in support, that many ectodermal cell types – neurons, sensory cells, nematocytes, gland cells, and so on – were derived from this "primeval cell."

Wolff's histological observations of *Tubularia* and *Hydra* were predicated on the use of Flemming's fixative (chromic/acetic/osmic acid solution) and Heidenhain's iron haematoxylin staining method. Although this technique was widely and successfully used at the time, in Wolff's hands it met with mixed results. By his own admission, he faulted his method for distorting the view of nerve elements in some of his preparations, so that his histological observations were met with suspicion by contemporaries. But that did not stop him from speculating on what he saw, including an endodermal nervous system. He conducted what he called physiological experiments on *Tubularia*, modelled after Zoja's own on *Hydra*, but what came out of these experiments amounted to anecdotal observations of dubious value. Yet he used these observations, combined with his neurohistology, to construct a theory according to which two reflex arcs preside over the motor reactions of these animals: a primary arc entirely contained within the imaginary neuromuscular cell, and a secondary arc mediated through a chain of cells (sensory to nerve to muscle cells). To his credit, however, Wolff appears to be the first among zoologists to have used the word "neuron" to describe nerve cells, more than a decade after Waldeyer introduced the term. As Haeckel would not have used the word, Wolff likely picked it up from his new boss in Berlin, Oskar Vogt, who as a neurologist would have kept pace with the jargon of medical physiology.

Sea Anemones

We saw in chapter 4 how the Hertwig brothers had used their superb observation skills to take in the distinctive morphology of actinians (sea anemones) and the impact of this morphology on the organization of their nervous system. Here was a clean departure from the jellyfish model. Sea anemones differ from jellyfish in that they live their entire life cycle as sessile polyps, anchored to a substrate on the sea bottom. They belong to the class Anthozoa which, like Hydrozoa, include colonial forms, known as soft or hard corals. What further contributions to the study of the nervous system of anthozoans were made by the successors of the Hertwig brothers? And how did these successors take up the challenge of advancing on the Hertwigs in sharpness of observation and the acquisition of new insights?

The furtherance of knowledge of actinian nervous systems started off on a bad foot. Lendenfeld, whom we met in the previous section, delivered yet another short communication from Melbourne, this time on the histology of Australian actinians (Lendenfeld, 1883). Scarcely three paragraphs are devoted to the nervous system of two actinians, *Adamsia* and a phyllactinid, probably *Phlyctenactis tuberculosa*. In the paper Lendenfeld merely mentions that a nerve layer (nerve net) is sandwiched between the supporting lamella (mesoglea) and the muscle-feet at the base of the ectoderm, adding very little to the Hertwigs' mammoth paper. In addition, he endows the cnidocil of the nettle cell with sensory capacity, noting that the process from this cell merges with the nerve layer. Nine years later Schneider, another familiar figure just discussed, revisited the sea anemone *Adamsia*, but only to assert brashly that its nervous system was no different than those of hydroid polyps (Schneider, 1892), despite reliable evidence to the contrary from the Hertwig brothers.

It was a full decade before another contribution on actinian nervous systems surfaced, but this time it was far less superficial than Lendenfeld's or Schneider's, and it added substantially to the solid work of the Hertwigs. The author of this next contribution was a Belgian, Jules Havet (1866–1948), born in the French-speaking city of Tournai. He studied medicine at the University of Louvain and soon practised psychiatry at the Gheel Asylum, near Antwerp. This was an institution off the beaten path, in fact a hospital surrounded by a commune of boarding rooms where the Gheel citizens cared for the patients in their abodes. Havet rose to become director of pathology at the asylum. It was, curiously, during his years there that he took an interest in invertebrate nervous systems. He maintained a link with his alma mater in order to facilitate his academic research and to keep his scholarship well honed. He was appointed *chargé de cours* in Louvain in 1900, and it was in his dual capacity of pathologist in Gheel and assistant teacher in Louvain that he produced his research on the histology of the nervous system of sea anemones at the turn of the century (Havet, 1901). He was thirty-six years old when he published this paper and, contrary to the majority of the protagonists in our story, did not enjoy the guidance of an authoritative luminary in conducting his research.

Havet's scientific maturity made him aware of the difficulties he faced. He noted the special histological challenge posed by sea anemones in comparison with jellyfish: the cells are small and compactly laid; the search for nerve elements requires thick tissue sections, but the opacity of the tissues makes this difficult; and its nerve cells are inextricably buried among other cell types from which they do not particularly stand out. The efforts of predecessors had been doomed by inadequate histological methods, except for the work of the Hertwigs who, in Havet's estimation, offset their flawed methods by their superb investigative skills. But he pointed out that other distinguished zoologists had not been as successful as the Hertwigs. The Swedes Oskar Carlgren and Gustav Retzius had failed in their attempts, and Ramón y Cajal's own brother, Pedro, also had proven unsuccessful at tackling sea anemones.

Havet used mainly two histological methods: the Golgi impregnation method as predicated by Ramón y Cajal, and methylene blue. In his hands these methods, especially the Golgi, had previously yielded highly satisfactory results with other invertebrates – gastropod molluscs (Havet, 1899) and annelid worms (Havet, 1900). Equipped with such unimpeachable expertise he approached the neurohistology of sea anemones with breezy confidence. He was rewarded with preparations that displayed nerve cells and their distribution in remarkable detail for the time and made his observations all the more reliable. In his original paper of 1901, *Metridium dianthus*, the sea anemone common on the shores of the North Sea, was the object of his study, but in a later paper (Havet, 1922) he added the species *Actinia equina*, *A. sulcata*, and *Heliactis bellis*, collected at the Station biologique de Roscoff in Brittany. He was careful to select young samples for best results. However, the results in the 1922 paper only confirmed the observations reported in 1901, adding no new information.

Havet described in the ectoderm of the body wall numerous sensory cells that make contact with a thin bundle of nerve fibres at the ectoderm-mesoglea interface (Fig. 7.7, below). The nerve cells constituting the basiectodermal bundle (nerve net) he called "cellules nerveuses motrices" (motor nerve cells); these are what other authors had better named ganglion cells, as they connect with other nerve cells as well as with muscle feet. In a new observation that had eluded even the Hertwig brothers, Havet noticed that the type and distribution of nerve elements in the endoderm are almost a mirror image of those in the ectoderm (Fig. 7.7, below). He also depicted in the mesoglea – which he calls mesoderm – large multipolar cells projecting processes

7.7 · (*Above left*) Jules Havet as professor at the Université catholique de Louvain. Courtesy Université catholique de Louvain. (*Below*) His rendition of a cross-section through the body wall of a sea anemone, showing sensory and nerve cells in the ectoderm and endoderm, and connecting mesogleal nerve cells. (*Above right*) Section through internal septa, showing endoderm richly supplied with sensory and nerve cells as well as mesogleal nerve cells. Drawings from Havet (1922), Figs. 3 and 5.

to the nerve net of both ectoderm and endoderm. He interpreted these as nerve cells bridging the two nerve nets, ensuring thereby the transmission of excitation from the ectoderm to endodermal muscles or vice-versa. However, in discussing these cells Havet cautiously warned that they might have been stellate connective tissue cells – known today as amoebocytes – which had been impregnated by the Golgi stain.

In the basal foot and in the oral disc he found even more nerve cells – sensory and ganglion cells combined – than in the column of the body wall. These he correlated with the greater sensitivity of these parts to stimulation. He also found that ganglion cells at the base of the tentacles send long processes to the longitudinal muscle of the tentacles, thus possibly mediating the contractions of the tentacles in response to touch or food signals. Probably his most original discovery was the dense assemblage of sensory and nerve cells in the septa, infoldings of the endoderm and accompanying mesoglea which traverse the cavity of the body wall between the outer ectoderm and the pharynx ectoderm (Fig. 7.7, above right). The muscles of these septa are associated with changing the diameter of the body column or with the crumpling behaviour of sea anemones.

Although Havet was keen to offer physiological or behavioural interpretations for his observations, he made no attempt to cast such observations in an evolutionary or phylogenetic mold as his contemporaries would be wont to do. At best, he pointed out that the nervous system of sea anemones originated not only in the ectoderm but also in the endoderm, and that it recalled to an extent that of flat worms (platyhelminthes). This interpretational slant can perhaps be explained by his deep religious feelings. The University of Louvain, where he studied and later worked, was run by clergy of the Catholic Church. His mentors there, Jean-Baptiste Carnoy and Eugène Gilson, were priests as well as biologists, and it is likely that teaching Darwinian evolution was anathema on the campus at the turn of the century.

The neurophysiology and anatomy of sea anemones (*Heliactis bellis*) was also the subject of a small section of Max Wolff"s 1904 dissertation. Wolff (see previous section) concludes from his experiments and those of others that a stimulus (tactile probing) on the oral disc does not propagate in all directions with the same ease. If you irritate the oral disc at the base of a tentacle it will contract at a given stimulus intensity – but only that tentacle. A stronger stimulus must be applied to cause the contraction of the neighbouring tentacles, and even a stronger one to bring the whole tentacle crown to contract. Wolff interprets this in terms of different thresholds that determine the spread of stimulus propagation. This was an important discovery of a phenomenon later known as synaptic facilitation of nerve net conduction (see chapter 10).

After discussing at length the behaviour and sensory capabilities of sea anemones, Wolff turned to histology to reveal a neuronal basis for his physiological results. He subjected his specimens to the very methods used by his immediate predecessor, Jules Havet, but with less success. Silver chromate precipitates prevented him from determining with certainty the presence of nerve cells in the endoderm that Havet had no difficulty revealing. Wolff would not have been aware of his technical flaws, however, for the simple reason that he was not even aware of Havet's contribution – it was not cited in his 1904 paper – which should have served as the comparison point. Similarly, Wolff failed to detect nerve cells in the ectoderm of the foot, where Havet had seen large concentrations of them. He saw a concentration of nerve cells akin to a nerve ring in the oral disc, whereas Havet never observed enough nerve cells there to justify such a designation. Wolff's illustrations give all the appearances of the poor workmanship that caused artifacts and distortions in his preparations. And yet, not knowing of Havet's work, he may have believed he had produced a highly original work.

Efforts to appraise the actinian nervous system were adeptly capped in 1909 by Pavel Grošelj (1883–1940), born in Ljubljana, the capital of Slovenia. In his formative years, Grošelj's interests vacillated between the natural sciences and poetry, but in the end he opted for a bachelor's degree in biology at the University of Vienna. He started his doctoral research at the same university in the fall of 1904 under the supervision of Berthold Hatschek, director of the Zoological Institute.

In the paper that resulted from his doctoral dissertation (Grošelj, 1909), Grošelj made no attempt to conceal that it was Hatschek who had suggested the research topic. Interestingly, he also revealed that he had sought advice from Karl Schneider, who had moved up the academic ladder since his work on coelenterate nervous systems in the early 1890s to become a professor at the Zoological Institute of the University of Vienna. To collect his sea anemones Grošelj shunned Naples, where most of his colleagues flocked, in favour of the Zoological Station of Trieste (northeastern Italy) on the Adriatic coast, just across the border from his native Slovenia. There, he sampled for histological processing specimens of the largest collection of sea anemones to date, to ensure that he could procure the best insights on their nervous system.

If Havet had found that the Golgi impregnation technique yielded his best results, Grošelj capitalized instead on Ehrlich's vital methylene blue method. He tinkered with the method until it gave highly satisfactory results on intact animals – no easy matter when considering the copious mucus secreted by living sea anemones and their readiness to close the mouth opening with a strong sphincter, thus making it difficult if not impossible for the stain to penetrate into the body cavity and reach the pharynx and septa. Fortunately, some of the species of sea anemones reacted to the irritation caused by the stain and the accompanying oxygenation by everting their pharynx, thereby allowing the stain to access the deepest recesses. Thanks to serendipity and to Grošelj's masterly handling of the stain, he was able not only to gain unparalleled cytological details of the nerve cells but also to enjoy a bird's eye view of the topological distrbution and density of the neurons.

Because the sensory nerve cells were more easily stained by methylene blue than ganglion cells, Grošelj chose to highlight them. He observed a large variety of sensory cells in the ectoderm of the tentacles (Fig. 7.8, right). The position of the cell body in the depth of the ectodermal layer varies greatly, and the outward processes from the cell bodies vary in length accordingly. Some of these processes reach the external surface, in which case they possess a terminal knob. The inwardly directed processes of these bipolar cells project in various directions and merge into a subectodemal nerve fibre layer. Many of these processes exhibit fine varicose swellings. In the endoderm of the same tentacles, sensory cells are stouter than those of the ectoderm, with more prominent end-knobs and varicose swellings. Ganglion cells, by contrast, took longer to stain than sensory cells and did so capriciously. When Grošelj was able to capture them in the pharyx wall, he saw some bipolar and many multipolar cells, the processes of which tangle with each other to form a plexus. Elsewhere in the body only a few ganglion cell bodies scattered here and there were successfully stained.

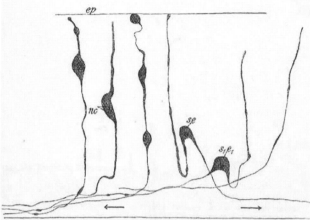

7.8 · (*Left*) Pavel Grošelj in the 1920s. Courtesy Digital Library of Slovenia (ref. no. BTN21W1F). (*Right*) A compendium of ectodermal sensory nerve cells in the tentacles of *Cerianthus membranaceus*. Drawing from Grošelj (1909), text-Fig. 1.

One interesting, if inaccurate, observation made by Grošelj concerns the capsules of the stinging cells. Instead of tracing an innervation to these cells as others before him had attempted, Grošelj actually depicted the capsule as a large organelle inside the cell body of a sensory cell, which otherwise looked like any other sensory cell with a distal knob and varicose processes. He found no evidence of innervated grand cells, however, contrary to claims by Wolff (1904). Muscle fibres, on the other hand, he found to be commonly innervated by cells that showed some analogy with the mesogleal nerve cells of Havet. Many of these cells send varicose processes directly or through branches to the muscle fibres, on which they run over a variable distance. In some cases the process ends on the muscle fibre in the form of a single, often triangular, neuromuscular ending. However, Grošelj refrained from calling them nerve cells outright, preferring to allude to them as merely "mesogleal cells," even though their connection to the muscle feet clearly evokes cells that fill a motor role.

Grošelj paid more attention to the distribution of nerve elements in sea anemones than Havet did. He noted a great concentration of sensory cells in tentacles, especially in the ectoderm, where these cells are in contact with the outside world. This led him to view tentacles as sensory organs in their own right. He is rather silent on the body wall, but his implicit message seems to be that sensory cells are more sparsely distributed there. In addition to numerous sensory cells, Grošelj found an increased density of the fibre layer (nerve net) in the oral disc, and even more so in the pharynx wall. His drawing clearly reveals the relatively stout sensory cells projecting their distal processes into the dense layer of varicose nerve fibres. This is a significant discovery that had escaped both the Hertwigs and Havet. It led Grošelj to consider that sea anemones do show an incipient centralization of the nervous system, but instead of seating it in the oral disc as the Hertwigs had done, he located it in the newly found pharyngeal ectodermal nerve net. Oddly, where he found ganglion cell bodies in the pharyngeal or oral disc nerve nets, he would notice here and there what appeared to him anastomoses between these cells.

In the conclusion to his paper, Grošelj reflected on the place of sea anemones as to the degree of organization of their nervous system in relation to hydrozoans and scyphozoans. Grošelj put the *Hydra* nervous system at the bottom of the scale, on the basis of Schneider's flawed 1890 paper. While he rated hydrozoan and scyphozoan jellyfishes above sea anemones in the development of their sensory centres, he considered the condensation of nerve fibres in the pharynx a cut above the nerve rings of jellyfish.

Colonial Anthozoans

Once an interest in colonial hydroids had developed in late nineteenth century, a similar curiosity drove some zoologists to the study of colonial anthozoans, with the objective of comparing their nervous systems to those of sea anemones. For some, the opportunity arose to ask the right (and uniquely important) question: is there a colonial nervous system separate from the polyp nervous systems? These colonial species include the hexacorallians minus sea anemones (stony corals) and octocorallians (soft corals and sea pens). Stony corals were neglected because their heavy calcareus skeleton made histological processing too difficult.

The first to provide significant information on the nervous system of soft corals was someone we are already acquainted with: the Russian Alexei Korotneff (1852–1914). While spending time as usual at the station of Villefranche-sur-Mer, Korotneff made an excursion further west along the coast, to Banyuls-sur-Mer near the Spanish border. There he found a ready supply of living sea pens (*Veretillum*). He published a short note (Korotneff, 1887) in which he described their nervous elements. Sea pens contain two types of individuals: feeding polyps and zooids. Feeding polyps are also reproductive individuals and are miniature look-alikes of sea anemones. In Korotneff's days the role of the diminutive zooids was poorly undertood, but they are now known to represent regressed polyps involved in siphoning water inside the colony chambers to keep the colony turgescent and oxygenated.

Korotneff found a plexus of ganglion cells between the epithelium and the longitudinal muscle in the ectoderm of the polyp tentacles and oral disc, an arrangement previously described in sea anemones. He asserted, mistakenly, that some of the neuron-like cells in the nerve net are light-emitting cells producing a "phosphorescence." Light-emitting cells with long bifurcating processes do exist in the polyps, but they are not nerve cells, and Korotneff seemed to be unaware that these cells are located in the endoderm, across the mesoglea. He saw fusiform, bipolar nerve cells closely associated with muscle layers, but he found no nervous elements in the endoderm.

If Korotneff had but a passing interest in sea pens, his successor made a career of studying coelenterates, and particularly octocorallians. Sydney J. Hickson (1859–1940), born to a prosperous London boot and shoe manufacturer, led a "distinguished life devoted to zoological research" and is said to have taken "an active part in the application of the principles of Darwinian evolution to morphology and embryology" (Gardiner, 1941). As an undergraduate, Hickson was mightily influenced by the lectures of two strong Darwinian advocates, Ray Lankester and Thomas Huxley. He went to Cambridge, where he developed research interest, first in the structure of invertebrate eyes, and then in corals and other sedentary coelenterates. He was appointed Beyer Professor of Zoology at Owens College, University of Manchester, in 1894, and the following year his paper on the morphology of the soft coral *Alcyonium* (Hickson, 1895) appeared. In it he devotes a small section to the nervous system, but his command of the literature was deficient. When he compares his observations on *Alcyonium* with those of sea anemones, his only reference point is the Hertwig brothers, and he seems to have been unaware of Korotneff's paper.

Hickson's description of the nerve plexus in the ectoderm of tentacles corresponds to that of Korotneff; in contrast to Korotneff, however, Hickson also observed a similarly organized nerve net in the endoderm. In addition, he saw a plexus of stellate cells with their interconnecting fibres throughout the mesoglea, but was reluctant to label them as nerve cells. The question of a mesogleal nerve net in anthozoans was controversial then, and remained so for a century to come. Havet was the first to report their presence. The fact that they were noticed in anthozoans but not in hydrozoans or scyphozoans certainly has to do with the much larger body mass filled by mesoglea in this cnidarian class. The controversy over the identity of the mesogleal cells was alive even within Hickson's own laboratory. Four years after the publication of his paper, one of his graduate students published a paper on the morphology of a new soft coral named after his supervisor, *Xenia hicksoni* (Ashworth, 1899).

James Hartley Ashworth, Hickson's graduate student, was born in 1874 and brought up in Bolton, Lancashire. He studied at University College, London, graduating in 1895, and did his doctoral thesis under Hickson at Owens College. Like his mentor, Ashworth was more interested in the general anatomy of octocorallians and in their spicules (units of the calcareous skeleton) and nettle cells. This objective was reflected in his choice of histological treatments: they were designed primarily to yield a general view of cellular organization, whereas Havet and Grošelj had targeted the nervous system above all. Consequently, Ashworth devoted a very small section of his paper to the nervous system, and its only topic was what he interpreted as a mesogleal nervous system (Fig. 7.9, right). In depicting it, he stresses its relationship with the other components of the polyp nervous system:

> The cells of the nervous system are exceedingly small, and usually fusiform, triradiate, or stellate in shape, the angles of the cell being produced into nerve fibrils.
> On tracing the fibrils outwards from the mesoglea into the ectoderm, they are seen to be in connection with small cells

7.9 · (*Left*) James Ashworth in 1924 at Woods Hole. Courtesy Marine Biological Laboratory Archives. (*Right*) Oblique section through the mesogleal nervous system of the soft coral *Xenia hicksoni* and its connections to the ectodermal and endodermal epithelia, according to Ashworth (1899).

which are situated n the deeper part of the ectoderm. Owing to the irregularity of the inner face of the ectoderm, and to the presence of spicules in the portion of the layer where the nerve-cells are situated, it is not possible to obtain a section which shows the ectodermic nerve plexus clearly, but small portions of it may be seen where the spicules are slightly less numerous.

In the case of the endodermic nerve plexus this difficulty does not exist, and an oblique section through the wall of a polyp shows that the plexus of fibrils in the mesoglea is connected with minute stellate cells situated upon the outer face of the muscle processes of the endoderm cells. (Ashworth, 1899)

This observation appeared to offer a persuasive argument for the existence of a mesogleal nerve net, but only five years later another of

Hickson's students demolished Ashworth's claim that the cells of this plexus resembled those of the nerve layer of the ectoderm.

Edith Mary Pratt was born in Dukinfield, Cheshire, on 5 September 1873. She obtained both her bachelor's (1897) and master's (1900) degrees at Owens College. She went on to make what must then have been considered, for a woman, a daring leap to doctoral studies. Pratt was among the few women training in science at the time, and it is to be assumed that Hickson showed an unusual measure of tolerance or liberalism by accepting her in his laboratory. She proved to be an astute experimenter. For her paper (Pratt, 1904) she gathered two types of evidence: high-power microscopic observations of living mesogleal cells over time in thick slices of polyps; and experiments designed to demonstrate the amoeboid nature of mesogleal cells. For the first set, she followed individual cells in time-lapses of thirty minutes and saw that they moved by pseudopodial thrusts. In the second, she exposed polyps to a suspension of carmine particles and observed that the coloured particles made their way into endodermal cells of the polyp's wall, which eventually developed pseudopods as they migrated to the mesoglea, eventually ending up as amoebocyte-looking cells. She concluded that "the amoeboid character of the mesogloeal cells affords substantial evidence that the so-called nerve cells of the mesogloea are endoderm cells which have become amoeboid and wandered into the mesogloea." Surprisingly, she did not dismiss their role as nerve cells out of hand, suggesting that "the amoeboid cells are nutritive and excretory as well as nervous in function, and may therefore be looked upon as neuro-phagocytes." But the issue was not entirely settled by Pratt because she considered only the large mesogleal cells. The status of Ashworth's smaller mesogleal cells remained uncertain and so the controversy lived on.

A more ambitious and comprehensive analysis of the nervous system of octocorallians than previously attempted appeared in 1908. This work was the product of Nicolai Kassianow, a Russian doctoral student of Otto Bütschli, the Heidelberg professor with whom we became acquainted earlier in reference to other protagonists in this story. We know next to nothing about Kassianow (1878–1948) apart from his association with Bütschli, who guided him in his dissertation on an obscure kind of hydrozoan, the stalked jellyfish (Kassianow, 1901). Kassianow felt no urge to return to Moscow, so when Bütschli suggested he stay in Heidelberg and take on the nervous system of octocorallians, he promptly obliged. Strangely, he concentrated his attention on the already much studied *Alcyonium*. He was aware of the contributions of Korotneff, Hickson, and Ashworth, but they did not deter him. He travelled to the Marine Biological Stations of Trieste and Villefranche, and to the Bergen Museum in Norway to collect his specimens (*Alcyonium digitatum* and *A. palmatum*). A preliminary account of his results appeared in the Bergens Museums aarbog (yearbook) of 1903 (Kassianow, 1903), but the definitive paper came out only in 1908 in a special issue of the *Zeitschrift für wissenschaftliche Zoologie* honouring Bütschli on the occasion of his sixtieth birthday (Kassianow, 1908a). By then Kassianow had returned to Moscow, where he wrote the article in 1907.

From the start Kassianow expressed two regrets: that he was unable to survey enough species to determine how representative the nervous system of *Alcyonium* was of octocorallians at large; and that, while he obtained a satisfactory picture of the polyp nervous system, he was unable to get a handle on the colonial nervous system. However, he took comfort in thinking that his knowledge of the polyp nervous system could be a springboard for eventual insights into the nature of the colonial nervous system. Kassianow approached methods conservatively, relying mostly on the maceration and osmium techniques employed by the Hertwigs. One significant departure, however, was the use of magnesium sulfate as a general anaesthetic for these animals. It allowed an unparalleled relaxation of the body and therefore a less crowded view of cellular layouts in the viewing field of the microscope. He picked up this technique at the suggestion of Professor Carl I. Cori, the director of the Trieste Biological Station, where the relaxant had become widely used. (Cori was the father of the winner of the 1947 Nobel Prize in Physiology or Medicine, Carl F. Cori.) Kassianow was a careful methodologist and observer but, if his plates of illustrations are any indication, his artistic talents left much to be desired.

From his observations Kassianow concluded that the nervous system of the polyps is primarily associated with the ectoderm, and this ectodermal nervous system forms a dense nerve plexus on the oral side of the tentacles, in the oral disc and in the ectoderm of the pharynx wall, thus mirroring Havet's and Grošelj's descriptions for sea anemones. It is composed of "multipolar and bipolar ganglion cells with long, extremely fine, varicose, branching extensions and spindle-shaped sensory cells, which protrude with their thin distal end beyond the epithelial surface." He found that this nervous system is best developed where the ectodermal muscle is strongest. He also noticed a concentration of nerve fibres where the oral disc met each of the eight septa separating the body cavity lengthwise. He saw large bipolar and multipolar cells with long processes in the ectoderm of the body wall, but stopped short of calling them nerve cells. He observed an endodermal nerve net composed of ganglion cells similar to their ectodermal counterparts, but to him it did not appear as developed as the one described in sea anemones.

Contrary to Havet and Ashworth, Kassianow could not identify any nerve cell in the mesoglea. He saw what he described as large, branched gelatinous cells in the mesoglea, which probably corresponded to Pratt's amoebocytes, but he thought it highly improbable that they would serve a nervous function. He was eager to find signs of a colonial nervous system, but came away disappointed. His physiological observations indicated a fair degree of independence among the polyps, thus obviating the need for a coordinating nervous system on behalf of the colony.

In the same *Festschrift* volume of the *Zeitschrift für wissenschaftliche Zoologie*, Kassianow also published an article comparing the nervous systems of octocorallians and hexacorallians (sea anemones) (Kassianow, 1908b). He stressed their similarities, such as the shared presence of endodermal and pharyngeal nervous systems, but also pointed out insightful differences. One difference is the distribution of sensory and ganglion cells in the tentacles. While these cells are distributed equally all around the tentacles of sea anemones, in soft corals they are more densely distributed on the oral side of the tentacles. Kassianow attributed this difference to the fact that soft corals, being limited in their number of tentacles – always only eight compared to multiples of six in sea anemones – must make optimal use of the sensory capabilities of these tentacles for feeding. The flattened aspect of soft coral tentacles and their feather-like appearance due to the branching-off of pinnules are designed to enlarge the surface area of sensory contact, especially in the oral direction as food is channelled to the mouth opening. Incidentally, from this and other observations, Kassianow concluded that the tentacles are specialized for chemical sensing and the oral disc for touch sensitivity.

In pursuing his comparative analysis, Kassianow touched again on the thorny issues of the mesogleal nerve net and the colonial nerve net. In assessing the likelihood of a mesogleal nerve plexus, which he failed to see in *Alcyonium*, he stressed that neither the Hertwigs nor Wolff found it in sea anemones. He would have welcomed persuasive evidence because such a presumptive nervous connection between ectoderm and endoderm might have helped explain how septal endodermal muscles contract when the ectodermal nervous system is stimulated. But for lack of such evidence he was inclined to think that what Havet saw in *Metridium* and Ashworth in *Xenia* were ectodermal ganglion cells that delaminated and sank in the mesoglea as a result of harsh histological treatment. As for the colonial nerve net, Kassianow saw no evidence of it in Hexacorallia, where very few colonial forms had been investigated; but, according to the literature on hydroids (Zoja, Wolff), nerve tracks seem to be present in the coenosarc (colonial binding tissue) and physiological evidence suggests that electrical stimulation of the coenosarc causes the contraction of polyps. Kassianow suspected that his methods may have been inadequate to allow observation of a colonial nervous system and that the Golgi impregnation method, which he was unable to work out in *Alcyonium*, might eventually settle the question. Seventy years later, when the colonial nervous system was revealed by the Golgi method in octocorals (Satterlie et al., 1980), his prediction came true.

By a quirk of historical symmetry, the flurry of studies on octocorallians ended on the eve of the First World War with the animal model it had begun with – the sea pen *Veretillum*. This time, the author of this study was not Russian, but Austrian. Albert Niedermeyer (1888–1957) was born in Vienna and studied natural sciences in Vienna and Breslau, where he earned his diploma of Doctor of Medicine as well as a PhD While in Breslau he fell under the spell of zoologist Willy Kükenthal (1861–1922), a former student of Ernst Haeckel who had moved on from a professorship in Jena to a post as professor of comparative anatomy and Zoology at the University of Breslau. His major research focus was the octocorals, of which he had a sizeable preserved collection in the natural history museum of the campus. So Kükenthal suggested

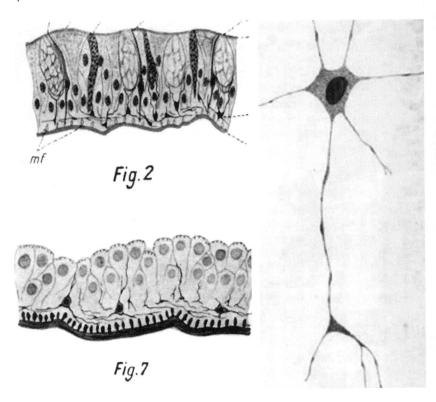

7.10 · (*Left*) Nerve elements of the sea pen *Veretillum*. Ectoderm of the rachis in which are seen sensory and ganglion cells (*top*) and endoderm of the pharynx wall where ganglion cells appear to send sensory process toward the surface (*bottom*). (*Right*) Details of a ganglion cell from a polyp tentacle. From Niedermeyer (1914), Figs. 2, 7, and 11.

that Niedermeyer work on *Veretillum* from preserved materials there that had originally been collected during the German deep-dea *Valdivia* expedition and from the Station zoologique d'Arcachon in France.

Niedermeyer used Apathy's gold chloride technique specifically to reveal nerve cells; otherwise, classical general stains such as haematoxylin-eosin prevailed. The paper that resulted from his investigations (Niedermeyer, 1914) dealt with the general anatomy, taxonomy, and phylogeny of the species, and only a few pages were given over to the nervous system. The quality of Niedermeyer's work could have suffered from the fact that he was handed material that had not been collected with paramount care for the preservation of nervous tissue, but surprisingly his illustrations suggests otherwise (Fig. 7.10). He lamented the impossibility of applying "proven" methods for nerve cells such as methylene blue or the maceration of fresh material, but the gold chloride method served him well for his purpose. As in sea anemones, neurons and other cells are small in octocorals, so Niedermeyer was forced to rely heavily on microscopic observations at high magnifying power, using oil immersion.

Niedermeyer expected the presence of a well-developed nervous system to supply the prominent muscles he was observing in both the polyps and the colonial tissue (rachis). He found nervous tissue in the usual locations reported by others: dense ectodermal nerve plexus in the tentacles, oral disc, and pharynx, and less dense in the body wall; loose endodermal nerve plexus in the tentacles and oral disc, and in septa. But where Niedermeyer broke from his predecessors, especially Kassianow, was in his discovery of nervous elements in the ectoderm of colonial tissue (Fig. 7.10, top left), in the endoderm of the pharynx lining (Fig. 7.10, bottom left), and in the endoderm of the cavities of the rachis (coenenchyme), where nerve cells appear to form a plexus overlying the well-developed longitudinal muscle layer. Sometimes he could make out the involved ganglion cells with clarity (as seen in Fig. 7.10, right); depicted here is a multipolar cell with a small cell body (6–7 microns) and a long varicose process that bifurcates at some point. The depiction gives the impression of a polarized cell, with what looks like a crown of dendrites and a single axon, which provocatively goes against the grain of the accepted dogma at the time; namely, that coelenterate nerve cells are non-polarized.

Niedermeyer, like Kassianow, could not detect the presence of mesogleal neurons. He criticized Kassianow who, because he saw few nerve cells in the endoderm, had strained credulity by speculating that ectodermal nerve cells could invade the mesoglea to help supply innervation to endodermal muscles. Niedermeyer, and rightly so, pointed out that the endodermal muscle sheets are too massive to be adequately supplied in this way; his finding of a rich nerve plexus in the endoderm closed the argument. Apart from this controversy, it is clear that Niedermeyer's major contribution was his discovery of a nerve layer in the colonial tissues, which suggested that the colonial organization of octocorals affected the nervous system, which adapted by neural re-arrangements designed to assist in the behavioural coordination of the individuals across the colony.

Jellyfish Revisited

No other animal of choice for coelenterate neurohistology would have been considered more daunting to the new generation of zoologists than the jellyfish, if only for the unsettling realization that giants such as Romanes, Schäfer, and Eimer had trodden there. What else was there to be discovered? And what feat was to be accomplished in order to make one's own mark? This was the challenge. How successfully was it met?

Only four years after Schäfer's landmark paper, Lendenfeld entered the fray with an article on the anatomy and histology of Australian scyphomedusae. We have seen how Lendenfeld, from down under, produced a short note on the nervous system of hydroid polyps. In the few years he spent in Australia and, to a lesser extent, in New Zealand, Lendenfeld undertook a prodigious scientific labour and produced innumerable papers on various aspects of the scyphomedusae, hydromedusae, and sponges, with an emphasis on taxonomy, classification, and anatomy. He discovered a new species of the scyphomedusa *Cyanea* that he named after his wife (*annaskala*). This species he studied in detail, including its nervous system (Lendenfeld, 1882). However, Lendenfeld did not add original information to what Schäfer and Eimer respectively had already contributed on *Aurelia* and *Cyanea*, except for two new discoveries.

The first of Lendenfeld's discoveries concerns stinging cells (cnidocytes), which are defining elements of cnidarians and were more popular objects of study than the nervous system. At the time a debate simmered among coelenterate zoologists about whether cnidocytes should be considered sensory cells, with the cnidocil as a sensing device. If so, then the cnidocyte would be expected to send a signal to a subepithelial ganglion cell via a process that contacted the ganglion cell. Lendenfeld thought otherwise. He believed the cnidocil was too short to effect a sensory function, and that the cnidocyte therefore has only a motor function; namely, to shoot the nematocyst micro-harpoon to a prey. If that were the case, Lendenfeld argued, it could only mean that ganglion cells make neuro-effector contacts with the clusters of cnidocytes found in *Cyanea* – which is what he observed.

His other discovery related to the development of the nervous system. Few if any investigators had thought of examining the nervous system at early stages. Lendenfeld, whose curiosity seemed boundless, sorted out ephyra stages of *Cyanea* and found ganglion cells inside the ectoderm epithelium, rather than ganglion cells forming a subepithelial nerve layer, as in the adult stage. This observation suggested that the nervous system appears first in a disorganized state, but Lendenfeld refrained from drawing evolutionary implications. Incidentally, he depicts *Cyanea*'s ganglion cell processes as anastomosed to each other, in contradiction with Schäfer's depiction, of which he became aware only shortly before his own paper was going to press.

Seven years after Lendenfeld's contribution, a paper came out dedicated to the morphological analysis of the eyes of scyphomedusae. Such a narrow specialization was uncommon at the time. The author of this study was Wladimir Schewiakoff (1859–1930), born in St Petersburg to a Russian merchant and a mother of Prussian roots (Fokin, 2000). Having fared rather poorly at school, he was denied entrance to university programs. He spent two years at the St Petersburg Mining Academy, all the while showing more interest in things zoological than geological. By 1880 he had become a serious amateur zoologist, even undertaking a solo expedition to the Caucasus to collect his pet animals, insects. Such dabbles convinced him that he should pursue a university education, which he did in 1881 thanks to special permission from the Ministry of Public Education (Fokin, 2000). However, after three years of zoological studies at the University of St Petersburg, it became painfully clear to Schewiakoff that the training he was receiving was mediocre by European standards. He determined to continue his education outside Russia. As S.I. Fokin, his biographer, explains:

He chose the University of Heidelberg, one of the best universities in Europe. Other Russians had graduated from this university before Schewiakoff, and had studied for the "preparation for the professor degree." He, of course, knew it to be one of the best places to study invertebrate zoology. The key element in this was that the head of the Zoological Institute in Heidelberg at that

time was the famous protozoologist, the eminent world authority, Professor Otto Bütschli (1848–1920). From the beginning of
his education there (1885), Schewiakoff became a student of this
remarkable person. (Fokin, 2000)

So Bütschli became the catalyst for yet another career in zoology
and of another significant investigation of primitive nervous systems.
The outcome of Schewiakoff's research under Bütschli was a doctoral
dissertation arcanely entitled (translated): "Where one wishes a precise
anatomical and histological examination of the marginal body of the
Medusa *Charybdea* with special reference to the eye and possibly taking
into account the eyes of related Medusae. Contribution to the knowledge of the eyes of Acalaphs [scyphomedusae]." The thesis earned
him the Gold Medal of the Philosophical Faculty at Heidelberg. The
paper based on his thesis was published the following year (Schewiakoff, 1889).

The species that Schewiakoff studied in detail, *Charybdea* (now *Caryb-
dea*) *marsupialis*, was then lumped in with the scyphomedusae, but is
now considered to belong to a distinct jellyfish group, the cubomedusae. These are in fact the much feared box jellyfish, among the most
venomous animals in the ocean, all too well known for their deadly
sting. No wonder Schewiakoff sheepishly admits that he could only
work from preserved specimens! *C. marsupialis* happens to possess the
most complex visual system of all jellyfish (Martin, 2002), so Schewiakoff's choice of animal model turned out to be most suitable for mining
new and original data. The specimens were already fixed in mixtures of
osmic, picric, and acetic acids; Schewiakoff completed the histological
treatment by embedding specimens in paraffin, cutting thin sections
which were then stained with haematoxylin or, for better singling out
of nerve cells, with the gold chloride method.

We are already familiar with the sensory complexes, the rhopalia, at
the bell margin from Romanes's experiments described in chapter 5.
There are only four rhopalia in cubomedusae, and Schewiakoff found
that in *C. marsupialis* (Fig. 7.11, right) each appeared club-shaped with
the stem connecting to the bell. In the rhopalia he numbered two
complex eyes (*Linsenaugen*), one large and the other smaller (Fig. 7.11,
right), equipped with lens, cornea, and a well-defined retina containing
ciliated photoreceptor cells. These photoreceptor cells, he wrote, send

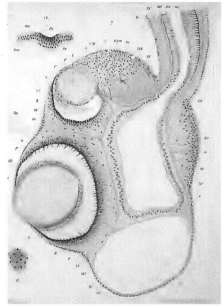

7.11 · (*Left*) Wladimir Schewiakoff at the Zoological Institute in Heidelberg in 1891, two years after the publication of his paper on the eyes of scyphomedusae. Courtesy Urban & Fischer Verlag (Elsevier GmbH). (*Right*) Section through a rhopalia of *Carybda marsupialis* showing the two complex eyes. From Schewiakoff (1889), Fig. 7.

their processes to a nerve layer that merges with the nerve ring of the bell margin. In addition, he spotted four simple "pigment eyespots" (ocelli), in which pigmented photoreceptor cells are found but no optical, image-forming structure.

Schewiakoff thought the photoreceptor cells of pigment eyespots were derived from ectodermal sensory cells, with little transformation. At the origin of eyecups, he speculated, were simple epithelial invaginations which then became filled with a vitreous body to protect them from surrounding detritus or excessive light; the vitreous body was later invaded by a lens to produce the camera-like complex eye. The latter design he found reminiscent of the eyes of higher animals such as molluscs, arachnids, and vertebrates, and compared them to the parietal eyes of lizards. He marvelled at the surprising sophistication of such eyes in so primitive an animal, and judged that the ocelli of scyphomedusae such as *Aurelia*, which, oddly, he assumed were of endodermal origin, pale in comparison.

The store of knowledge on the nervous system of scyphomedusae was enlarged a few years later by a detailed description of the nervous system and sense organs of *Rhizostoma*. The young German investigator who undertook this work, Richard Hesse (1868–1944), was a student of Theodor Eimer. Eimer, as we saw in chapter 5, had gone on to pursue scientifically questionable ideas and had abandoned jellyfish. What, then, possessed him to suggest a jellyfish project to Hesse is a mystery. Hesse had already completed his habilitation thesis in 1894 – on the comparative anatomy of earthworms – but he was obviously in no hurry to leave Tübingen when Eimer proposed the project. The young and ambitious Hesse, now promoted to the rank of *Privatdozent* in zoology and employed as assistant at the Zoological Institute, wasted no time; he began in March and April of 1895 with a sojourn at the Naples Zoological station to collect his specimens, and completed the histological work back in Tübingen in May and June. This whirlwind pace was maintained with the publication of his results the same year (Hesse, 1895).

If the content of Hesse's paper is any guide, the research was perhaps too hastily done. It contains no original information, providing at best confirmatory observations on what Schäfer and Eimer had found in other scyphomedusae. His pedestrian use of conventional histological techniques was hardly conducive to landmark contributions. One interesting observation stands out, however. He found a highly developed nerve tract at the base of the marginal body (rhopalia). The marginal body, he points out, has two sensory pits. Nerve fibres from the "outer sensory pit pass through the mesogleal layer to a subendodermal tract surrounding the epithelium of the marginal body canal." According to Hesse, the nerve fibres then pass from this region to the nervous epithelium of the "inner sensory pit" lying underneath the base of the marginal body, which contains a rich supply of ganglion cells and which he considers to be the nervous centre of *Rhizostoma* (*Nervencentrum*), for want of a proper nerve ring as in hydromedusae.

He offered an interesting reflection on the significance of Eimer's and Romanes's jellyfish experiments in the light of his own neuroanatomical observations in *Rhizosoma*. Hesse realized, for instance, that

his mentor, Eimer, had been wrong in thinking that removing the rhopalia was tantamount to removing the nerve centre; the nerve ring in
Rhizostoma is a few millimetres away and can still access contractile
zones downstream via the bipolar cell nerve plexus. In contrast, when
Eimer caused paralysis by removing all eight "contractile zones" from
the swimming bell on the exumbrella side, Hesse suspected that in this
case the nerve centre was effectively compromised. He reiterated the
notion that the organization of the nervous system – the radial arrangement of the sensory centres and the nerve ring, and the diffuse umbrel-
lar nerve plexus – easily explains the power of motor recuperation of
jellyfish from all manner of surgical cuts.

Bringing the Eye into Focus

The next contributions on the topic came from across the Atlantic
and picked up the cubomedusae where Schewiakoff had left off. Two
young and promising Americans were involved: Franklin Story Conant
(1870–1897) and E.W. Berger (1869–1944). Conant was born in Boston
and earned his bachelor's degree at Williams College, Williamstown,
Massachusetts, in 1893. He entered the PhD program of Johns Hopkins University in 1894 and, like William Conn before him (see earlier
in this chapter), he fell under the influence of Professor W.K. Brooks,
who had transformed the Zoology Department at Johns Hopkins into
a dynamic and influential centre of post-Darwinian zoology. After mas-
tering the zoological literature Conant took on research projects at
the field station of Johns Hopkins in Beaufort, South Carolina, focusing on chaetognaths and crustaceans, and producing papers from his
labours. Brooks then suggested that Conant tackle a comprehensive
investigation of the cubomedusae.

So far in this narrative, for lack of records, we have had no opportunity to report on the feelings or assessments of students by their
teachers. This opportunity now arises as we read what Brooks had to
say about his young charge in the preface to Conant's published PhD
dissertation (Conant, 1897): "I myself felt confident that the career on
which he had entered would be full of usefulness and honor. I was de-
lighted when he was appointed to the Adam T. Bruce Fellowship, for
I had discovered that he was rapidly becoming an inspiring influence

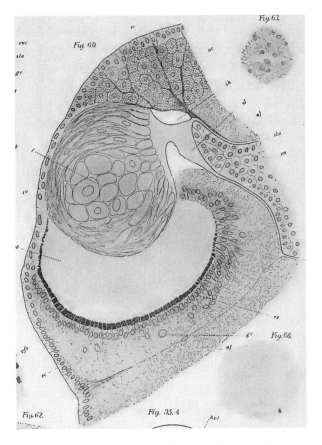

7.12 · Vertical section through the smaller complex
eye of *C. xaymacana*. From Conant (1897), Fig. 69.
Courtesy Johns Hopkins University Press.

among his fellow students in the laboratory, and I had hoped that we
might have him among us for many years, and that we might enjoy and
profit by the riper fruits of his more mature labors."

For his fieldwork, Conant in June 1896 joined a group of scholars
who worked out of Port Antonio in Jamaica, where Brooks had recently organized a tropical field station for his Johns Hopkins brood.
There Conant discovered two new species of box jellyfish, *Charybdea
xaymacana* and *Tripedalia cystophora*, and these served as the basis for
his anatomical investigations. He used his time wisely and was able
to expedite his thesis examination a year later, in June 1897. He de-

voted a sizeable portion of his thesis to the nervous system "because of the high degree of development attained by their nervous system" and for the reason "that here among the lowly-organized Coelenterates we find an animal with eyes composed of a cellular lens contained in a pigmented retinal cup, in its essentials analogous to the vertebrate structure" (Fig. 7.12).

In fact, Conant believed the nervous system of cubomedusae to be the most developed among all jellyfishes. It shared with hydromedusae the presence of a nerve ring. The latter is clearly visible at the surface of the subumbrella, where it forms a complex bundle of more or less packed nerve fibres overlaid by ganglion cell bodies. Here the ring is seen to interrupt the course of the circular muscle, which must curve its path to bypass the ring and continue its course around the bell. Strangely, this peculiarity led Conant to the statement: "It is evident that the tissues which elsewhere on the subumbrella were differentiated into muscle epithelium and muscle fibre have here become nerve epithelium and nerve fibre, a point that has not been remarked upon before, so far as I remember, and that may be of interest in connection with the neuro-muscular theory." This appears to be a rare lapse in scientific judgment on the part of a young scholar who "enriched all that he undertook by sound and valuable observations and reflections," as it smacked of free-wheeling speculation on a theory that had already lost its credibility on scientific grounds. In addition to the nerve ring, an assemblage of ganglion cells and their nerve fibres run radially on the surface of the subumbrella facing the cavity of the bell. The nerve ring connects to the sensory clubs (rhopalia) by sending two roots through the mesoglea ("gelatine") that reach the conical base of the stalk of the clubs.

In the four sensory clubs Conant observed the "special sensory organs" previously reported by Schewiakoff: the four simple eyes, seen as "simple invaginations of the surface epithelium," the two complex eyes, and the crystalline sac containing a large "concretion" (statolith). Here Conant describes the retina, which he saw as composed of pigment cells and visual cells. The visual cells project "rod-like processes" into the vitreous body that separates the retina from the lens. These processes are the long light-sensitive cilia of the visual cells.

Although Conant had satisfied the requirements for his doctorate, he resolved to return to Jamaica to further his knowledge of the development and physiology of the sensory organs of the box jellyfish (Conant, 1897, preface). So he arrived in Port Antonio in June 1897 and spent three months making experiments and taking notes. But his work was suddenly interrupted by a breakout of yellow fever that took many lives, including that of the leader of the Johns Hopkins field expedition, Professor James E. Humphrey. While helping to care for his fellow staff struck with the fever, Conant himself fell ill. As his condition worsened, he was transported *in extremis* back to his native Boston, where he died a few days later, on 13 September 1897, shortly before his twenty-seventh birthday. A promising career was tragically cut short.

Brooks was left with the incomplete notes of Conant's project but decided they contained enough promising material to warrant the enlistment of a new graduate student to pursue the investigation. The student he chose, Edward W. Berger, was born in Berea, Ohio, in 1869. He received his bachelor's degree in 1891 from the local Methodist institution, Baldwin-Wallace College, and an unspecified PhD degree in 1894 from the same institution (Denmark, 1995). Berger's aspirations must have shifted toward world-class zoology and his ambitions must have set him on a search for a higher-rated institution, for he ended up in Baltimore and there he was pulled into the circle of Brooks's Biological Laboratory. Berger started work with celerity on the box jellyfish project. Equipped with the material and notes left by Conant, mainly on the sensory physiology of *Tripedalia*, Berger spent time analysing and expanding on the experiments carried out by Conant – using the hydromedusan *Olindias* caught in Woods Hole for comparison – and conducting histological work on the sensory structures from *Charybdea* material preserved by Conant. He first published a preliminary note in which he reported the important discovery that the retina of complex eyes undergoes photomechanical changes (Berger, 1898). During daylight, pigment granules in the pigment cells move inside the cell's finger-like process extending into the vitreous body, thus screening the visual cells from excessive light, and during the night phase the pigment granules are retracted into the pigment cell body.

Berger's 1899 dissertation based on his and Conant's research appeared in published form the following year (Berger, 1900). The salient histological contribution made by Berger relates to the structure of the retina of the eyes in the sensory clubs. He provided the most detailed histological description so far of these structures. Where his predeces-

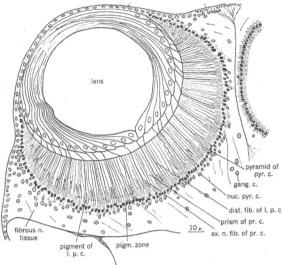

7.13 · (*Left*) E.W. Berger in the 1910s. From Denmark (1995). Courtesy Division of Plant Industry, Florida Department of Agriculture and Consumer Services. (*Right*) Sagittal section through the distal complex eye of *C. xaymacana*, showing the retina with its visual cells and their ciliary outer segments, and "pigment cells" with their extension into the capsule of the round lens. Also shown are the axial nerve fibre of prism cell (ax. n/d. fib. of pr. c.), distal fibre of the long pigment cells (dist. fib. of l. p. c.), fibrous nerve (n.) tissue, ganglion cell (gang. c.), nucleus of pyramid cell (nuc. pyr. c.), pigment (pigm.) zone, pigment of long pigment cells (l. p. c.), prism of prism cells (pr. c.), and pyramid of pyramid cells (pyr. c.). Adapted from Berger (1900), Fig. 7, by Bullock and Horridge (1965). Courtesy of Professor G. Adrian Horridge, with permission.

sors saw only one type of visual cell, Berger distinguished two in the distal complex eye: prism and pyramid cells (Fig. 7.13, right). In addition, he noted the presence of pigment cells sending long processes as far as the lens capsule, which suggested to him "that these cells may serve to conduct impulses to the lens, and that the latter is adjustable." We know today that he mistook a type of visual cells for pigment cells, and that all visual cells are pigmented and capable of migrating their pigment. He also followed what he calls the axial fibre (axon) of all these cell types converging "centrad" toward the nerve ring via the stalk of the sensory club. The eyecups of simple eyes, Berger observed, are packed with a single type of pigmented cell. The nuclei of these cells are straddled by a thick outer layer, from which basal bodies extend into inner "flagella" (probably photoreceptor cilia) toward the lumen of the eyecup. The nuclei give rise to the nerve fibres on their outer pole, and these fibres join the nerve ring in the same way as those of the complex eye retina. Like Conant, Berger discussed to a limited extent the functional implications of his observations but made no effort to interpret them in a phylogenetic context.

In short order the simple eyes (ocelli) of hydromedusan jellyfishes came under new microscopic scrutiny. This was due to the sponsorship of Wladimir Schewiakoff who, as we saw earlier, had produced a seminal paper on the eyes of scyphomedusan jellyfishes. Schewiakoff had continued with his career (Fokin, 2000), and it was in his capacity as professor of zoology and dean of the Physical-Mathematical Faculty of the University of St Petersburg that he persuaded a young Russian investigator, Alexander K. Linko (1872–1912), to adopt this project. Linko was born in Kargopol, near St Petersburg. His interest in natural sciences was awakened at an early age, and a friend of the family of German descent, A.K. Gunther, soon detected and promoted this interest. After Linko graduated from the Imperial University of St Petersburg, he was accepted as a doctoral student in 1898, likely under the supervision of Schewiakoff. His paper on hydromedusan eyes appeared as part of his dissertation, in the publishing organ of the Imperial Academy of Sciences in 1900.

In this paper (Linko, 1900), Linko noted that the Hertwig brothers had neglected the eyes in favour of the "hearing organ" of hydromedusae, and that his goal was to remedy this neglect. He collected his material in the Arctic Circle: the White Sea and the Murmansk area of the Barents Sea. It was certainly a harsher environment for fieldwork than the cosy stations of the Mediterranean like Naples and Villefranche preferred by his fellow Russian zoologists! He used only standard histological techniques, as he was unsuccessful in revealing nerve cells with either the methylene blue or the Golgi method.

Linko described eyecups reminiscent of those seen in cubomedusae, but also the simplest of light-sensitive organs: the eyespots present at the bell margin where the tentacles emerge. These pigmented eyespots, Linko found, were associated with what he called a sensory pad (*Sinnespolster*), which was separated from the eyespot by a vitreous body. The eyespot contains pigment cells and photoreceptor cells, the latter sending fibres to the back of the eyespot where they form a nerve fibre layer. Linko saw a gradation of complexity among the species he examined. The simplest ocellus is found in *Catablema*, and that of *Oceania* represents a transition form toward the eyecup design. A nascent form of eyecup is betrayed by the presence of a vitreous body, such as in *Hippocrene*, *Staurostoma*, and *Lizzia*. The most developed eyecup is found in *Tiaropsis* which, contrary to the other forms, originates in the endoderm. In this design, Linko surmised, the pigment cells protect the light-sensitive segment of the photoreceptor cells from excessive or harmful light. The level of sophistication reached by the ocellus of *Tiaropsis* reminds Linko of the ocellar designs found in the scyphomedusa *Aurelia* and the cubomedusa *Charybdea*.

Linko envisioned the evolution of these simple eyes as a balancing act between the need for flexibility, where eyes have room to rotate to scan different visual fields, and the need for protection, where the eyes are recessed and form an eyecup.

Women Pitch In: Their Contribution to Studies of Hydromedusae

The authors of the last two contributions pertaining to this section concerning the nervous system of hydromedusae happen to be women. It was rare for doors to be opened to women in European universities of the time, so the accomplishments of Ida Henrietta Hyde and Sophie Krasinska, are all the more remarkable. Both women studied under Otto Bütschli, who, in addition to welcoming Russian students as we have noted and as is well documented elsewhere (Sergey Fokin, St Petersburg State Unversity, "Otto Bütschli and His Russian Students"), had a liberal atttude to receiving women in his laboratory.

Ida H. Hyde (1857–1945) was born in Davenport, Iowa, to German immigrant parents – the original last name was Heidenheimer. The following account of her early life is based on a biographical sketch in a book about women biologists (Creese, 1997). After her father vanished from home Hyde's mother moved the family to Chicago, where she started a business and prospered. However, the Great Fire of Chicago in 1871 destroyed everything they possessed and economic necessity forced Ida to leave school at sixteen. She worked as a milliner, but managed by sheer determination to catch up with her education. One day she happened on a copy of *Views of Nature* by Alexander von Humboldt, and she became hooked on natural history. She earned her Bachelor of Science degree at Cornell University in 1891. Then she moved to Bryn Mawr College, where she caught the attention of up-and-coming biologists Jacques Loeb and Thomas Hunt Morgan, who agreed to supervise her graduate research on the development of jellyfish, which she completed in 1893.

Her research attracted attention in Europe because it settled an issue of contention between Alexander Goette in Strasburg and Carl Claus in Vienna over the emergence of tentaculocysts in the ephyra developmental stage of jellyfish. Her work supported Goette's view, so he invited her to work in his laboratory, which she did after receiving a scholarship to study in Europe. Neither Goette nor Hyde and her mentors were aware of the obstacles she was to face in European universities over the simple fact of her being a woman. The relatively open climate in academic America had not prepared her for what unfolded in Strasburg, as she recounted somewhat jocularly in an autobiographical article:

It was not until I had worked many days in the splendid laboratory assigned to my private use that it dawned upon me that I was occupying a unique position, and that I was regarded by the students, faculty members and their wives as a curiosity. In the university circle the news quickly spread that an American "woman's rights" freak, a blue stocking and what not, had had the boldness and audacity to force entrance into the college halls. At *Kaffee Klatchen* she was served for gossip and dissection. It was not unusual for a professor, student, or *diener* [dissection assistant}, seemingly by mistake, to open the laboratory door, look frightened, and quickly retreat. Or students would congregate at the windows of the botanical building opposite the laboratory, and from sheer curiosity stare at my windows, greatly to the annoyance of the professors in both buildings. (Hyde, 1938)

Despite the efforts of Goette and others, Hyde was refused permission to take the examination for a doctoral degree. She then heard that the University of Heidelberg "was somewhat more liberal toward women students, having accepted several as auditors since 1891" (Creese, 1997). So she approached Otto Bütschli, who welcomed her and let her complete her research and other academic requirements for the doctorate at his Zoological Institute. The only drawback, given that physiology was her favourite field of study, was that the dean of the Medical School and professor of physiology, Wilhelm Kühne, gave his cooperation only grudgingly. Armed with the PhD granted her in 1896 when she was already thirty-eight, Hyde spent a few months at the Naples Zoological Station working on the physiology of the salivary glands of *Octopus*, and at the University of Bern in Switzerland on problems of muscle physiology. She returned to the United States in 1897 and undertook postdoctoral training in the laboratory of physiologist Henry P. Bowditch at Harvard Medical School. In 1898 she was appointed assistant professor of zoology at the recently endowed University of Kansas in Lawrence, where she was promoted to associate professor of physiology the following year.

Hyde had been a regular summer visitor to the Woods Hole Marine Biological Laboratory since her days at Bryn Mawr, when the laboratory was only a few years old. She resumed the practice even from her new base in faraway Kansas. In the summer of 1902 she went there to do physiological work on the jellyfish *Gonionema* (=*Gonionemus*) *murbachii*, but soon realized that it was "important that I should know the distribution of its nervous system which at the time had not been described or known with certainty" (Hyde, 1902). She used Hans Bethe's methylene blue technique without sectioning her material, just cutting pieces and mounting them whole on the platinum of the microscope. Alternating between immersion oil and lower magnification, and drawing with the aid of *camera lucida* images, she made arresting and vivid reconstructions of the nervous elements.

As a result the nerve cells contributing processes into the lower and upper nerve rings are excellently rendered, as well as the multipolar nerve cells of the subumbrellar nerve plexus (Fig. 7.14, left). Hyde observed also that some of the ganglion cells of the nerve ring send processes to innervate muscle fibres, and that some processes of ganglion cells cross the space between the two nerve rings. Hyde provides just as precise an account of the radial nervous elements. There we see the striking variety of nerve cells contributing processes to the radial nerve fibre bundles. Some of these nerve cells are sandwiched between the radial muscle and the endodermal epithelium (Fig. 7.14, right, L) and others are located just beneath the ectoderm of the subumbrella (Fig. 7.14, right, R). "The manubrium," she noticed, "also has its network

7.14 · (*Left*) Ida Hyde in Bütschli's
Heidelberg laboratory, 1896 (from
The Physiologist 24, 6, 1981). Courtesy
of the American Physiological
Society. (*Opposite, left*) Schematic
view of a piece of the marginal
nerve rings. From Hyde (1902), Fig.
1. (*Opposite, right*) Schematic view
of a piece of the radial nerve tissue.
From Hyde (1902), Fig. 2. Courtesy
Biological Bulletin, Marine Biological
Laboratory, Woods Hole.

of cells and fibres, and along the margin of the manubrium is a row
of large multipolar cells that send fibres to the periphery where they
join special sensory cells that lie among the epithelial cells" (Hyde,
1902). She provided very few functional interpretations for her observations, and no comparative analysis at all. She rationalized that hers
was a preliminary report to be expanded at a later date, which failed
to materialize.

We know next to nothing of Sophie Krasinska, the woman who
picked up the work on the hydromedusan nervous system in the wake
of Hyde's contribution. All we know is that Krasinska, whose name
clearly suggests a Polish heritage, was accepted by Bütschli – not surprisingly, given his almost unique liberal attitude to women – for doctoral studies around 1910, and that Bütschli himself suggested she take
a fresh look at the histological anatomy of both hydromedusan and
scyphomedusan jellyfishes. She did all her fieldwork at the Russian
Zoological Station in Villefranche-sur-Mer, where she was accommodated by its director, Mikhail M. Davidoff, himself a former student of
Bütschli. She produced her inaugural thesis dissertation in April 1913
and it was published the following year.

Krasinska's specimens included the hydromedusae *Aequorea forskalea*,
Carmarina hastata, *Neoturris pileata*, and the schyphomedusa *Pelagia
noctulica*. She followed the maceration and staining technique of the

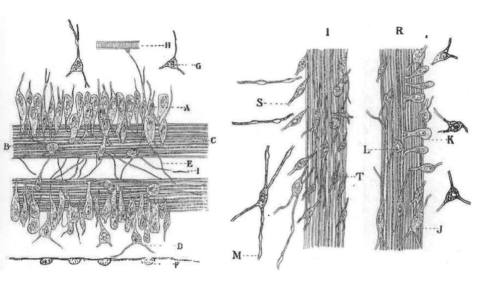

Hertwigs, but also used the newer formaldehyde-based fixation and paraffin sections stained with haematoxylin-eosin and other staining mixtures. She expressed disappointment that she could not make the nerve-specific staining methods work for her. The emphasis of Krasinska's paper (Krasinska, 1914) was clearly on the musculature, and the nervous system took a back-seat. She can arguably be credited with the most thorough exposé of the jellyfish muscle system produced so far, and is particularly noted for her description of striations in the circular musculature of the subumbrella of the hydromedusan jellyfish. At the time, striated muscles were thought to be more advanced and more performing than smooth muscle, and therefore to be the prerogative of more advanced animals such as vertebrates. Their presence in hydromedusan jellyfish and their association with jellyfish swimming did away with such thinking.

Without the advantage of methylene blue staining, Krasinska could not hope to command the depth of field of whole-mounts that helped Hyde obtain a more complete picture of the organization of the nervous system. Incidentally, Krasinska was unaware of the existence of Hyde's paper, even though the latter had been published twelve years earlier and Hyde had been one of Bütschli's students. This is symptomatic of the neglect the American scientific literature was suffering in Europe at the time; inversely, the Americans, who looked up to the

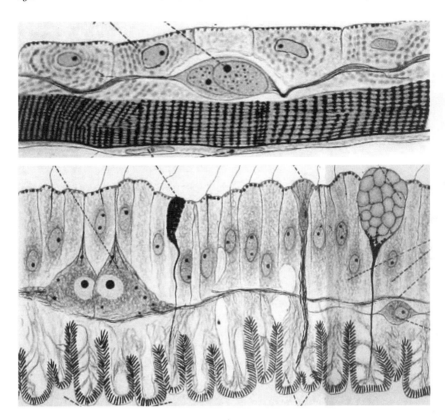

7.15 · Views of sections of the subumbrellar ectoderm in *Carmarina hastata* (*top*) and in *Pelagia noctulica* (*bottom*), highlighting the large ganglion cells sandwiched between the epithelium and the muscle sheet. From Krasinska (1914), Figs. 2 and 10.

Europeans, were more scrupulous in citing them. However, because her predecessors had repeatedly put emphasis on the "central nervous system" (nerve rings), Krasinska judiciously decided to place hers on the peripheral nervous system and the sensory cells.

Krasinska contrasted the ectodermal nervous system in hydrome-dusae (Fig. 7.15, top) with that of the scyphomedusae *Pelagia* (Fig. 7.15, bottom). In both, the ganglion cells and their nerve fibres were found inside the ectoderm, sandwiched between the other ectodermal cell bodies and the muscle layer, the latter striated in hydromedusae and smooth in the scyphomedusae. She saw only one type of ganglion cell in the hydromedusae, but distinguished large and small ganglion cells in *Pelagia*. She failed to observe direct contacts of ganglion cells with the subjacent muscle fibres, except in the tentacles where she suspected their presence.

The originality of Krasinska's paper rests largely in her depictions of sensory cells. She was the first to report the presence of a complex sensory cell in hydromedusae, which is composed superficially of a "basal body" surrounding a "sensory hair." What she saw is known today as the ciliary cone complex, seen also in sea anemones, which in fact involves a central sensory cell with its stereocilia and microvilli from adjoining cells (Fautin and Mariscal, 1991). Krasinska believed that such a complex sensory arrangement was comparable to similar attainments in higher metazoans. She also described in *Pelagia* a type of sensory neuron "which has numerous short stiff bristlelike projections" (Bullock and Horridge, 1965); and lastly, Krasinska shook the notion that ganglion cells do not fulfill a direct sensory role by depicting in *Pelagia* large ganglion cells sending a process to the surface (Fig. 7.15, bottom).

Krasinska concluded her paper with a phylogenetic statement. The notion that epithelio-muscular cells are directly irritable, she points out, was weakened by the discovery of the coelenterate nervous system by Carl F.W. Claus and the Hertwigs. She is adamant that the existence of a complement of sensory cells connecting with the nerve plexus and with the epithelio-muscular cells leaves no more doubt. Krasinska believed that the ability of coelenterate muscles to be directly stimulable, as opposed to being activated through the input of sensory and ganglon cells, is no greater than in higher metazoans. The implicit meaning here is that in her estimation the coelenterate nervous system was more sophisticated than anticipated from animals otherwise simply organized.

A Crowded Field of Neurohistologists

It is a great thing – science. There is everything: pleasure, relaxation and away from reality ...

Letter of W.T. Schewiakoff to his son Alexander (1917)

Looking back, it is striking that Romanes, Eimer, Schäfer, and the Hertwigs were the only investigators of significance in their generation to address the issue of the origin of nervous systems through their examination of coelenterate nervous systems. The irony, of course, is that despite the small numbers of players in the field they managed to step on each other's toes as they vied for priority of discovery. For the following generation – those who did their work between the mid-eighties and the eve of the Great War – the earlier generation was a hard act to follow and the pressure was on to perform more original work and to introduce new approaches. And yet, despite the daunting nature of the challenge, it is equally striking that so many were attracted by it. How did this new generation fit into the picture? What did they add to the contributions of the previous generation? How did they tackle the evolutionary issue? And what became of their careers?

Of the twenty-eight investigators discussed in this chapter, four were established academics, eighteen were doctoral students, and six floated in-between when they worked on their coelenterate projects. Few of them initiated their project on their own volition. Among those who did, the established investigators of course come to mind (Eimer, Richard Hertwig, Hickson, Hyde), but some of the floaters did also (Korotneff, Lendenfeld, and Havet). The remaining floaters (Schneider, Kassianow, Hesse) were still too dependent on their mentors to voice their own preferences or aspirations; and all the doctoral students deferred to their mentors for selection of dissertation topics. Some of these students descended from the lineage of the early protagonists of this narrative: Haeckel mentored Schaeppi and Wolff; Richard Hertwig – himself a student of Haeckel – mentored Bethe and Schneider; and Eimer mentored Hesse. But Bütschli, who mentored the largest number of students (Samassa, Jickeli, Schewiakoff, Kassianow, and Krasinska), made his reputation as a pioneer protozoologist and never personally invested himself in coelenterate research. In the English-speaking academic world, a single mentor supervised either the two British students (Hickson for Ashworth and Pratt) or the three American students (Brooks for Conn, Conant, and Berger), and both mentors had personally invested in coelenterate research.

This swarming of doctoral students to the topic of coelenterate anatomy and neurohistology is a reflection of the demographics in European universities during that period. K.H. Jarausch provides an eloquent quantitative demonstration of this trend by enumerating the educated professionals in different countries:

The rudimentary numbers that are available suggest an impressive expansion: in England the size of eight professions rose by half, from 127,354 to 191,384 between 1880 and 1911; in France the number of liberal professionals and intellectuals similarly swelled from 83,359 to 121,257 between 1876 and 1906; in Germany the number of those educated multiplied from about 63,000 (31,418 in Prussia) in 1852 to 335,252 in 1933; finally, in Russia higher education leavers increased from 133,660 to roughly to 233,000 between 1897 and 1927. (Jarausch, 2004)

The steep rise in graduation numbers was not smooth, howerver. Jarausch also noted cycles of overproduction of graduates in relation to employment opportunities in Germany and Austria, the countries where the large majority of our protagonists earned their doctorates. The expansion of German universities following the reforms instigated by Wilhelm von Humboldt halted in the 1830s. By the 1860s the resulting stagnation in enrolment spurred a new demand, which led to a new expansion and renewed overcrowding of classrooms in the 1880s – known historically as the *Qualifikationskrise*. This pattern suggests that the doctoral students of the 1880s and early 1890s portrayed in this chapter (Chun, Jickeli, Korotneff, Schneider, Zoja, Samassa, Schewiakoff) contributed to the crowding of the research laboratories as well. After a lull, an enrolment explosion returned in the last decade before the war, in which many of the protagonists in this chapter who operated in the early 1900s participated (Kassianow, Pratt, Wolff, Grošelj, Hadzi, Niedermeyer, Krasinska).

Behind these statistics loomed societal changes reflected in new conceptions of the intellectual mission of universities. The academic framework had altered since the student days of the patriarchs (Haeckel, the Hertwigs, Eimer, Romanes, Schäfer). A shift of emphasis from theology and law to the sciences and the learned professions led to a tightening of rules for the examination of students' knowledge and skills, so as to raise the bar of academic excellence. This new outlook also permeated graduate studies. Here is how Jarausch put it:

In accordance with more stringent scientific training, the examination requirements for admission to a university career also increased from a sometimes perfunctory dissertation to a second, extensive piece of original research, called *thèse d'état* in France, *Habilitationschrift* in Germany ...

In the long run this professional conception of scholarship proved irresistible because of its enormous success in promoting empirical discovery and a secular scientific world-view. Hence it was passed on to students who entered bureaucracy, academic occupations or the general public ...

The new research ethos also endowed professors with a higher mission than before and demanded a different kind of inner-wordly asceticism, no longer based on religion but on secular enlightenment. Finally, a dense network of scholarly associations promoted a new gospel of scientific discovery within a bewildering and ever-increasing variety of new disciplinary specialities. (Jarausch, 2004)

This was the new climate in which professors coached the students mentioned in this chapter. Disciplinary specialties, of which zoology was an example, sprang up on all fronts, and this new reality was soon reflected in academic structures, as Walter Rüegg explains:

Before World War I, mathematics and the natural sciences had their own departments in Germany only in Tübingen (1869), Strasburg (1872), Heidelberg (1890), and Frankfurt-am-Main (1914). The conviction that philosophy would guarantee – on an institutional level – the intellectual unity of the humanities, natural sciences and social sciences was so strong that Kiel, Cologne and Marburg maintained an undivided philosophy department until the 1960s, as did Graz and Vienna until 1975. (Rüegg, 2004)

Where natural sciences could not nestle in their own academic department, the next best thing was a zoological institute. This was the structure in vogue with leading zoologists and the one under which most young investigators worked. What did they accomplish? They made their most substantive contributions in territories unexplored by the previous generation. In particular, the nervous system of siphonophores and other colonial cnidarians was discovered, the organization of the ctenophore nervous system was shown to depart from that of the other coelenterates, and *Hydra* was found to possess a nervous system like all other cnidarians. The remaining contributions, for the most part, added relatively minor new information.

For a generation largely trained in Darwinian evolutionary thinking, it seems surprising that views on the origin of nerve cells and nervous systems showed little progress in the years between 1880 and 1914 – if and when the protagonists adopted a position at all. It seems that the earlier views of the Hertwigs, Romanes, and Eimer remained unchallenged. Some young researchers compared specific features of nervous systems among different coelenterate groups or between coelenterates and higher invertebrates, but none came up with a comprehensive interpretation of their own with regard to the process by which nerve cells emerged from "indifferent" cells and organized themselves into an integrated nerve cell complex. While there were a few forays into behavioural observations and experiments that led to functional interpretations of nervous system organization for both sensory input and motor output, the emphasis of those contributions was anatomy. The overall impression is that functional considerations had greater appeal among these students than an evolutionary outlook. Why was it so?

A straightforward answer is that scientific rigour, an increasingly ingrained currency of intellectual life and scholarship from the 1870s on, demanded caution against mis- or over-interpretation of observations. For these students, who for the most part were undertaking their first serious research project and were eager to please their supervisors, it was safest to take the path of caution and steer clear of speculation. Engaging in evolutionary interpretations would seem riskier than offering functional interpretations, the latter having the advantage that intuitive and teleological tools can be brought to bear on anatomical observations sometimes supported by experimental observations. More daring hypothesis-making of the kind needed to vent opinions on the origin of nerve cells and nervous systems was best postponed until one's academic career was secured and one could submit new ideas with more authority. But as the great majority of this cohort did not continue with coelenterates in their ensuing research careers, little opportunity arose to engage in such debates.

A subtext to this explanation concerns the authorship practice in the period of interest. The reluctance of students to offer consensus-changing interpretations of their observations in their scientific papers, even when there was some support from these observations, was encouraged by the common practice of letting students stand as single authors of their papers. The student's supervisor proposed the research topic and offered suggestions along the way, but the student took responsibility for his or her choice of methods and courses of action in moving the project forward, and this responsibility extended to the writing of the scientific report, in which the style and personality of the student were laid bare. The professor tended to remain at arm's length from these proceedings. This process contrasts with the now-current practice of multi-authorship – which I have experienced throughout my career – whereby co-authors, and especially the supervisor, have a say in the content and direction of the text and press students to discuss the results fully and spare nothing, short of undue speculation, in interpreting every angle of the results. Students of the late nineteenth and early twentieth centuries felt no such pressure, and as a result missed opportunities to discuss the evolutionary implications their papers warranted.

Career Sketches of the Turn-of-the-Century Generation

To the question of what happened to these promising students after completing their doctoral years, the answer depends on the category of career outcome one is looking for. If we ask how many spent their professional career studying coelenterates, the answer is simple: none. Hickson in England kept on publishing papers on cnidarians, especially octocorallians, throughout his entire career; but he was already a tenured professor when he penned his paper in which the nervous system of alcyonians is described. Others pursued research interests that had some relation to their thesis topic. Chun, for instance, went on to investigate planktonic and deep-sea animals; Lendenfeld, who was introduced to sponges by his mentor, Schultze, adopted them as his pet research animals; Korotneff, who as director of the Station zoologique de Villefranche had easy access to planktonic animals, moved away from coelenterates but remained an invertebrate specialist; Hesse, who had included jellyfish sense organs in his thesis work, made the comparative study of visual organs one of his lifetime pursuits; and Hadzi, alone among them, developed and maintained an interest in phylogenetic studies, culminating in a controversial book on the origin of metazoans (Hadži, 1963).

Who among the students secured a satisfying, if not brillant, academic career? Carl Chun, Herbert W. Conn, Robert von Lendenfeld, Alexei Korotneff, Wladimir Schewiakoff, Albrecht Bethe, and Richard Hesse answer this roll call. Chun became a professor at his alma mater, the University of Leipzig, in 1892, and in 1898 he spearheaded the *Valdivia* deep-sea expedition, which explored the sub-Antarctic seas and the waters around the coasts of South America and produced numerous scientific reports of note. He is famous for his discovery of the deep-sea vampire squid (*Vampyroteuthys infernalis*). He died in Leipzig on 11 April 1914 at sixty-one.

Herbert Conn was appointed professor at Wesleyan University in 1884, the year of his doctoral graduation, and the following year he married Hattie Herrick, whom he had courted during his Johns Hopkins days. On becoming chair of the Department of Natural History in 1891, he moved quickly to modernize it into a Department of Biology. Much of what we know about him we owe to a biography by his son (Conn, 1948). His research interests at Wesleyan focused at first on invertebrate zoology and embryology, interspersed with reflections on evolution in several articles and books, but after the 1890s he converted to bacteriology, of which he became an American pioneer. Conn started experiencing heart troubles in his fifties, and although he was forced to curtail his activities, he found it hard to stick to the new regime. As his son points out, "His death came suddenly, although we had known for years it might come at any time." He died on 18 April 1917.

Robert von Lendenfeld returned to Europe from his explorations in Australia and New Zealand in 1886, and wrote a book about the experience (Menzel, 2004). In the same year he was hired as an assistant to Ray Lankester at the British Museum, where he worked on sponges and coelenterates. He later transferred to the continent, taking short-term assignments at the universities of Innsbruck (Austria) and Czernowitz (in today's western Ukraine). He finally settled as a professor at Charles University in Prague, where he enjoyed a meteoric rise as both a scholar and an administrator, as Randolf Menzel, his great-grandson, attests:

> For several years around 1910 he was rector of this famous
> university, founded by Charles the Fifth as the first university
> north of the Alps. With his group, he worked on the Porifera
> [sponges] from the Challenger and the Valdivia expeditions and

the catch from the "Albatross," an American research ship. R. von Lendenfeld's many and mostly quite long books are full of spectacular, often multi-colored lithographies, which show the sponges' habits, histology, and, most importantly, their needles. (Menzel, 2004)

Using preserved collections from these great oceanographic expeditions, Lendenfeld's industry paid off in the form of beautiful monographs on sponges, coelenterates, and deep-sea fishes from around the world. He even studied the light-emitting organs of fishes. Privately, he was an original and colourful personality:

> R. von Lendenfeld was a remarkable character. In his youth he was extremely strong and fit; in his later years he was a massive and imposing figure, which helped to reinforce his authority ... He was a strong, short-tempered personality ... For example, when he was rector of the Charles University, the German-language university, and Czech students attacked his office, he successfully rescued the university's centuries-old official seal. He gave his lectures in English and kept in contact with his colleagues all over the world with research trips to marine research stations, especially on the Mediterranean. (Menzel, 2004)

Lendenfeld died prematurely at fifty of unknown causes, after only one year as university rector.

Alexei Korotneff was appointed professor at St Wladimir University in Kiev, Ukraine, in 1886, just as he was completing his study of the nervous system of the sea pen *Veretillum*. He remained in this position until 1912. Besides his key role in creating the Station zoologique de Villefranche, which he directed between 1886 and the year of his death (1915), he was a tireless traveller who visited, and published on, local invertebrate faunas. In his later years Korotneff became increasingly settled on the Riviera and exercised cultural as well as scientific influences there:

> The Station's scientists and, first of all, Korotneff himself were a centre for the cultural community in Villefranche and Nice, a favourite resort of Russians on the French Riviera in the early 20th

Century. In the years before his death, A.A. Korotneff often lived in his own villa in the suburbs of Nice, not far from Villafranca, where he was visited by many prominent men of science and culture. One of the reasons was a good private collection of fine art established by Korotneff. (Fokin, 2008)

The other Russian who had a successful academic career was Wladimir Schewiakoff. After completing his doctorate in Heidelberg under Bütschli, Schewiakoff decided to dedicate his reseach career to unicellulars (protists), at the very core of Bütschli's interests, no doubt in emulation of his esteemed mentor. He started off in 1889 by undertaking an extended trip around the world under the auspices of the Moscow Imperial Society of Naturalists and of the St Petersburg Imperial Geographical Society, during which he collected and studied freshwater protists in America, the Pacific islands, New Zealand, and Australia (Fokin, 2000). These investigations resulted in two monographs that helped establish his scientific reputation. After taking temporary assistantships in Karlsruhe and Heidelberg, where he rose to the rank of *Privatdozent* in 1893, Schewiakoff obtained an assistant professorship in St Petersburg, but only after preparing for a Russian doctorate; strangely, his diploma from Heidelberg, more distinguished than any Russian university at the time, was no entrance ticket to an academic position.

In 1895, he married Lydia Kowalevsky, the second daughter of the famous zoologist and embryologist A.O. Kowalevsky (1840–1901), who had conducted groundbreaking studies on the protochordates and their evolutionary link to vertebrates. Kowalevsky had been instrumental in advancing Schewiakoff's career in Russia. Schewiakoff quickly rose through the ranks, first to full professor in 1899 and soon afterward to dean of the Physical-Mathematical Faculty. After many years of productive research in protozoology, he took up administrative duties of the first order in the prewar years:

His administrative duties began to take up more and more of his time. Very often, his closest co-workers, V.A. Dogiel and N.M. Rimsky-Korsakov (1873–1951), elder son of the famous Russian composer N.A. Rimsky-Korsakov (1844–1908), helped him in the department. In 1908, he was elected as a Corresponding Member of the Imperial Academy of Science. It looked as if this was

the highest point of his career, but two years later Schewiakoff
received a proposal from the Ministry of Public Education to
become the head of the "industrial colleges" division of the
ministry. After 11 months he became vice-minister in January 1911.
For six years, Schewiakoff abandoned his scientific work and left
the university forever; but he kept the position of professor in
the Women's Pedagogical Institute. Keeping this position proved
to be a wise decision as after the 1917 revolution it saved his life.
(Fokin, 2000)

It was the worst of times in Russia. With his elder son, Alexander, at
the front and his daughter Tatiana serving as a hospital nurse, Schewiakoff and his wife did their share by helping to organize a military hospital in St Petersburg. But he became ill in 1916 and after his recovery
the revolution had changed the city, now called Petrograd, into a very
disquieting place. By 1918 he was installed in Perm in the Ural, unable
to return to Petrograd, as the civil war and the "red terror" were raging.
He finally settled in Siberia, obtaining a professorship at the University
of Irkutsk and painstakingly reshaping his life, while avoiding political repression by keeping his past as a tsarist vice-minister and the
enrolment of his two sons in the White Army under wraps. In this way
he was able to resume research on protozoans, especially radiolarians,
until his death in October 1930.

Just as Bütschli and Kowalevsky helped Schewiakoff along, Richard
Hertwig's academic prestige helped propel Albrecht Bethe's career.
Bethe took his first professorship (*Privatdozent*) at the University of
Strasburg (1899). In Strasburg he married Anna Kuhn, the daughter
of a Strasburg professor, and their only child, Hans, the future Nobel
Prize winner for Physics, was born. Bethe also honed his skills as a
physiologist and wrote a highly influential book on the comparative
anatomy and physiology of the nervous system (Bethe 1903), in which
he also discussed the coelenterate nervous system. In fact, the influence of this book was such that Bethe's conviction that nerve fibres
anastomose led to the misguided thinking of two generations of neuroanatomists and caused some to recant the discontinuity theory. After
an appointment at the University of Kiel (1912), Bethe was called to
Frankfurt-am-Main in 1915 to build a physiology department in the
fledgling new university. He was appointed editor-in-chief of the presti-

gious *Pflügers Archiv für Physiologie* as his career flourished in Frankfurt, where he remained until his death in 1954.

Similarly, Richard Hesse was helped along by Eimer, who secured him a lecturer's position in his Tübingen institute in 1894. Hesse went on to replace his deceased mentor in 1901 as associate professor in the Zoological Institute at the University of Tübingen. In 1909 he was appointed professor of zoology at the Agricultural College of the Friedrich-Wilhelms-Universität, Berlin. Presumably feeling the confines of an agricultural college as too narrow for his ambitions, Hesse moved to the University of Bonn in 1914. He returned to the Berlin university in 1926 as professor and director of its Zoological Institute, where he remained until his retirement, forced by the Nazis for political reasons in 1935. He died in a Berlin devastated by bomb raids in December 1944.

Hesse's scientific legacy was significant and far-reaching. His comparative studies of invertebrate and vertebrate eyes led him to propose hypotheses on the phylogenetic sequence of eye evolution. These works inspired the career of his students, particularly the German zoologist Hansjochem Autrum, who went on to become a leader in the field of invertebrate visual physiology. But Hesse's impact was felt most strongly in the emerging field of ecology. It began with a book inspired by his observations of land animals at the Berlin Agricultural College (Hesse, 1914) in which he argued that the mode of life of an animal is informed by its structural organization. And it culminated in a groundbreaking book (Hesse, 1924) that gave birth to the field of animal biogeography. Its influence was further felt with its translated and augmented 1937 edition in the United States, entitled *Ecological Animal Geography*.

In contrast to these leading scholars, the following had rather lacklustre careers: Karl Schneider, James Ashworth, Jules Havet, Max Wolff, Pavel Grošelj, and Jovan Hadži. Schneider started his career promisingly, only for it to take a bizarre and tragic turn later on. He was appointed *Privatdozent* at the University of Giessen in 1897, but the next year found him at the Zoological Institute of the University of Vienna where he stayed for the rest of his career. His progress was slow, as he remained assistant professor for many years, working under Professor Hatschek from 1905 to 1911. He was finally promoted to full professorship in 1911. His publications on coelenterates culminated with a textbook on the comparative histology of animals in 1902. He moved from zoological interests to animal psychology interwoven with vitalism at the turn of the century. Dabbling of this sort probably caused a few frowns among his colleagues and might help explain the sluggish advancement of his career. With the security of his tenure in 1911, he was emboldened to shift his research emphasis toward the psychology of the occult and to parapsychology. His zeal for off-base subjects attracted him to the wayward notion of sending pro-war tracts, in which he justified war cruelties through his ideas on animal psychology.

Schneider sent some tracts to the pacifist Albert Einstein, who replied with this sarcastic letter dated 24 February 1918, from Berlin (Schulmann et al., 1998):

> Esteemed Sir,
> You gave me great pleasure by sending me some of your little tracts. I admire the versatility of your interests and knowledge, as well as your fine and amusing style. I have read almost all of it. The "Academy" with my Mr. Double amused me especially as well; this character is somewhat less precise in his statements than the original, but for it considerably more comical and palatable. Only your strong-arm attitude, sailing under the Germanic flag, goes very much against the grain for me. I prefer to hold with my countryman Jesus Christ, whom you and your mind-mates consider irretrievably obsolete. Suffering really is more preferable to me than exerting force. History may instruct us where this mentality, which you and so many contemporaries in this country extol, is going to lead us; who can know? But moving beyond all designs: De gustibus non est disputandum; my taste is otherwise.
> Very respectfully,
> A. Einstein

When Schneider began lecturing on the parapsychology of occultism at the University of Vienna in 1926, there were clashes with Dean Othenio Abel about the extent of his teaching assignment. Matters grew worse, and at the funeral of the botanist Richard Wettstein in 1931 Schneider perpetrated a gunshot attack on Abel, by then rector of the university. Fortunately, the attack failed and Schneider avoided a criminal conviction on grounds of having a mental disorder at the time of his action. Needless to say, his academic career abruptly ended in 1932.

After several months of hospital treatment, he left Vienna, and lived on his brother's poultry farm until his death in March 1943.

James Ashworth and Jules Havet had far less controversial careers. Ashworth followed up his studies on soft corals with several sojourns at the Naples Zoological Station. He specialized in the taxonomy of invertebrates, especially chaetopods, and was appointed professor of zoology at the University of Edinburgh. He remained in that city until his death in 1936 at sixty-two (obituary in *Nature* 137:304–5, 1936).

Havet's career peaked quite early. In 1902, the year after the publication of his substantial work on the sea anemone nervous system, he was appointed assistant professor (*professeur extraordinaire*) at the Université catholique de Louvain, and a promotion to Chair of histology and embryology followed in 1906. Havet's research interests shifted to the cellular pathology of mental diseases and to histological pursuits beyond the nervous system. He lost his wife, with whom he had had three daughters, during the war, and this loss somehow led him to enter the priesthood in 1918, which he did secretly. Even his eulogists had to admit that the quality and impact of his early researches on invertebrate nervous systems were never surpassed in the mainstream of his career. He retired in 1936 and died in July 1948.

As for Max Wolff, he became an obscure and shadowy academic figure after his doctoral thesis of 1903 on the nervous system of cnidarian polyps. He stagnated as an assistant for the next decade, first under Oskar Vogt at the University of Berlin and under Ernst Stahl in Jena, where he specialized in plant diseases, then at the Zoological Institute of the University of Halle and finally, from 1906 to 1914, at the Institute for Plant Diseases of the Academy of Agriculture in Bromberg. In 1914, he was at last appointed professor of zoology at the Eberswalde Forestry Academy, where he worked on plant pests, forest insects, and pest control. His academic record was marred by his joining the National Socialist party in 1933 and signing the Loyalty Oath of professors at German universities and colleges to Adolf Hitler. After his retirement in 1941 Wolff set up a private laboratory in Naumburg, where he died in November 1963.

Grošelj and Hadži were born in different states of the Balkans but both spent their academic life at the University of Ljubljana in Slovenia. After Grošelj's 1909 publication on the nervous system of sea anemones, he became a lecturer (1910) and docent (1923) in biology at the Ljubljana Medical School. He divided his interests between literary studies and scientific writings in obscure local journals. At the time of his premature death in January 1940 he had almost finished a work on the nervous system of jellyfish, which was never published.

Hadźi started his career in Zagreb after the publication of his study of the *Hydra* nervous system in 1914, but in 1920 he moved to the Zoological Institute of the University of Ljubljana, of which he later became the head. He is best known for his bold and peculiar views on animal classification and metazoan evolution, which were popularized in a widely read book (Hadźi 1963). Hadźi's ideas were considered far-fetched by many of his contemporaries, however, and never gained acceptance in the zoological community at large. His original research offered descriptions of invertebrate animals dwelling in caves and mountains, many of which were newly discovered species. He died in December 1972 at the age of eighty-eight.

Carl Jickeli, Theodor Schaeppi, Edward Berger, and Albert Niedermeyer, four of the neurohistologists profiled in this chapter, were led by choice or serendipity into respected career paths outside academia. Jickeli, after his publications on the histology of hydroid polyps, contented himself with returning to his birthplace (Hermannstadt), where he became curator of the Siebenbürgen Museum and resumed the studies of molluscs that had attracted his attention before his doctoral studies under Bütschli. He died at the age of seventy-five in 1925. Schaeppi returned to Zurich, where he continued publishing on invertebrates until 1906. He later turned to medical research and practice, again in Zurich, until his death around 1967.

Across the Atlantic, Berger also returned to his hometown after completing his PhD at Johns Hopkins, to teach at his alma mater, Baldwin Wallace College (1899–1906). But he soon grew restless. Having developed an interest in insects, he took a position as an entomologist at the Experimental Station of the University of Florida between 1906 and 1911. Later he became chief entomologist of the State Plant Board, Florida Department of Agriculture, and therefore in charge of pest control. According to biographical notes that appeared in the *Florida Entomologist*: "He retired [June 1943] due to failing eyesight and a chronic disease, diabetes. Dr. Berger was admitted to the hospital with an infected foot and died 3 days later on 24 August 1944."

The fate of Niedermeyer contains elements of both fascination and sympathy. He was the only member of the cohort of this chapter to have immediately abandoned zoology for medical practice after his doctorate. He already held a medical degree, but over the following years he trained in gynecology (1918) and in legal medicine jurisprudence (1924). He established his practice in Görlitz, Germany, and as a Catholic he became a strong and vocal defender of prenatal life. He demonstrated fortitude while opposing Nazi eugenic policies, and he refused to participate in the implementation of the sterilization law (Kater, 1989). In protest, Niedermeyer emigrated to his native Austria in 1934 and worked at the Familienamt (Family Clinic) of Vienna. But after the Anschluss in 1938 he was arrested and deported to the Sachsenhausen camp. Thanks to the intervention of colleagues, he was released and practised as a civil contract physician until the end of the Second World War. Amazingly, at fifty-seven, he decided to retool himself in gastroenterology and became head of the Berlin Institut für Postoralmedizin between 1948 and 1955. He died in 1957.

An astounding third of the cohort had their careers cut short for different reasons. Some of them lost their lives abruptly. Just as Conant had died prematurely in the throes of yellow fever, Rafaello Zoja also died tragically in his twenties. Zoja and his younger brother Alfonso, a student of Camillo Golgi, were montaineering aficionados of the same ilk as Lendenfeld. In September 1896, only five years after Rafaello published his papers on *Hydra*, he, Alfonso, and a medical doctor close to the Zoja family, Dr Filippo de Filippi, went on a climbing expedition to Mount Limidario (Gridone Rock) in the Italian Alps near Switzerland (Mosso, 1898). As they neared the summit at 2,100 metres, de Filippi noticed that "both [brothers] were pale, their teeth were chattering, they complained of nausea and of slight headache, were apathetic, irresponsive alike to my jokes and entreaties" (Mosso, 1898). Both brothers died of altitude sickness, and only Dr de Filippi survived to tell the tale.

The life of Edith Pratt, a promising woman scholar, was also marred by tragedy. After her 1904 doctoral thesis on digestion and mesogleal cells in soft corals, Pratt found it hard to follow up on her leading achievement for women in Great Britain. Her doctoral diploma did not open the doors of academia and yet she wanted to continue scientific research. Hickson accommodated by giving her the position of honorary researcher in his laboratory at the Victoria University of Manchester, as they now called Owens College. In this capacity she continued to generate papers. In 1909 she married Stanley Musgrave, who was likely a friend of James Ashworth, as he also came from Bolton, Lancashire. She unfortunately died within a year of an unknown cause, possibly a complication of pregnancy or typhoid-type fever. The Edith Mary Pratt Musgrave Fund was later created for women graduates by Cambridge University "for the furtherance of research on the anatomy, physiology, or life history of the Alcyonaria, corals, or related organisms." It is still in existence today.

Alexander Linko lost his life shortly after his career had taken off. Having completed his doctoral work on jellyfish eyes, he continued work on coelenterates but also expanded his interests to all aspects of the plankton of Northern Seas. He became senior staff scientist at the Zoological Museum of the Imperial Academy of Sciences in St Petersburg as well as assistant in the Department of Zoology and Comparative Anatomy at the Imperial University. However, Linko suffered from an incurable kidney disease, and the man who was said to have "a soft and modest character" and who "gained the sympathy of everybody by his frankness, simplicity and mildness" died at forty on 4 September 1912.

The careers of both Nicolai Kassianow and Sophie Krasinska appear to have been stillborn because of the advent of the First World War. After finishing his work on the nervous system of soft corals, Kassianow returned to Moscow, but his academic affiliation there is unknown. A paper by him on the evolutionary emergence of arachnoids (spiders and scorpions) in the light of the *Limulus* theory appeared in 1914 in the *Biologische Zentralblatt* (Leipzig), and then nothing. It turned out that he fled Russia during the Revolution of 1917 and exiled himself in Switzerland (Fokin, on Bütschli's Russian students). From there Kassianow published a pamphlet on Germany's geopolitical designs on Russia (Kassianow, 1918). It is evident that his scientific career was definitively scuttled. He eventually emigrated to the United States and died in Alameda, California, in 1948.

As for Krasinska, not a single paper of hers shows up in the scientific literature after her 1914 paper based on her thesis on jellyfish. At the outbreak of the war, activities at the Villefranche station where Krasinska worked were substantially curtailed (Fokin, 2008). Bütschli's female students tended to develop a scientific career in one form or another, as examplified by Ida Hyde, who had a distinguished career as a professor of physiology at the University of Kansas and in the 1930s invented the extracellular microelectrode, which greatly extended the reach of physiological inquiry. Krasinska, in contrast, vanished from the record after 1914.

Paul Samassa, the remaining student in the cohort, made a deliberate decision to move away from an academic career in 1897, stating health reasons (Schödl, 1988). He re-invented himself as a journalist and politician, writing books on historical and political analyses of places like South Africa and his native Austria, published in the early 1900s. Although there is no direct evidence, it is likely that Samassa's career was derailed by his scientific blunder – the denial of a nervous system in ctenophores – and by his stand against better scholars and better-connected men than himself.

The shadow of the giants was indeed long, and their footsteps hard to fill.

CHAPTER EIGHT

George H. Parker and the Broad View of the Elementary Nervous System

> In Europe I worked for the most part on the structure and the function of the nervous system, with special reference to these states in the lower animals. In this way I proposed to gain as broad a view as possible of the field in which I planned investigations. Little did I think at that period that this problem would last me most of my lifetime.
>
> George Parker, 1946

Between 1891 and 1893, while neurohistologists of the coelenterate nervous system were busy with their sporadic investigations all over Europe, a young and eager American visited several European laboratories to sit at the feet of the luminaries of disciplines of zoology and physiology and to acquire hands-on experience at their laboratory benches. This young man returned to the United States with great dreams of research plans, and he would eventually become the leader in his field. George Parker, a central figure in our story, embarked on a series of investigations lasting over a quarter of a century which entirely shifted the approach to the problem of the origins of nervous systems by emphasizing experimental models and integrated sets of scientific inquiries. His intellectual journey raised reflections on the origins of nervous systems to a higher level and provided a script for the emergence of nerve cells and nervous systems that would prove satisfying to the biological community for decades to come.

A Boy with a Consuming Interest in Natural History

We know at least as much about Parker as about Agassiz or Romanes, thanks to his delightful autobiography (Parker, 1946), obituaries (Romer, 1967; Beebe-Center, 1955), and his personal papers stored at the Library Archives of Harvard University. I have relied on these documents to provide a narrative of his life.

George Howard Parker (1864–1955) was born in Philadelphia on 23 December 1864, exactly four months before Abraham Lincoln's hearse passed through the streets of Philadelphia with baby George and his mother among the attending crowds, and six months before the Civil War was declared officially over. His father was a moderately prosperous businessman of unspecified trade. Although his parents were participants in the First Unitarian Church, the paternal grandfather, who lived in his son's home, was a Quaker and pressed the family to send George to a Friends' School. Its curriculum was truly inspired for the period. "Besides the ordinary English studies," Parker wrote, "we were schooled in Latin, in French or German, in drawing, in modern geography, and even in natural history. We also attended lectures on anatomy and physiology, physics and chemistry, subjects not commonly met with in the school curricula of those days" (Parker, 1946). In such academic surroundings a penchant for natural history was readily fostered.

The first inkling of Parker's infatuation with the world of nature came when he was six years old and the family, in what became an annual event, escaped the summer heat of Philadelphia by retreating to the seaside community of Cape May, New Jersey. The boy's collection of shelled animals was only discovered when his parents sought out the source of the awful stench in his bedroom closet. He learned the hard way that only empty and cleaned shells will do for indoor purposes! His father responded to his interest by taking him to the Philadelphia Academy of Natural Sciences, where George beheld wide-eyed what a professional collection of specimens looked like. By the age of seven or eight he was struck by the morphological appearance of embryos compared with adults, and by naturalists' tendency to classify animals and plants in a systematic fashion and give species Latin names.

When the worldwide financial panic of the mid-1870s hit home. George's father lost his business and the family house had to be sold.

They moved to a humbler abode and had to scratch out a living. This meant George had to leave the Friends' School at fifteen to help replenish the family's coffers. He recalled his deflated spirit as he moved to help support his folks:

> But for me the streets of Philadelphia were a poor place to begin picking up a living, for there were times when I lacked even such rolls of bread as Benjamin Franklin once carried up Market Street under his arms. My father had business associates in the publishing house of J.B Lippincott & Co. and through their consideration I was enabled to act as a local sales agent for a new edition of Thackeray's novels which they were then publishing. It was my job to obtain subscribers to the set of volumes and then, as one number after another appeared, I would deliver them to the persons concerned and collect the price. (Parker, 1946)

A year later, at sixteen, George seized an opportunity to extricate himself from his predicament by winning a Jessup Fund fellowship at his beloved Philadelphia Academy of Natural Sciences. "Each beneficiary," he explained, "appointed for a term of two years, received a monthly stipend of twenty dollars, in return for which he worked one half day in arranging some part of the Academy's collections and the other half in any kind of natural history study that was to his liking." He was now truly in his element. He made the most of it, working hard and endearing himself to the staff. His industry and charm were effective enough to land him an extension of his fellowship in 1882 to act as a caretaker of the butterfly collection.

That same year, George's father decided to create a way for his son to further his education at college level even though his business prospects had not improved greatly – in the census of 1880 he had listed stockbroker as his occupation. He advised no less than Harvard to his son, because it "was an institution where a liberal education could be had and it was a center where Natural History had been highly developed, especially under Louis Agassiz." But Parker, having been away from formal schooling for three years, first had some catching up to do. So, with the help of volunteer tutors, he prepared to write all the admission examinations by June 1883. He was admitted to the Lawrence Scientific School of Harvard College "with conditions in Latin, Algebra, and

Trigonometry." To finance his first year at college, his father tapped a past business colleague who owed him a favour. To George's relief the donor "advanced me two hundred dollars without security and to be returned to him at my convenience. What a Providence!"

Burdened by meagre resources and academic "conditions," Parker began the fall semester of 1883 with a determination to prove his intellectual valour so that more financial assistance might come his way. The strategy worked: he was awarded a scholarship, followed by an assistantship in zoology. Being a quick learner and a born organizer, he found time to supplement his income with tutoring jobs. One of his students was William Randolph Hearst, the future newspaper publisher and tycoon. His account of Hearst's performance ably illustrates what men like Hearst and Parker could accomplish in the permissible Harvard environment of those days:

> Professor Davis [the Geology Professor] questioned me about Hearst's work and I assured him that it was done by Hearst himself and with an expedition and economy that was almost unbelievable. I also told him that I had never before tutored a man who had as quick and clear a mind, and as retentive a memory as Hearst had. It was a joy to work with such a fellow ... To a man of Hearst's mental capacity, the college work of those days was a mere bagatelle. He could do it and still spend half his college year in other pursuits ... The system allowed the real student an almost ideal opportunity for his work, and consequently it bred such men as George Santayana ... It likewise allowed the sluggard or the stupid one to slump on for a year or two and then to die in the traces, but to die for the most part unmolested. (Parker, 1946)

Much of that college culture was owed to the administrative style of President Charles W. Eliot, under whose aegis Parker blossomed both as student and lecturer. Described by Parker as "an austere and forbidding personality," Eliot nevertheless ran things with a light rod. As long as students paid their tuition and faculty resigned themselves to their measly pay cheques – after all, it was often stressed, Harvard dons are paid in prestige – he was happy to offer the luxury of attractive and authoritative courses to select from. What you did with your studies was up to you, and if anything, the teachers had a freer hand in developing their course curriculum than at other American universities. It worked well for Parker, who promptly earned his Bachelor of Science in Natural History in June 1887.

Parker was too steeped in his long-standing field of interest to even waffle over the wisdom of engaging in graduate studies. Focusing on his ambition to devote his life to zoology, he enrolled under the only Harvard professor competent to train him in the discipline. Edward Laurens Mark (1847–1946) had succeeded Louis Agassiz after the latter's death (1873) as Hersey Professor of Anatomy and director of the Zoological Laboratory – Agassiz's son, Alexander, replaced his father as head of the Museum of Comparative Zoology. Mark is better remembered today for having instituted in 1881 the "Harvard system" – the method of citing references by author and year, usually in parentheses – than for his scientific work (Chernin, 1988). Parker thrived under his mentorship: "From the time you took your research problem under him till the last punctuation point marking the completion of your printed thesis you were under Dr. Mark's eye. The result of this personal attention was that he brought up the most devoted body of students that I have ever met." His fellow graduate students in the Zoological Laboratory were Herbert H. Field and Charles B. Davenport, who eventually had good careers and became lifelong friends, even becoming related by alliance in one case, as will soon transpire.

During his graduate years (1897–91) Parker divided his time between an instructorship in zoology and his research to fulfill his PhD requirements. It is fair to say that zoology at Harvard had either a small following or a poor standing with President Eliot, for Mark and Parker constituted the entire teaching staff. Alexander Agassiz, Parker recalled, was reported to have remarked tongue-in-cheek: "zoology at Harvard is ... in the hands of a man and a boy" (Parker, 1946). The "boy," however, went about his research projects with genial precocity and deliberate zest, working out the structure of the eyes of scorpions as an undergraduate project, before taking on the structure and development of lobster eyes as well as an overview of compound eyes for his doctoral dissertation. His choice of dissertation topic, though, was not particularly original; invertebrate morphology, as previous chapters have illustrated, was the rage among zoologists of Europe, and it had spread to America. Parker was engulfed in it like so many others.

But as a sideline he also studied the regressed eyes of cave-dwelling fishes. This pattern of rapid productivity lasted his whole career.

It is perhaps timely to ask what had led him to the choice of sensory organs for his thesis in the first place. Soon after he began his undergraduate studies, Parker became acquainted with William James, the famous professor of psychology at Harvard and brother of novelist Henry James. James was the first to introduce psychology courses in the United States, and his *Principles of Psychology* would be published two years after he first met Parker. It is clear from his autobiography that Parker felt strongly the catalytic role that James played in the nascent development of his research program. Parker recalled where their conversations led them:

> Professor James and I drifted into my object in coming to
> Harvard to study biology, and I told him that my chief interest was in the evolution of the nervous system, the steps of
> which I thought might be discovered in the nervous structures
> and responses of the lower animals. The idea, which seemed
> to Professor James a novel approach to a problem that was of
> interest to him, attracted his attention and, after I had disclosed
> to him some of the lines of research which it opened up, he gave
> it his growing approval. From time to time during my undergraduate days I had the opportunity of talking over my immature
> plans with Professor James, and always with increasing clarity
> and security; till finally, in my later college life and just before
> graduation, I had formulated a program which continued to
> mean more and more to me as a biologist whose chief interest was
> the steps by which the nervous activities of animals had evolved.
> (Parker, 1946)

There is no evidence that any of the protagonists of this book, with the possible exception of Romanes, had the scientific maturity and embracing vision of a research career path that Parker displayed. James must have been struck by the exceptional intellect, personality, and self-will before him, so full of promise. But for now Parker's plan had materialized only as far as tackling the part of the nervous system that monitors the outside world, sensory structures; and even then he limited himself to visual organs and to invertebrates (crustaceans) that hardly represented the lower animals. It was a good start, however, and a very good performance for a PhD student.

As Parker pondered his next career move after completion of his thesis, events of a personal nature crept into his life. In 1888 he had met a cousin of his lab mate Herbert H. Field, Louise Merritt Stabler from Brooklyn, on the occasion of the Harvard commencement. He had other brief encounters with her in the course of his graduate studies, and by the time he was contemplating a postdoctoral tour of European laboratories to round out his scientific training she was much in his mind. In fact, according to his autobiography, he had decided that she was the woman best suited to share the rest of his life, especially as he pondered the dedication to science that his future life entailed. If his mind was irrevocably made up on this point, hers was not. He would have to keep wooing her before gaining her consent.

After receiving his doctoral degree in 1891, Parker won a travel scholarship from Harvard and resigned his instructor job. To many young American scholars of the period, European, and especially German, scientists were the best in the world and the leaders to look up to for up-to-date training. Besides, where else but in Europe could he get exposure to neurohistologists and physiologists? So Parker planned to spend half a year in Leipzig in Rudolf Leuckart's laboratory, where his own mentor, Mark, had done his doctoral studies; and to spend the ensuing half year in Berlin under Franz Schulze and in Freiburg in the laboratory of the comparative anatomist Robert Wiedersheim. He would conclude with a stay at the indispensable Naples Zoological Station.

Parker spent his first two months in the Harz Mountains immersing himself in conversational German and reached the University of Leipzig for the start of classes late in September of 1891. There he spent a forlorn winter, for he "had been requested not to write to Miss Stabler." He transferred to Berlin in the spring of 1892. His longing for Louise was requited in the summer when he joined her and her family on their vacation trip along the Rhine River. This time Louise responded more positively to George's entreaty, but urged him to keep their engagement secret until she completed her education at Barnard College. Parker went back to Berlin more hopeful and elated, now that the ban on letter-writing was lifted. As the Harvard archives demonstrate, he made up for lost time by turning out two or three letters to her weekly.

His mood is well encapsulated in a letter dated 30 July 1892: "Oh, that I could be with you one short moment to tell you how precious you are to me, how truly I love you."

His letters reveal intriguing exchanges over delicate subjects. One was over religion, as Louise came from a dedicated Quaker family. To her query about his religious feelings he answered ambiguously, in a letter dated 20 August 1892, that they evolved with the growth of his mind, but also that he was then neither Friend (member of the Quaker faith) nor Unitarian (the faith of his parents). In another letter, dated 9 August 1892, George urges Louise, in a somewhat paternalistic tone, "to know what the cell theory and Darwinism is in the organic sciences." In the same letter he entreats her to observe accurately, to express herself clearly, and to be sympathetic toward the arts. Obviously, he believed she had a steep leaning curve to accomplish before declaring her an accomplished woman. But a refreshing sense of humour sometimes crept into their exchanges. To Louise's query about snoring, George responded with a certificate issued in French by his lab mate and roommate in Naples – Victor Willem, a young Belgian zoologist from the University of Gand – according to which Parker had been tested and proven snore-free. What a relief for the prospective Mrs Parker!

Banter aside, Parker assiduously attended the lectures of the great German luminaries of zoology and physiology, such as Rudolf Leuckart and Carl Ludwig, while pursuing his own investigations on the nervous system. However, his European travels were his foremost learning experiences: touristic, cultural, and scientific. He was under no illusion that, however seasoned a researcher he thought he was, his lab investigations in Germany were anything more than child's play designed to bounce off human sounding boards – the distinguished German professors – if only to reinforce his self-confidence by the approval of his elders.

Parker returned to the United States in July 1893 to resume his post as instructor at the Zoological Laboratory and to make arrangements for his wedding. Harvard had just suffered a budgetary deficit and President Eliot, to whom a deficit was anathema, went about forcing the resignation of some professors and "turning out gas-lights to save expense" (Parker, 1946). So Parker was hugely fortunate to keep his job. His good fortune extended to his private life when he finally wed

Louise on 15 June 1894. His happiness knew no bounds. One can almost hear him giggle when he recounts how patrons at a boarding house in Nantucket, where they spent their honeymoon, learned that George and Louise were newly wed by overhearing Louise say that she did not eat fish when George passed her the plate. Or when Louise "said that one of her greatest surprises was to find that when I prepared for bed I did not take off my beard" (Parker, 1946). On returning to Cambridge, however, his elation was dampened by the news of the sudden illness and death of the younger of his two sisters, Mary. But now he settled into a routine of teaching assignments and scientific research that he maintained for the rest of his life. He was twenty-nine years old.

The Bumpy Road to Discovery and Synthesis

In spite of his avowed desire to act on a plan to investigate the evolution of the nervous system, Parker was slow to implement it. He was perhaps compelled to allow more time to mature his ideas and devise a cohesive strategy. His publication record shows an eclectic array of research topics once he resumed academic life at the Zoological Laboratory: papers on the retina and brain visual centres of the crayfish, on methods of preservation of mammalian brains, on pigment screens in the shrimp eye, on variations in the morphology of the vertebral column of the mudpuppy (*Necturus*). Finally a paper came out in 1896 that hints at an embryonic positioning of his research plan.

The paper in question addressed the behaviour of the sea anemone *Metridium* when presented with food (Parker, 1896). The genesis of this investigation may be traced to Parker's sojourn in Naples in 1893. Victor Willem (1866–1952), his closest colleague at the Zoological Station (and who had declared him snore-free), had published several papers on the eyes of molluscs and arthropods, so it was inevitable that Parker, whose PhD research was on that subject, was attracted to him. But Willem also worked on the digestive process in sea anemones and, when he and Parker were colleagues in Naples, had just published a paper on food absorption in sea anemones (Willem, 1893). Parker's acquaintance with Willem's work may have been a factor in his decision to select sea anemones as the starting point of his research programme on the evolution of the nervous system. In keeping with his long-term goal, however, Parker chose to approach the problem of food intake from the angle of behavioural physiology rather than probing the digestive function.

Feeding in sessile animals such as sea anemones was known to involve secretion of mucus for the adhesion of food particles, tentacle movements to collect and direct food to the mouth, and opening of the oral sphincter to circulate the food in the pharynx. Parker found that the pharynx is lined with cilia, which as a default beat their strokes upward but in the presence of food reverse their beating to push the food downwards (Parker, 1896). He assumed that this reversal was triggered by a chemical cue in the food, but others later found that mechanical (tactile) stimulation by the food particles was the major factor. He also assumed, correctly this time, that ciliary reversal was not mediated by the nervous system.

In the same paper Parker reported on experiments revealing the kind of habituation noted before by Romanes for jellyfish swimming (see the second section of chapter 3). Here is how Lulu F. Allabach summarized them:

In Parker's experiments alternate pieces of meat and filter paper (soaked in meat juice) were given to the tentacles of one side of the [oral] disk of *Metridium*. It was found that while the meat was swallowed each time with equal readiness, the time taken in swallowing the paper increased, and after three or four trials the animal no longer ingested the paper, though the latter contained each time the same amount of meat juice as at first. After reaching this result with the right side of the disk, the same series of experiments was performed on the opposite side of the disk of the same specimen. It was found that the left side had not become modified by the experience of the right side. It at first took the paper, then by the same gradual change seen previously on the right side, it came to refuse the paper. (Allabach, 1905)

The experiments uncovered a discrimination between tactile (meat pieces) and chemical (soaked paper) stimulation, showing that only chemoreception displayed what Parker called "sensory fatigue," or what is described today as habituation, involving modulation of neurotransmission. They also revealed that the habituation is localized in the stimulated area, and is not transmissible to unstimulated parts by nervous means. As Parker put it, the experiments illustrate "the extreme looseness, or even independence, of the nervous activities of the two sides of the animal."

After this first foray into the workings of the nervous system of a lower animal, Parker fell silent for over a decade. What happened? It was not for lack of inspiration or industry in carrying out research and publishing papers, for he published forty-nine papers of original research between 1897 and 1910, the year he returned to experiments relating to the problem of the origin of nervous systems. He continued to publish on sea anemones, but none of these few papers dealt with topics relevant to the nervous system. There is the suspicion that, no matter how unshakeable his determination had appeared when he confided his thoughts to William James, the pursuit of the origins of nervous systems amounted to a dream eventually to be realized but for which he had not really prepared in any pragmatic way. So as his career unfolded he chose the path of least resistance and let other research projects take precedence. He worked especially on sensory systems of both vertebrates and invertebrates.

Other tasks distracted him from his grand scheme. Upon his return from Europe Parker had resumed his teaching and spent a great deal of time honing his lectures for the freshman course Zoology I. He was a very conscientious teacher; his lecture notes, which are preserved in the Harvard Archives, are written with a steady, smooth hand, and portray a man of great organizational skill. His attachment to teaching is remarked upon by Alfred Romer:

> Unlike many research workers, who consider elementary teaching as beneath their dignity, Parker considered undergraduate teaching a matter of basic importance, and taught large elementary courses to the end of his professorial career. He was a forceful and inspiring lecturer. His effect on the students was heightened by their knowledge that here was a research worker who knew firsthand the things he taught, in contrast to the type of pedagogue who may lecture glibly but whose knowledge does not extend far beyond the text assigned. (Romer, 1964)

He worked hard to be deserving of the promotions that President Eliot granted very sparingly. He finally rose to the position of professor of zoology in 1906, thirteen years after his return from Europe and sixteen years after starting out as an instructor. Some tasks of a more sporadic nature also made heavy demands on his time. A prime example was his appointment as chairman of the organizing committee of the 1907 International Zoological Congress, which took place in Boston. According to a letter from Parker to Alexander Agassiz, dated 11 March 1907 (Museum of Comparative Zoology, Ernst Mayr Library Archives), 248 American and 177 foreign participants were expected to attend. From the spring of 1906 Parker spent long hours on all matters related to organizing such a large meeting. He was assisted primarily by Charles S. Minot, a distinguished professor of anatomy and embryology at Harvard Medical School, who served as fundraiser, and by Henry Fairfield Osborn, professor of paleontology at Columbia University and curator of vertebrate paleontology at the American Museum of Natural History, who acted as secretary. The latter intervened on Parker's behalf to petition Eliot to grant Parker a sabbatical year so that he could attend to congress business in calmer conditions, but Eliot refused. Parker had to trudge on, dealing with innumerable letters that demanded response. Minot complained that Philadelphia, ostensibly the second richest city in the country, was not contributing more substantial funds for the congress, and that Stanford University, which claimed to be the richest such institution, was contributing a measly twenty-five dollars to science! In a letter of 29 May 1906 Osborn forwarded to Parker Anton Dohrn's suggestion that inviting Theodore Roosevelt – a Harvard College alumnus and then American president – to the congress would have a beneficial influence on the German Emperor and on the Hamburg-American Line – potential carriers of European attendees – but also noted that Alexander Agassiz questioned the wisdom of such an invitation. In fact Agassiz, presiding over Parker in the matter, informed the latter a week later that his decision was reached: Roosevelt would not be invited. Such were the matters, petty and grand, that occupied Parker and kept him from his lab.

When he discussed the nervous system during that "dry" period, Parker used the channel of review articles, presumably in an attempt to establish himself as an authority on the subject outside the academic community despite his slim credentials so far. In the first of the two articles, which was based on a lecture delivered at the New York Academy of Sciences (Parker, 1900), he discussed the challenges to Waldeyer's neuron doctrine posed by Stefan von Apathy and Albrecht Bethe after their discovery of neurofibrils (see chapter 6).

The second review article, based on lectures delivered at the University of Illinois (Parker, 1909), was more ambitious and addressed the issue of the origins head-on. Published in instalments in a magazine read by the lay public, its tone is strikingly didactic. More important, it gave Parker the chance to review what his predecessors had accomplished and to articulate his own thoughts on the matter a year before he was to strike out again with an important research paper. He began to formulate a few principles founded on the notion that the basic operational unit of the nervous system is the reflex. The latter was understood as "the chain of consequences that begins with the reception of a stimulus on the surface of the animal and, leading through the central nervous organs, ends in the excitation of a reaction by some such organ as a muscle."

But even as Parker defined the circuit in terms of neural conduits, he acknowledged that the components of the circuit – receptors, adjustors, effectors – need not be construed as nervous and muscular in nature. To achieve this, he hypothesized, "a physiologically continuous thread of protoplasm must connect the two extremes [receptor and effector]." Parker observed such a mechanism operating in ciliated protozoans to explain how these unicellulars can respond to touch by adjusting the ciliary beat responsible for their movements. However, the substrates to achieve reflex action in the single cell "are so remote from those of a true nervous mechanism that, interesting and significant as they are, they had better be termed neuroid than nervous."

Parker opined that the neurally based reflex circuit did not emerge fully formed in the course of the evolution of multicellular animals, but on the contrary had occurred gradually:

> The first stage was that of the independent effector, the muscle which was brought into action by the direct influence of environmental changes as seen in the pore sphincter of sponges. The second stage was that of the combined receptor and effector in which the receptors, in the form of diffuse sensory epithelia or specialized sense-organs, served as delicate triggers to set the muscles in action and thereby render the effectors responsive to a wider range of stimuli than they would be under independent stimulation. Finally, the third stage is seen in the complete neuromuscular mechanism in which a central nervous organ or adjustor has developed between the receptors and the effectors. This

adjustor serves as a switchboard for nervous transmission and a repository for the effects of nervous activity. (Parker, 1946)

None of Parker's predecessors had reached that depth of understanding of the issues involved, or had produced as reasoned an evolutionary scenario of the emergence of nervous systems, as he reached in his 1909 synthesis. He was helped in this by his physiological bent, viewing the problem as one of functional integration of the parts, whereas his colleagues had merely seen anatomical units. Now it was time for him to design experiments that could throw light on how the scenario unfolded.

The Thorny Issue of the Sponges

Parker correctly suspected that sponges, being considered early representatives of evolutionarily emergent multicellular animals, were key to detecting the trail that would lead to the first signs of the formation of nerve cells and nervous systems. Coelenterates, especially since the presence of nerve cells in *Hydra* had been confirmed, were important actors in the story, but they already possessed fully formed, albeit "simple," nervous systems. So rather than continue his long-halted physiological studies of sea anemones, he embarked in 1909 on an ambitious investigation of sponges.

When Parker entered the fray, sponge biology had become a murky field. Sponges disoriented investigators because the usual signposts of the histological organization of animals were largely missing. For a start, there was controversy among investigators as to the very presence in sponges of epithelia, defined here as an outer sheet of tightly abutting cells with membrane boundaries. There were those who flatly denied the existence of true epithelia and substituted for them a "membrane" in which cells were loosely organized and their boundaries poorly defined (forming a syncytium). At the other extreme, represented by Franz Schulze and his pupil Robert von Lendenfeld (see chapter 7), not only did epithelia exist, but the three classical tissue layers of triploblastic metazoans – ectoderm, mesoderm, endoderm – were identifiable even in sponges. For Schulze and Lendenfeld, the ectoderm represented the border epithelia, the endoderm the epithelia of the aquiferous canals, and the mesoderm (or mesohyl for some), contrary to the mesoglea of coelenterates, contained the largest variety of cell types. Scholars with less extreme views acknowledged the exist-

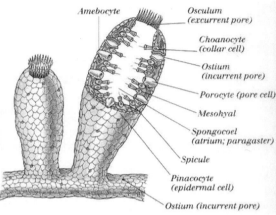

8.1 · (*Left*) Constantin Merejkowsky. From entry "Konstantin Mereschkowski" in ru.wikipedia. (*Right*) A modern depiction of sponge organization and cellular constituents. From an image in "Sponges: the original animal house," posted by Jennifer Frazer (17 Nov. 2011) in blogs.scientificamerican.com.

ence of "dermal membranes," meaning true cellular epithelia but a kind peculiar to sponges and sometimes called epithelioid (epithelia-like).

Observers of sponge behaviour were at a loss to account for the ability of these animals to respond to stimuli in a cohesive fashion. In coelenterates, such responses could be attributed to the underlying organization of the nervous system, but the tissue organization of sponges was such as to dispel the notion of a nervous system orchestrating these reflexes.

One of the first to have offered a rigorous and lucid description of the reactions of sponges was Konstantin Sergeevich Merejkowsky (1855–1921), who became famous for his theory of evolution based on cellular symbiosis. He was a graduate student in St Petersburg when he conducted his research on the sponges of the White Sea in the summer of 1877. His results were published in French early the following year (Merejkowsky, 1878). Although the paper dealt principally with morphological descriptions, it reported experiments on the osculum, a conical or cylindrical opening through which water is expelled after entering in the spongocoel via small pores (Fig. 8.1, right). The osculum was said to contain "muscle cells" (myocytes) believed to be implicated in the closing mechanism of the oscular opening. Merejkowsky noted that the osculum of the sponge *Rinalda arctica* reacted to having its edge probed with a needle by slowly closing the opening. The extent of the closure is proportional to stimulus strength, a strong enough needle stroke causing complete closure. In this case it took ten seconds for the closing process to even start after needle contact, and another twenty seconds for closure to reach completion. The osculum remained closed for two and a half minutes before gradually reopening over the next two minutes. This reflex is thus much slower than any recorded in coelenterates. Merejkowsky also noticed that prodding of any part other than the rim region of the osculum failed to induce the closure of the opening.

The sphincter-like reflex action is not the only contractile activity that Merejkowsky observed. If the tip of the osculum is raised above water level, the osculum will bend until it is fully immersed again. Of course, he believed that myocytes were involved in these contractions, but he offered little else to explain what happened in the sponge tissues to implement the reflexes. He assumed, like the majority of his contemporaries, that the time frame of these events precluded any nervous activity, but that some kind of transmission of excitability by protoplasmic channels must be involved, however reluctant he was to characterize them.

But only seven years later Lendenfeld published a paper in which he claimed to have discovered sensory and ganglion cells in sponges (Lendenfeld, 1885). Although Lendenfeld had concentrated his investigations on coelenterates in Australia (see chapter 7), he was already following in the steps of his mentor, Schulze, by studying Australian sponges. In contrast to most zoologists, Lendenfeld viewed sponges as coelenterates that differ from cnidarians in their uniquely well-developed cellular mesoderm. Consequently, he classified sponges as Coelenterata Mesodermalia, to distinguish them from the Coelenterata Epithelaria (cnidarians), in which the ectoderm is paramount. Not surprisingly, since according to him the mesoderm in sponges assumed the role that the ectoderm plays in cnidarians, it followed that nerve cells, if they existed at all in sponges, should be found in the mesoderm. In his 1885 paper Lendenfeld's descriptions of sensory and ganglion cells are tentative and unsupported by illustrations. But in a comprehensive monograph on Australian sponges published four years later (Lendenfeld, 1889a), in which he writes of New South Wales and Sydney as "the Eldorado of the spongiologist," he gave a more minute descrip-

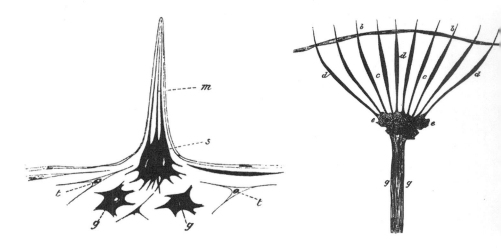

8.2 · Lendenfeld's depictions of sensory and nerve cells in sponges.
(*Left*) The "synocil" (m) of *Sycandra compressa* in which sensory cells (s) project
fine processes, with ganglion cells underneath (g). (*Right*) The "nerve ring" of
ganglion cells (e) of *Hippospongia canaliculata* intimately associated with the
slender sensory cells (c) irradiating above them and the "muscular membrane"
(g). From Lendenfeld (1889a), text-Figs. 7 and 8.

tion of these cells, accompanied by drawings (Fig. 8.2). He explained
the basis for his identification of sponge nerve cells as follows: "In a
great number of sponges cells are observed which have a spherical or
oval nucleus and highly granular, readily-stained protoplasm, which
assumes, when treated with osmic acid, the same aspect as the protoplasm of the sensitive and ganglion cells of higher Coelenterates assumes when similarly treated."

Lendenfeld went on to describe what he interpreted as a sensory
structure similar to the sensory cone of actinians, then called "palpocil." As the palpocil characteristically contains the process of one
sensory cell and the sponge equivalent contains processes from multiple sensory cells, he called it a "synocil" (Fig. 8.2, left). He observed
what he took for stellate ganglion cells proximal to the sensory cell
bodies. He also described nervous structures associated with so-called
muscle elements which for him also had a mesodermal origin. "The
muscular membranes," as he called them, "which divide the lacunose
from the more solid parts of the sponge in *Hippospongia canaliculata*,
do not reach to the surface, but appear crowned by a zone of spindle-

shaped sense cells" (Fig. 8.2, right). He perceived a nerve ring inserted between the muscle membranes and the sensory cells. Finally he described what appeared to him bundles of sensory hair cells and affiliated ganglion cells in inhalent pores.

Lendenfeld's claim of a nervous system in sponges met with his colleagues' acute skepticism as much as his lumping of sponges with coelenterates or his diagnosis of triploblasty for the sponges. As W. Clifford Jones put it (Jones, 1962), one of these critics (Bidder, 1898) stated that the synocils are "nothing other than the remains of slender monaxons ("hair-spicules") together with their pair of formative calcoblasts." In other words, they were part of the skeletal components and Lendenfeld's delusion stemmed from his lack of appreciation of the artifacts that result from histological processing of difficult tissues such as those of sponges. Critics also pointed out that his "ganglion cells" never seem to be interconnected or to connect with "sensory cells." Jones also suggested that Lendenfeld's sensory hair-cells were in reality contracted or transdifferentiating pinacocytes. The pinacocytes are the constituent cells of the "ectoderm" of the sponge basket – known as the pinacoderm – and are characterized as flat, polygonal cells forming a single layer. In Lendenfeld's time, as reported by Emile Topsent (Topsent, 1887) these cells had already been observed in diverse forms and it was soon discovered that these cells, not mesodermal myocytes – which are in fact connective fibres – are the actual contractile cells of sponges. In Topsent's own words, these cells "are capable of strong amoeboid movements and are found in all locations where contraction-relaxation cycles are necessary" (my translation). No wonder, then, that these chameleon-like cells confused Lendenfeld and led him to see sensory cells where effector (motor) cells in fact lodged. Five years after the publication of Topsent's monograph the contractile nature of the "ectodermal cells" (pinacocytes) was confirmed independently in England (Minchin, 1892).

From then on the consensus of sponge specialists was that the pinacocytes respond directly to adverse stimuli, without the intervention of a nervous system or other transmissive means, to cause the closure of the osculum. Such was the status of the problem when Parker began his investigations. His own reading of the literature made him side with Ghaltherus C.J. Vosmaer and Cornelis A. Pekelharing, who stated that sponge cells "are not connected in a way so as to enable them to conduct stimuli from one cell to another," and that it therefore followed that sponges were "destitute of the principle, the significance of which culminates in nervous tissue" (Vosmaer and Pekelharing, 1898). But Parker kept an open mind and made it his mission to test physiologically the hypothesis that sponges possess a nervous system.

To do so, he first scouted around for a seaside spot where large numbers of sponges could easily be collected and where lab facilities were available. He settled for the U.S. Bureau of Fisheries in Beaufort, North Carolina, and moved there early in the summer of 1909. In letters to his wife, Louise, who had stayed at home in Cambridge, except for one visit (Harvard Archives), he reported on his progress. On the first of July he wrote: "The sponges are behaving well and I have got a good set of results today. I am following my attempted habit of working on sponges in the morning and reading, writing, etc. in the afternoon (evening)." He was unhappy with the quality of local assistants: "a rather shiftless lot ... I have been offered one as a helper but declined." Consequently, he took it upon himself to arrange "a system of tanks to try drugs on sponges." On 5 July it appeared that snags were preventing Parker from fully implementing his plans: "The sponge work goes on well and I see the end in view though I shall have to omit the question of the relation of the lateral pores [ostia] and cilia over which I broke my head several days in vain." But three days later things looked up and his confidence returned as he followed a new trail:

Work comes on well and I have finally succeeded in getting a means of separating the action of the choanocytes [ciliated cell] and lateral pores which when thee [you, Louise] was there I despaired of. I was looking at a sponge fragment today and noticed a peculiar circulation and this gave me a hint that I followed up to the end and here I am. Consequently I have much to do between now and Sunday to go on my work in a new way but with past experience I can do it quickly. (Parker Papers at Harvard University Library Archives)

Parker's good fortune continued. A triumphant letter announced the completion of his work on 11 July:

I put in a long hard day yesterday and was at sponges this morning at 5:30 am. Now 3:30 pm. I have finished the job much better than I expected when thee was here through my last turn of good luck … Most of yesterday and today has been spent repeating experiments till I felt sure of results and I now [feel] I have a pretty good answer to the question about sponge reactions that I entered Beaufort with. I have not written up results but I shall expect to do that at once and get the paper off soon for publication. (Parker Papers at Harvard University Library Archives)

And he did write the paper with such dispatch that it appeared in January 1910, amounting to forty-one pages in print (Parker, 1910a). Parker identified the sponge he worked on as *Stylotella heliophila*, but the accepted name today is *Hymeniacidon heliophila*. He found that circulation in the sponge begins with the water current entering the ostia, and continues in large subdermal cavities, connecting in turn via incurrent canals to flagellated chambers, and thence to outcurrent canals from which water flows to the gastric cavity and out through the osculum (see Fig. 8.1, right). His experiments determined that the oscula close in still seawater and when exposed to air or injured in adjacent parts, and that they are unaffected by low temperature or light. Like the oscula, ostia close when injured in adjacent parts, but in contrast they open in deoxygenated water and at 35°C. The currents created by the flagella of choanocytes are stopped in high temperatures (40–45°C), slow down at low temperatures (9–10°C) and at first speed up in deoxygenated water. As implied by these results and further brought out by specific experiments, the ostia, oscula, and choanocytes act independently of each other – a hint of this is found in the above citation of the 8 July letter to Louise. He concluded that no integration of functions brought about by conducting pathways occurs in sponges, and that activities could only be elicited by direct stimulation.

Parker expressed the significance of his experiments for the question of the existence of neuromuscular functions in sponges in the following manner:

In the ostia, oscula, and flesh contraction is accomplished by spindle-shaped cells, the myocytes, which resemble primitive, smooth muscle-fibres.

The body of Stylotella is almost without transmission and such transmission as is present is so sluggish in character and so slight in range as to resemble transmission in muscles and not in nerves. It is probable that Stylotella possesses no organs that can reasonably be called nervous. (Parker, 1910a)

Parker's reference to muscles needs to be addressed because he speaks of them as the alternative means of signal conduction in the absence of nerve pathways. The "myocytes" he found in the flesh and in the "inner wall" of oscula and ostia, forming a sphincter responsible for their opening and closing, were not seen by his American colleague and sponge specialist Henry V. Wilson, who worked at Beaufort in the summer of 1909 after Parker had left the station. In a publication that appeared shortly after Parker's (Wilson, 1910), Wilson, who respected Parker the physiologist, nevertheless challenged Parker's histology, claiming that what Parker saw as myocytes were "the same elongated epithelioid cells which I have described as lining the large efferent canals." According to Jones (1962), the rugose surface assumed by Parker's sponges when the flesh contracts is due to the contraction of the pinacoderm of the deep-seated canals. Wilson's epithelioid cells are none other than Topsent's (1887) contractile pinacocytes, which are involved in the closing of the ostia and oscula as well as the general contraction of the fleshy parts of the body. Both Parker and Wilson seemed unaware of Topsent's seminal monograph, despite the fact that it was published in a prominent and well-circulated French scientific journal.

It was more than Parker's histology that met with criticism. Parker's 1910 paper also included numerous experiments with drugs, as alluded to in letters to Louise from Beaufort. Before him, Lendenfeld had conducted similar experiments with drugs (curare, veratrin, atropine, cocaine, strychnine, ether, chloroform) designed to bolster his case for the presence of nerve cells and neuromuscular transmission in sponges (Lendenfeld, 1889b). Jones (1962) argued persuasively, however, that these experiments were flawed for lack of adequate controls, and that no significance could therefore be attributed to Lendenfeld's observations. Similarly, Jones was less than impressed by the scientific quality of Parker's pharmacological experiments:

These results can be criticized in that Parker did not completely disentangle the influence of the drugs on the oscula from their action on the dermal ostia. When investigating their effect on the latter, he did remove the osculum, so that there can be no doubt that the effect he looked for, namely the cessation of ostial currents, did in fact entail the closure of the pores. For his investigations on the oscula, however, he employed whole sponges and since he states that the ostia closed more quickly than the oscula in the ether and chloroform solutions, it is possible that the closure of the oscula was induced by the cessation of the outflow caused by the prior closure of the ostia, and not directly by the action of the drugs on the oscula. Similarly, the slower closure of the oscula in 1:15,000 strychnine can perhaps be correlated with the gradual closure of the pores in this medium, while the inhibitory effect of 1:10,000 cocaine and 1:1000 atropine on the closure of the osculum in still water may have been due to the opening of some previously closed ostia, which may have resulted in an augmented current. (Jones, 1962)

To Parker, the effects of drugs on sponges resembled those obtained on vertebrate smooth muscles, thus reinforcing his conviction that the contractile cells of sponges were primitive forms of smooth muscle. But the suspicion that some of his drug experiments were flawed, and his unfortunate comparison of drug effects on a tissue which by his own admission is nerve-free with drug effects on vertebrate tissue in which smooth muscles and nerves are intermingled, rendered his argument very shaky.

In the end, the majority of Parker's conclusions were on the whole valid insofar as the design of his experiments allowed. On the question of how nervous systems emerged, Parker's conclusion from the results of his sponge inquiry was markedly at odds with those of his predecessors. Contrary to the views of Kleinenberg and the Hertwigs, he believed that the muscular and nervous systems did not develop simultaneously in the course of evolution; nor did he think that they developed independently, as Claus and Chun assumed. Instead, Parker was adamant that "muscles, independent effectors, as represented by the sphincters of sponges, were the first of the neuromuscular organs to appear and these formed centers around which the first truly nervous organs, receptors, in the form of sense-cells developed giving rise to a condition such as is seen in the coelenterates today." This was as plausible a scenario as it was possible to produce in 1910. Decades later other investigators would claim the presence of a nervous system in sponges, only to retract themselves eventually, and the nature of the signal conduction mechanism in cellular sponges is still a matter of debate today (Leys, 2007). So the essence of Parker's views on the integrative abilities of sponges and on their position in the development of "neuroid" conduction in multicellular animals has endured pretty much to this day.

There is evidence that Parker resumed work on sponges as early as the summer of 1910, for in a letter to Louise from Woods Hole, dated 14 July 1910, he writes: "I never saw so much dirt in a sponge. Clean it out often but it seems to have much more in it." Presumably he had started to work on the physiology of water circulation inside sponges, which led to a couple of publications on the subject in 1914, but in which no more was said on the question of the origin of nervous systems.

Sea Anemones Revisited

Parker's first acquaintance with Woods Hole, where much of his research on sea anemones was conducted, dates back to 1888 when he was a graduate student. This was the year of the foundation in Woods Hole of the Marine Biological Laboratory (MBL), which was fashioned after the zoological stations of Europe, particularly the one in Naples that Parker had visited. His attraction to the Cape Cod location increased over the following years; unlike Beaufort it was close to home – just 120 km from Boston – and the local marine fauna were rich enough to satisfy every researcher's fancy. He became a regular summer fixture there, so much so that he was appointed a trustee of the MBL from 1907 until his retirement.

The earliest evidence we have that Parker had rekindled his interest in sea anemones comes from a letter to Louise from Woods Hole, dated 15 July 1910, in which he writes: "I have been getting interesting results on the anemones and [did] work last night till about 10^{30}." The next logs of his progress appeared two years later, also in letters written from Woods Hole in July and August (1912), in which he expressed his pleasure with the outcome of his experiments and mentioned his projected experiments on the locomotion of anemones. But in August one of his legs became lame from an accident and the research slowed down. The letters also reveal that Parker was without a salary at the time – perhaps because of a new budgetary crisis at Harvard – and Louise was worried that they would not be able to afford a planned trip to Nova Scotia. All this must have put a damper on his work at Woods Hole. His research must have progressed haltingly because the fruition of his efforts did not appear in full papers until 1916, by which time he had reached the ripe age of fifty-two.

Parker continued the pattern set for his study of sponges. It was as if he had an irrepressible need to keep cultivating the intellectual ground that impelled his research and to broadcast the results of his reflections through the medium of reviews or "popular articles." While his research on sea anemones was still in its early phases, between 1910 and 1914, Parker produced no fewer than four such articles. In the first (Parker, 1910b) he argued that the "muscle" system of sponges, as an independent, nerve-free unit, responds directly to stimuli, and that the first step in the formation of a nervous system was the appearance of sensory cells that connected directly to myocytes – the receptor-effector stage – and which provided a more sensitive, lower-threshold "trigger" for effector responses than accomplished by independent effectors. He added that "coelenterates usually show more than a simple receptor-effector, for the fine branches from their sense-cells not only reach their muscle-cells but also anastomose with one another and form a nervous net." The next stage was adding ganglion cells to these diffuse nerve nets, as seen in sea anemones. The centralizing trend in the evolution of nervous systems, he proposed, was a response to the aggregation of sensory structures, such as occurred in jellyfish, or to the anterior sense organs of higher, bilateral animals which led to the development of true brains. Brain nerve cells, he claimed, depart from the coelenterate model of syncytially connected nerve net cells to become connected by Sherrington's synapses. Obviously, Parker, who was unaware of Edward Schäfer's work on jellyfish, could not countenance the notion that the neuron doctrine applied universally, including to coelenterates.

In the second paper (Parker, 1911) Parker stressed that effector systems are an integral part of the nervous system to the same extent as receptors and adjustors, and consequently he called the receptor-adjustor-effector chain the "neuromuscular mechanism," even though it may sound odd that he referred to the evolutionary stage of independent effectors as a neuromuscular mechanism. He offered this view of the gradation of nervous system organization in coelenterates: "Judging

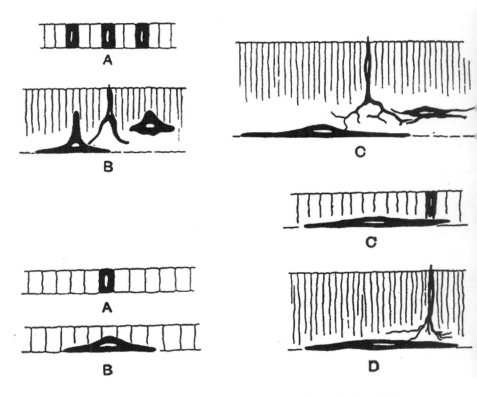

8.3 · Sketches contrasting the Hertwigs' (*top*, A–C) and Parker's (*bottom*, A–D) views of the origin of nerve and muscle cells. From Parker (1911), Figs. 2 and 3. Courtesy of the American Philosophical Society.

from the more recent work on the nervous system of these animals, centralization can scarcely be said to the present at all in hydra; it is but little more pronounced in the sea-anemone; and, though most marked in the jellyfishes, it does not rise even here to a grade that entitles it to comparison with what is seen in such foms as the earthworm."

Parker's investigation of sponges led to his reconsideration of the Hertwigs' theory of the origin of nerve and muscle (Fig. 8.3). In the Hertwigs' view, the initially "indifferent" epithelial cells simultaneously differentiate into sensory, ganglion, and muscle cells, followed by the sprouting of processes by sensory and ganglion cells to make contacts between themselves and with muscle cells (Fig. 8.3, top, A–C). In Parker's rendition (Fig. 8.3, bottom, A–D) the same indifferent epithelial cells give rise first to muscle cells (sponge stage), and then to partially differentiated nerve cells proximal to fully differentiated muscle cells, to arrive finally at the early coelenterate stage of sensory nerve cells connecting to muscle cells.

Parker's brooding thoughts in the years preceding his output of sea anemone papers extended beyond nervous system evolution. As an experimentalist he grew increasingly self-conscious about his way of dealing with the lab environment for conducting physiological research and training aspiring zoophysiologists. He couched his thoughts on the subject in a short paper published in *Science* (Parker, 1914). In it he calls for a better preparation in chemistry and physics for future experimentalists. He even advocates knowledge of psychology for those wishing to study the "mental states" of lower animal forms. "These problems are psychological," he adds, "and I should, therefore, regret to see a prospective zoologist omit from his preliminary training a reasonable grounding in this field of investigation." The new mindset of experimentation was slow in coming, and Parker expresses this in the following caricaturesque frame: "The student of thirty years ago was concerned with methods of preserving animals and he never felt safe until his catch was in the alcohol jar; the modern student is all alert to keep his stock alive and he consigns it to preservatives with funereal rites. This change of attitude is part and parcel of the new growth and is working a slow and steady revolution in the equipment of our laboratories."

However, the *Science* paper made it clear that to Parker, being enamoured with lab apparatus is an exercise in vanity. He reminds us that "many of the pioneers in the new [experimentalist] movement have already demonstrated to us fundamental results by means as strikingly simple. To [Jacques] Loeb the problem of the universe is soluble in a finger-bowl; to [Thomas Hunt] Morgan in a milk-jar; and we must never forget that the importance of a result is often inversely proportional to the complication of the apparatus by which it was attained." One wonders if he was a bit conceited here, and whether, in reality, the simplicity of his lab apparatus was proportional to the financial support he was receiving for research. His salary at Harvard was modest, but even more so his research funds. There was no organized government funding for academic resarch in Parker's day. Under President Eliot, Harvard was transformed into a research university, but research

8.4 · (*Left*) The Agassiz Hall of the Museum of Comparative Zoology where Parker worked, as it looks today. Photographed by author. (*Right*) Parker in his lab at Woods Hole, around the early 1920s. Courtesy Marine Biological Laboratory Archives.

expenses were met by monies from the operating budget of the institution, unlike the case in German universities, which had a tradition of receiving grants for research equipment from governments or philanthropic foundations. A German zoologist of Parker's stature would have used the kymograph, invented by Leipzig physiologist Carl Ludwig fifty years earlier, to record the amplitude of muscle contraction or movements of sea anemones over time. The kymograph was not considered a luxury by European standards and it yielded precise measurements and insights that simpler methods could not provide. But even if Parker had shown interest in such instruments, he would have been told that his department could not afford them; surely the piggybank was already broken to make the purchase of microscopes and histological necessities!

For all Parker's preaching about the experimental approach, it is confounding that he started his series of papers on sea anemones with a histological study (Parker and Titus, 1916). The histology of *Metridium* had been well studied previously and its neuroanatomy known since the 1879 monograph by the Hertwig brothers. So why did Parker bother? Probably to acquaint the reader with the species (*M. marginatum*) on which his experiments were to be conducted. But Parker's securing of a co-author who was not one of his students suggests a lack of confidence in his own histological skills. Titus, the co-author, is acknowledged in the paper as the designer of the neurohistological method used – a variant of the Golgi impregnation technique.

After describing the various parts of the anemone, including the different sets of muscles, Parker comes to the matter of the nervous system. He describes what he calls neurofibrils, many of which lie in the mesoglea, but no cell body to link them to; from past acquaintance with previous neurohistological papers in this book, these fibrils look suspiciously like connective fibres. The illustrations that support the histology look shoddy, not at all up to the standards to which many predecessors had accustomed us. Parker conducted some experiments in an attempt to bolster the neurohistology. Stimulating the surface of the body column triggered a contraction of deep mesenteric muscles, and Parker sensibly explained this by invoking a connection between ectodermal sensory cells in the column and a nerve net in the endo-dermal mesentery across the mesoglea. He thought, albeit wrongly, that the mesogleal "neurofibrils" could constitute the bridging con-nection to account for the experimental result. He also wrongly de-duced that nerve cells were missing from certain parts of the anemone body because they went undetected by Titus's silver nitrate impregna-tion method or because some muscles were refractory to anaesthesia with magnesium sulphate. Parker should have known that a negative result, especially one based on the use of a single histological method, does not signify the absence of the cells looked for. And the appar-ent resistance to anaesthesia may be due to differences in threshold for effective action or to failure of the anaesthetics to penetrate deeply enough into the tissue. From these flawed premises Parker concluded that four different neuromuscular organizations exist in *Metridium*: one akin to the sponge-like independent effectors; a second where circular muscles are directly excitable yet loosely controlled by sensory cells; a third where sensory cells tightly innervate ectodermal tentacle muscles; and a fourth in the mesenteries, where sensory cells act through a con-ducting nerve net to activate longitudinal and transverse muscles. The gradation in complexity, here, makes one wonder if Parker was intending to show that sea anemones embody a recapitulation of evolutionary stages in the management of muscle activity and body movements.

In his next paper (Parker, 1916) Parker takes stock of four types of effector systems available to the sea anemone: mucus cells, motor cilia, nematocysts, and muscles. Here he endeavours to show that of all four effectors only muscles are under nervous control. He musters up several lines of circumstantial evidence to build his case, but the one that he relies upon most heavily is anaesthesia with chloretone or magnesium sulphate. He found that anaesthesia failed to stop mucus secretion, ciliary beat reversal, and nematocyst discharge, while blocking muscle contractions. He ran counter to the Hertwigs, who had reported that nematocytes are innervated. But amazingly for someone who was performing poorly in his own neurohistological work, he believed more in the meaning of his anaesthesia experiments than in the neurohistological evidence of the Hertwigs, whose credentials seemed unassailable. At the heart of the controversy was Parker's black-and-white approach to the role of nerves; for him nerves were there only to trigger effector events, so if anaesthesia was ineffective in blocking a response to stimulation, one could only conclude that nerves were absent in the effector's vicinity. It never occurred to him that the effector might be innervated for the purpose of fine-tuning – that is, to modulate the sensitivity threshold of the effector's response to stimulation – and that in this case no form of anaesthesia could betray the presence of nerves. This is what he missed in his appraisal of the nematocyst discharge.

Turning his attention to muscles, Parker was awed by the diversity of muscle development in sea anemones compared to other coelenterates, a condition to which he attached great physiological significance. Muscles are involved in sphincter activities, body column shortening, tentacle writhing, oesophagus expansion, foot retraction, and more. This variety allowed sea anemones to enrich their behavioural repertoire. In addition, Parker saw the importance to these animals of rhythmic, peristaltic activities in relation to feeding, and he could not help drawing an analogy with the peristalsis of the vertebrate gut muscles under the influence of the myenteric nerve plexus. To him these activities overrode reflexes in importance, and he therefore became convinced that "there is not the least doubt that some of the neuromuscular responses in

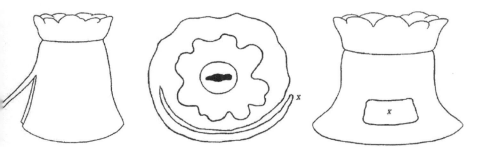

8.5 · Sketches illustrating some of the experiments by Parker (1917a). Explanations in text. Area of stimulation marked by "x." Courtesy John Wiley & Sons.

Metridium are true reflexes, though the majority of such operations are more usually exhibitions of excessive tonicity or of rhythmic motion."

Parker's papers of 1916 were followed by a prodigious output in 1917, when he produced no fewer than five full papers on sea anemones. The first of these, previewed in a short commumication to the *Proceedings of the National Academy of Science*, describes a series of surgical experiments in the style of Romanes to support the involvement of nerve net transmission in sea anemones (Parker, 1917a). Here Parker focused on the effects of various cutting procedures on the retraction of the oral disc by contraction of the longitudinal mesenteric muscles. The retraction "could be called forth by mechanical stimulation of varying intensity" on the surface of the column, especially near the sphincter and the foot. Basically, the experiments were designed to test the path of nervous transmission inferred by the Hertwig brothers on the basis of their neurohistology. In Parker's own words, "impulses from the ectoderm of the animal were supposed to pass to the oral disc, thence down the ectoderm of the oesophagus to the mesenteric filaments and thus to the entodermic musculature, such as the longitudinal muscles of the mesenteries."

Parker found that removing the oral disc entirely had no effect on the mesenteric contraction, no matter where the stimulus was applied, thereby refuting the Hertwigs' contention that the oral disc and oesophagus were obligated paths for transmission. After carving off a tongue of the column from the foot, as seen in Fig. 8.5 (left), he observed that oral disc retraction was unimpeded upon stimulation of the tongue extremity. Producing this incision in reverse, from the oral disc region down, gave similar results. If a similar incision from the foot region is practised transversely, as seen in Fig. 8.5 (middle), once again stimulation of the tip sends the retraction to the rest of the body. Extending the incision around the circumference of the body, however, impedes retraction considerably. In a critical experiment Parker carved off a segment of the column with the underlying mesentery as the only attachment left to the segment (Fig. 8.5, right). Stimulating that segment still induced the total retraction of the oral disc.

From these and other experiments Parker concluded that diffuse neural pathways were available to ensure communication between the ectoderm and the endoderm. "Nervous transmission in actinians," he asserted, "is accomplished in large part over tracts that penetrate the mesoglea and is in no sense limited to the basal part of the ectoderm and the entoderm." In cases of incisions resulting in muscle response delays significantly longer than one second, Parker assumed that no nervous transmission was involved. This was his implicit assumption from believing that the nerve net was syncytial, so that little slowing of impulse conduction was expected. However, if one assumed, as did Schäfer, a synaptic-like nerve net, a significant slowing down of transmission due to extensive incisions should be expected, owing to time delays in crossing multiple synaptic barriers along sinewy paths.

In the next paper (Parker, 1917b), Parker took great pains to find experimental evidence that much of the activity of sea anemone tentacles can be accounted for without recourse to a coordinating nerve net. He describes a series of clever experiments designed to show that what suggested the involvement of nervous transmission was in fact a simple interplay of physical forces, such as internal hydrostatic pressure and muscle tonus directly adjusting to these forces. He finds little difference between the spontaneous or elicited responses of intact and excised tentacles, and to him it constitutes an argument in favour of the lack of involvement of the oral disc nerve net. Viewing tentacles as a form of limbs, he suggests that sea anemones represent an early stage in limb development, stating that the actinian tentacle, "in contradistinction to such appendages as those of the arthropods or the vertebrates, contains a complete neuromuscular mechanism by which its responses can be carried out quite independently of the rest of the polyp." But Parker provides no compelling evidence that the tentacles, despite showing evidence of "neuromuscular" independence when excised, are not subjected to coordinating nerve input from the nerve-rich oral disc when attached to the animal. In this paper, as in others by Parker, he tends to disparage claims of the presence of nervous tissue by other researchers if they clash with his own 1916 neurohistology paper, even if his own status as a neurohistologist was shaky by comparison. One cannot help suspecting that the drive for some of these studies is ideological, as Parker appears insistent on proving the hypothesis contained in his essays of 1910 and 1911; namely, that sea anemones represent an intermediate step between the independent effectors of sponges and the neurally controlled effectors of higher animals. To take this view, one has to posit that in sea anemones one encounters a mixture of independent effectors and neurally controlled effectors.

Following strict logic, Parker's next investigation pertained to another "limb" of sea anemones, the pedal disc (Parker, 1917c). As the name implies, the pedal disc of the anemone is akin to a foot: it anchors the sessile animal to the substrate at sea bottom, but it can also loosen itself and by undulating movements carry the animal away to a different location and substrate to glue itself anew thanks to special muscles and slime-secreting glands. The bulk of Parker's paper is devoted to the orientation and mechanics of pedal locomotion, and little is said of the role of nervous tissue in controlling the circular, basilar, and longitudinal muscles of the mesenteries that are involved in it. Seemingly anxious to debunk the Hertwigs' attribution of a central role for the dense nerve net of the oral disc (see chapter 4), Parker conducted cutting experiments that showed that the oral disc nerve net was no more involved in pedal locomotion than in tentacle movements. But as in the paper on tentacle movements, Parker abstained from discussing the local innervation of the pedal disc, apart from a reference to the mesogleal "neurofibrils," which are no help in explaining anything, as we found earlier.

Parker now felt the need to bring out all his findings to date and draw general conclusions on the behavioural repertoire of sea anemones. In his own words, he "planned to examine some of the more complex activities of these forms with the view of gaining a clearer insight into their elements and into the relation of these elements to the animal as a whole" (Parker, 1917d). He threw in a few additional experiments to clarify points that he or other researchers had raised in papers. One point of contention was that others had sought to interpret the apparent complexity of actinian behaviour as evidence that these animals had reached a certain capacity for consciousness and will. Parker had no patience with such far-flung speculations, and he comments no further on the subject of sea anemone intelligence except to state that one could "see in Darwin's natural selection one means at least of explaining adaptations without recourse to such a factor." He also refers to his friend and Woods Hole mate Jacques Loeb (1859–1924), the famous German-born physiologist and discoverer of artificial parthenogenesis, as the authority on animal psychology who was most instrumental in introducing mechanistic alternatives to theories of animal consciousness (Loeb, 1900). Loeb, who had himself conducted experiments on feeding reactions of sea anemones, in the fourth chapter of his book compared the reactions of sea anemones to the behaviour of insect-eating plants. He implied that relatively complex behaviours such as those of sea anemones need not be explained by complex nervous activity seated in a central nervous system, but can be accounted for simply by "a sensitive and quick protoplasmic conductor," whether of a nervous or non-nervous nature. One can see how Parker was deeply influenced by Loeb's ideas and how these thoughts may have contributed to his research strategy for sea anemones.

In his paper on actinian behaviour, Parker concentrated on feeding and the retraction-expansion cycle of the oral disc and body column. Turning to feeding, he was struck, as others before him, by the declining response of sea anemone tentacles to food that is repeatedly offered. Some had attributed this to an adaptation related to satiety and others to muscular fatigue. Parker's originality is that he was the first to make the distinction between muscular and sensory fatigue, and he forcefully argued that sensory fatigue was responsible for the decline in the feeding response. The astonishing observation by Herbert Spencer Jennings (1868–1947), a student of Parker's postgraduate colleague Charles Davenport and best known for his work on the behaviour of protozoa, that the desensitization of tentacles in contact with food was also observed in tentacles not exposed to food (Jennings, 1905) intrigued Parker. Jennings's observation suggested that "the animal reacts as a unit"; that is, there is a coordinated response, which was something that ran counter to Parker's precepts. So Parker conducted further experiments that led him to conclude that the sensory fatigue experienced by the tentacles untouched by food was the result of the presence of food in the oesophagus. This conclusion comforted Parker in his ongoing view of "the relative independence of parts" and in his rejection of nervous coordination. It is surprising that Parker, who was aware of Havet's and Grošelj's descriptions of a dense nerve net in the lining of the oesophagus, did not envisage that the presence of food in the oesophagus would trigger a nerve net–mediated feedback inhibition in the tentacle nerve net, and that this was a more parsimonious interpretation of the results than sensory fatigue.

In the second part of the paper Parker revisits the retraction-expansion behaviour of sea anemones. Retraction is a protective behaviour, whereas expansion allows anemones to resume normal physiological activities such as feeding and respiration. They expand at night and Parker found that, in *Metridium*, bright light causes closure of the sphincter and partial retraction. Strangely for a keen investigator of the nervous system, Parker never discusses how light affects the behaviour, whether there are light-sensitive cells involved and, if so, whether these cells act directly on muscle cells or through the nerve net. He is entirely silent on such questions, as if the thought never occurred to him.

Finally, the paper addresses the issue of the "psychological development" of sea anemones, asking: "Is the actinian an organism that responds as the vertebrate heart does, or does it necessarily include in its activities elements of a psychic order?" To Parker the more an animal acts as an organismal unit – that is, in a coordinated fashion – the higher its psychic order. We have seen that Parker found little of unity of action in sea anemones. He conceded, however, that the creeping locomotion of sea anemones has all the trappings of an activity coordinated by the nervous system. And the phenomenon labelled by him "the modifiability of the animal's responses" is also characteristic of a higher psychic order. This concept was particularly developed by Jennings (1905), for whom the previous history of an organism determines its subsequent responses. The adaptability of responses, the decline in the amplitude of succeeding responses, was variously attributed by Jennings to metabolic or other changes in the responding cells as a means of adjusting to changing internal conditions. What was observable to Jennings and Parker is known today as the phenomenon of habituation or desensitization, a form of elementary nonassociative learning best studied at the nerve cell level in the sea slug *Aplysia* (Castellucci et al., 1970). For Parker that was the highest psychic order reached by sea anemones, however humble an evolutionary achievement it represented. All his attempts to experimentally demonstrate a higher form of learning – learning by association – in sea anemones, met with failure.

However reluctant he was to grant the nervous system a commanding role in sea anemone activities, Parker had to admit that his observations depicted a much more complex picture of the functional capabilities of these coelenterates than he had anticipated, and that sea anemones harboured the incipient stages of the neuromuscular complexity found in higher animals. Now he asked himself whether an intermediate stage of evolutionary development existed between the organizational simplicity of sponges and the complexity of sea anemones. Wishing to keep the comparison on a level ground, and since sea anemones are solitary polyps, he turned to hydrozoan polyps for an answer. Hydrozoan polyps are as a rule very small and therefore intractable to experimentation. From his friend William E. Ritter (1856–1944), Parker had heard of one species in the Pacific waters of the Californian coast that was large enough for conducting the kind of experiments he routinely practised on sea anemones.

After earning his PhD at Harvard in 1893, Ritter had taken a position in the Department of Zoology of the University of California at Berkeley, where he had risen to chairman. Looking for a good seaside location to establish a marine laboratory, Ritter happened on La Jolla, near San Diego, and there managed to set down roots in circumstances described here by Parker (1946): "This laboratory was established through the efforts of my old friend and associate of Harvard graduate days, Dr. William E. Ritter, and by funds contributed by Miss Ellen Scripps and her brother, Mr. Edward W. Scripps, of newspaper fame. The Scripps family did much for La Jolla, but nothing so considerable as the founding of its celebrated marine laboratory."

Parker had first visited La Jolla with Louise as a tourist and Ritter's guest in 1914, but he, and especially Louise, were so enraptured by the scenic place that they had determined to return for longer visits. The occasion arose two years later with Parker's resolve to conduct experiments on the large hydrozoan solitary polyp, *Corymorpha palma*, a species abundant in the mud flats of the area. The results of this work appeared in the fifth paper of importance on coelenterates to be published in 1917 (Parker, 1917e). When Parker set to work on *Corymorpha*, he was building on a knowledge base of several papers published on the species by one of Ritter's master's students, then on staff at the Scripps Marine Laboratory, Harry B. Torrey (1873–1970). Between 1902 and 1912 Torrey had published nine papers on different aspects of the biology of *Corymorpha*, and he was considered a leading authority on hydroids (Calder, 2013). So Parker felt the need to step cautiously lest he tread on Torrey's toes, which probably stood close to his own in the La Jolla lab. In his paper Parker strove to justify himself by using as diplomatic a tone as could assuage any resentment Torrey may have harboured: "It might seem superfluous after the very able work that Torrey ... has done on Corymorpha to undertake a study of the reactions of this animal, but my objects were somewhat different from those that Torrey had in view and, though I went over much of the ground that he covered and confirmed most of what he did, I believe I have brought to light some new facts that add to our knowledge of the natural history of this most interesting species."

Some of the new facts that Parker uncovered blatantly challenged Torrey's own findings. Torrey had reported the presence of cilia in *Corymorpha*, but Parker found that neither mucous glands nor cilia could be counted among the effectors of this hydroid. Thus, contrary to sea anemones, only nettle cells and muscles remained as effectors in *Corymorpha*, and Parker reiterated his oft-repeated interpretation that only the muscles are under nervous control. The longitudinal muscles of the stalk and proboscis (mouth appendage), but not the circular muscles, require neural input for their actions. However, he made an exception for the circular muscle involved in peristalsis of the proboscis, apparently unaware of the contradiction in having an innervated muscle for peristalsis but not for other actvities, as if the innervation could retract when peristalsis was turned off! Parker designed skillful experiments that clearly demonstrated the workings of a diffuse, ectodermal nerve net, although a longitudinally oriented nerve net conduction prevailed in the stalk. Presupposing that only a central nervous system can bring about coordinated behaviour – a nerve net is prevented from doing so by its diffuse nature – he failed to see that some of his experiments pointed to mechanisms of coordination of motor activities. So he again put forward the opinion he had expounded for sea anemones; namely, that *Corymorpha* is constituted of independent or semi-independent parts.

The stalk of *Corymorpha*, no matter what position Parker forced it into, invariably righted itself back to a vertical position; this he identified as negative geotropism. By dint of cutting experiments and anaesthesia with chloretone, Parker demonstrated that, contrary to Torrey's interpretation, this geotropic response was dependent on neuromuscular mechanisms. Similarly, feeding appears to be under nervous control, showing a pattern of coordinated activities involving stalk bending and oral flexion of tentacles. But strangely Parker scuttles his own observations by rating the performance of *Corymorpha* as inferior to that of sea anemones. To an uninformed reader, *Corymorpha* and sea anemones differ in feeding methods, but nothing in Parker's descriptions would lead one to conclude that coordinating capabilitiy is tipped in favour of sea anemones. And yet Parker failed to use cautious judgment in this case.

In the end Parker had to admit that *Corymorpha*, after all, proved inadequate as an evolutionary intermediate in the development of the neuromuscular system. The latter in *Corymorpha*, Parker concedes, "is not intermediate between that of the receptor-effector system of actinians and the independent effectors of sponges. It resembles a reduced actinian system rather than a primitive state from which such a system could be derived." On this rather disappointing note, the hectic pace of Parker's investigations of coelenterate behaviour and underlying neuromuscular functions came to a close.

The Crowning Achievement

The cnidarian nervous system: elementary my dear Parker?
George Mackie, 1990

No documentation was found in the Parker Archives at Harvard which traces the genesis of Parker's epochal book on the early evolution of the nervous system (Parker, 1919a), and his autobiography is silent on the subject. We are left with conjectures based on bits and pieces of incidental information. It is likely that Jacques Loeb or his good friend and colleague at Harvard, Winthrop J.V. Osterhout, invited him to write a book for their newly created series of Monographs on Experimental Biology. Loeb, who had moved in 1910 from Berkeley to the recently created Rockefeller Institute for Medical Research in New York, was advocating a departure from a purely zoological basis for the field of experimental biology, toward an approach rooted in the physico-chemical make-up of living cells. In Loeb's view, "experimental biology – the experimental biology of the cell – will have to form the basis not only of Physiology but also of General Pathology and Therapeutics" (letter to Abraham Flexner cited in Osterhout, 1930). To this end Loeb in 1918 created two instruments aimed at broadcasting his ideas: the *Journal of General Physiology* and the Monographs on Experimental Biology, co-edited with Osterhout – a noted plant physiologist – and the geneticist T.H. Morgan.

Parker was to partake of both ventures. He wrote a paper in the second issue (November) of the first volume of the journal founded by Loeb, in which he reported experiments that allowed an accurate measurement of the speed of propagation of impulses in the nerve net of his pet sea anemone, *Metridium marginatum* (Parker, 1918). The paper is of interest not only for its intrinsic scientific value and contribution to the understanding of primitive nerve net activity, but also for offering a rare glimpse of a Parker abandoning his distaste for complex lab instruments and mindful of the importance of quantitative measurements. The work was conducted at Woods Hole in the summer of 1918, so it is not clear if the kymograph he used for recording the waves of contractility of the sea anemone preparations had been purchased by him or, more likely, belonged to a staff member of the Marine Biological Laboratory. At any rate, this was a happy time for Parker, who was enjoying the presence of his wife's relatives vacationing in the Woods Hole area. George and Louise had no children of their own, so they showered their affections on their nephews and nieces. The affectionate tone of his relationship with his sixteen-year-old niece, Anna B. Stabler (1901–1991), is best illustrated in her short letter of 11 September 1918, when Parker had returned to Cambridge ahead of his relatives:

Dear Uncle Howard,
Thank you so much for the delicious candy you sent us. We haven't had any since you were here. We are now looking for a good place to put the boat for the winter. I think we shall have to take it out of the water tomorrow or next day. Saturday we had

an exciting boat race. The "Scout" came in second. Ben Webster won, and Josephine Choate came third. J.C. has a new jib for her boat but it didn't seem to do her much good this time.

We are coming home on the fifteenth and I shall be very glad to see all the people in Cambridge again.

Give my love to Aunt Louise and keep plenty for yourself.

Anna.

(Parker Papers at Harvard University Library Archives)

Notwithstanding family distractions, Parker that summer managed to accomplish a satisfying feat – being the first to record the speed of conduction of nerve net excitation somewhat directly, in contrast to his predecessors, Romanes (1877) and Harvey (1912), who had inferred the speed indirectly, from waves of muscle contraction in jellyfish. The matter was of urgent interest to Parker because comparing conduction in the coelenterate nerve net with nerve conduction in highly developed nervous systems should give a symbolic measure of the evolutionary distance separating the two levels of nervous system organization. Parker's method of recording this speed of conduction was very ingenious, like many of his experimental designs. He found that "the pedal edge of the column of *Metridium* may be partly cut off as a long tongue of tissue, one end of which can be left still attached to the animal [Fig. 8.6, left]. When the free end of this tongue, which may be 10 to 15 cm. in length, is stimulated, the animal responds in the characteristic way [contraction of mesenteric muscle] but without exhibiting muscular contraction on the length of the tongue itself. This portion of the animal contains a nerve net and hence may serve as a region in which to measure the rate of transmission in that type of nervous tissue."

Once the preparation was set up on the basis of this reasoning, electrical stimulation was applied to the tip of the tongue (Fig. 8.6, left, *2*) and the contractile response in the column recorded on a kymograph (Fig. 8.6, top right), allowing the measurement of the time between stimulus application and the response. Next, the stimulus was applied at the base of the tongue (*3*) and the response was recorded as before (Fig. 8.6, bottom right). The difference of stimulus-to-response time delay between the two recordings gave as accurate an estimate of the rate of transmission of impulses through the nerve net as could be obtained in those years. In this way he obtained values between 12 and 15

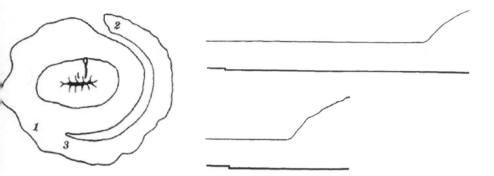

8.6 · (*Left*) Sketch of the column of *Metridium* viewed from above, showing the incision from near the pedal disc. (*Right*) Kymograph recordings showing the time of stimulation (deflection in lower trace) and the time of onset of the contraction (rise in upper trace). The trace that marks time intervals is omitted. See text for further explanations. From Parker (1918), Figs. 1, 3, and 4. Courtesy Rockefeller University Press.

cm per second. Not surprisingly, these values paled in comparison even to the lowest transmission speeds of nerve fascicles in higher invertebrates and vertebrates, but Parker's interpretation of the difference is rather lame and self-evident, stating that "the nerve net has generally been supposed to be a less rapid and efficient organ of transmission than the nerve trunk." But because Parker assumed that the nerve net was syncytial, he should have anticipated that transmission speed would be similar to the lowest values found in higher metazoans. Had he espoused the view that the nerve cells of the net are discontinuous and, therefore, constitute a synaptic nerve net, the discrepancy in transmission speeds would have made more sense, as crossing barriers between cells would slow down transmission further.

Now, as this paper was being published in the fall of 1918, Parker was deep into the writing of his book gathering together his findings of many years and exposing his encompassing vision of the nature of the primordial nervous system and how it emerged. *The Elementary Nervous System* (Parker, 1919) was the second book published in the Monographs in Experimental Biology series, just after Loeb had inaugurated it with his own contribution the previous year (Loeb, 1918). It was followed shortly by the third monograph written by no less than the pioneering geneticist and future Nobel Prize winner Thomas Hunt Morgan

(Morgan, 1919). The company Parker kept here testifies to the status he had reached in American biology and to the importance his colleagues attached to his research.

In the book preface Parker states that, by using the experimental approach, "an attempt has been made in this volume to portray the elementary nervous system as it exists in the simpler animals and in the simpler parts of the complex forms." By simpler animals he means sponges, cnidarians (his coelenterates), and ctenophores. By simpler parts of more complex animals he means, for example, the nerve net–like submucous plexus and myenteric plexus of the mammalian intestinal wall. Thus, to him coelenterates and vertebrates share the presence of an autonomous nervous system, the latter constituting the dominant, if not the only coelenterate nervous system in the form of nerve nets, but a subsidiary one in animals endowed with a well-developed central nervous system.

The first section of his book covers the sponges as representatives of the pre-nervous evolutionary phase when animals managed with "independent effectors" such as cilia and contractile cells, and coordination was achieved by "neuroid transmission." The section goes on to develop the theme that independent effectors and neuroid transmission were not lost with the advent of nervous systems, but remained physiologically relevant, even if more or less marginally, in higher animals. In the second and largest section, Parker discusses the emergence of what he calls receptor-effector systems of coelenterates, contrasting the neuromuscular systems of polyps such as sea anemones and of medusae such as jellyfishes. He here tends to minimize the role of ganglion cells as go-betweens linking sensory cells to muscle fibres, and favours direct connections between sensory and muscle cells as more rapid means to effect responses to environmental stimuli. He concedes, however, that the role of ganglion cells may be more important in jellyfish, on the basis of critical experiments by his friend, the physiologist Albrecht Bethe, which led to the conclusion that stimulation of the marginal bodies is more likely to activate muscles through the nerve net under natural conditions (Bethe, 1909).

In the same section Parker discusses at length the nature of the nerve net, its mode of impulse transmission, and its role in complex activities such as feeding. He makes it perfectly clear that implicit in his definition of nerve net is its syncytial character. He had wavered over this for years, weighing the evidence and arguments of those siding with the protoplasmic continuity theory versus those pushing the neuron doctrine (see chapter 6). In his book he is unabashed in his full commitment to the protoplasmic continuity of the nerve net, to the point of extending its syncytial nature to the intestinal "nerve nets" of higher metazoans. His respect and admiration for Bethe, a forceful militant for the continuity cause, may have influenced him in this regard. Their closeness as colleagues is suggested by documents in the Parker Archives at Harvard testifying to epistolar exchanges and mutual laboratory visits. Their intimacy is further demonstrated by a postcard, dated 18 September 1933 from Frankfurt, announcing proudly the birth of Bethe's granddaughter.

Now, what exactly was Parker buying into when he supported the network hypothesis? We have seen in chapter 6 how Ramón y Cajal had done away, apparently persuasively, with the continuity theory. But near the turn of the century Bethe and the Hungarian histologist Stefan Apathy (1863–1922) published results that breathed a second life into the theory (Frixione, 2009). Apathy had developed a staining method based on a mixture of gold chloride, formic acid, and methylene blue that revealed the presence of a dense meshwork of "neurofibrils" in the ganglion cells and nerve fibres of annelid worms (Apathy, 1897). More important, Apathy thought that neurofibrils form a continuous network from nerve cell to nerve cell, as if nerve cells "in fact were only roadways over which the neurofibrils travelled" (Parker, 1929). Apathy's work was done at the Naples Zoological Station, and when Bethe came to the station to work on the nervous system of crabs in 1896, Apathy showed him his histological slides. At first skeptical, Bethe was bowled over by what he saw: "What has been shown to me then, however, was of such convincing clarity, that I was forced, after some pretended objections, to relinquish my opposition. What happened to me was experienced by many others, and nobody of normal vision can elude the convincing impression of Apathy's preparations unless his eye is beclouded with envy or injured vanity" (quoted in Florey, 1985).

Bethe thus came to share Apathy's conviction that neurofibrils were the conduits through which impulses were conducted through the nervous system (see Fig. 8.7). He designed what became known in the academic community of the period as the "Bethe experiment,"

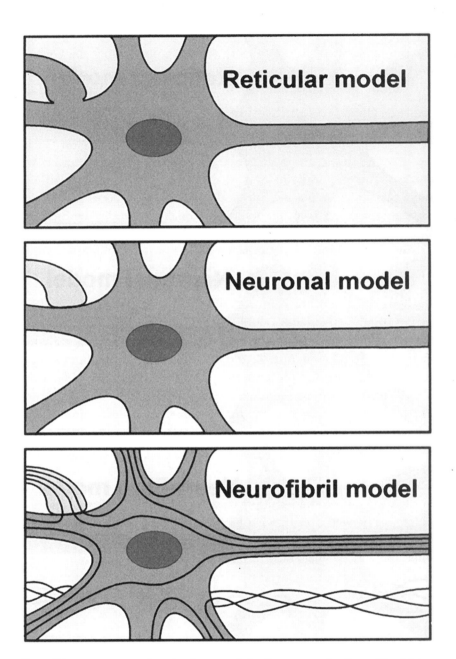

8.7 · Diagrams contrasting the three models of nerve conduction with which Parker struggled before proposing his own theory. From Frixione (2009). Courtesy Elsevier ScienceDirect.

which consisted in removing the ganglion cells from the mass of fibres supplying the second antennae of the crabs under study, and cutting the connectives and commissures to other ganglia. As this treatment failed to affect the muscle tonus and reflex of these antennae, Bethe concluded that ganglion cells are not required for nerve transmission and that neurofibrils are both necessary and sufficient to account for these actions (Bethe, 1898). His experiment convinced Bethe that "we must stop considering the neuron as a physiological unit and must admit that one and the same neuron is capable of many diverse actions, depending on which fibrillar tract is in operation" (quoted in Florey, 1985). Between Apathy's showcase neurohistology and Bethe's experiment, many neurohistologists were impelled to discard the neuron doctrine. Among them was the well-respected neuropathologist Max Bielschowsky (1869–1940), whose silver preparations of neurofibrils Parker had seen with his own eyes and found convincing, as he admitted in his masterly historical review of the neurofibril controversy (Parker, 1929).

In that review paper Parker also admitted the influence that Charles Sherrington's paradigm-changing book on the integrative action of the nervous system (Sherrington, 1906) had exerted on his views of the coelenterate nervous system expressed in 1919. In that book Sherrington, not surprisingly for the man who introduced the synapse concept, had come down strongly on the side of the advocates of the neuron doctrine, but made an exception for coelenterates. Sherrington, who along with Parker respected Bethe the fellow physiologist, had relied on Bethe's work (1903) to state that the "nerve-net of Medusa appears an unbroken retiform continuum from end to end." Parker, wavering over the merits of the two theories, found solace in Sherrington's approval of Bethe and seized the compromise solution that Sherrington offered.

The diffuse mode of signal transmission of nerve nets, Parker admits, is such that the existence of reflexes as understood in vertebrates is highly questionable in coelenterates. And yet, as Parker's experimental work on sea anemones shows, "some of these animals may have involved in their nerve-nets special tracts that enable them to carry out simple but obvious reflexes. Such conditions," concludes Parker, "indicate the steps by which a nerve-net may be converted into that type of central nervous organ that is characteristic of the higher animals." To the question of whether the coelenterate nervous system is capable of delivering the kind of central integration seen in "the higher nervous functions of the most complex animals," Parker, on the basis of his studies of feeding, locomotion, and defensive retraction in sea anemones, answers with a resounding NO. Even these complex responses, he says, "prove on examination to consist of operations none of which necessitate the assumption of activities other than those consistent with the nature of the nerve-net." In other words, whatever loose integration of activities coelenterates are capable of, they can only be those compatible with the functioning of diffuse nerve nets.

The third and last section of the book deals briefly with Parker's hypothesis of an evolutionary passage from the elementary nervous system to the central nervous systems of higher animals. As we saw earlier in this chapter, Parker had developed his ideas on the subject in the form of sketches as early as 1911. He envisioned first a passage from a system based on sensory cells as the only nerve elements, whose processes connect directly with the muscle cell layer at the base of the ectoderm, to a system seen in many coelenterates, in which nerve cells receive connections from sensory cells and connect theirs to muscle cells. Such intervening nerve cells were called ganglion cells by the Hertwigs and motor nerve cells by Havet (1901). Parker found both expressions inappropriate: "ganglion cells" because they are not constituent cells of a ganglion in coelenterates, and "motor cells" because these nerve cells may have other functions, such as forming by their connections with each other a net through which impulses can be propagated over long distances. Parker prefers to call them "protoneurones," and their emergence, in his opinion, brings us "a step nearer a centralized nervous system of the higher animals than one in which the protoneurone is not present." But the presence of protoneurones is not in itself really conducive to complex or higher functions because the nervous system remains diffuse, uncentralized. What allows complex responses is the diversity of effectors; the greater the diversity the more complex are the animal's activities, as seen in sea anemones but to a lesser extent in jellyfish.

In essence, Parker's network of protoneurones constitutes the first manifestation of the adjustor in his functional chain of command:

receptor-adjustor-effector. Changes in this primitive, epithelial adjustor are what led to centralized nervous systems:

> Its growth is associated with an inward migration whereby it retreats from the surface of the animal to a deeper situation and comes thus to gain what is significant of its growing importance, a certain degree of protection. In the coelenterates the nervous elements are mostly contained in the epithelial layers of the body and especially in the external epithelium, the ectoderm. In the worms, where the body has gained greatly in thickness and solidity as compared with the coelenterates, this inward migration is clearly seen. These animals no longer possess the diffuse system of the lower forms, but have a definitely centralized band of tissue extending the length of the body. (Parker, 1919a)

But probably the most intriguing aspect of Parker's musings on the evolutionary development of the nervous system concerns synaptic contacts. In the concluding pages of his book he stresses the advent of synapses as a key step in the evolutionary transition from nerve nets to central nervous systems. If you assume, as he does, that the nerve net is syncytial, it follows that synapses cannot be part of it – that their presence in it would be an absurdity. That is because a syncytial nerve net is a continuum of protoplasm (or neurofibrils) best described as a diffuse, unpolarized version of a long axon. As the nerve cells condensed together to form ganglia in a post-coelenterate step, a polarity developed, with the dendritic processes at one end of the neuronal soma and the axon at the other. To Parker this polarization and the ensuing directedness of signal propagation were key selective pressures for the emergence of synapses.

In developing this construct Parker took a huge gamble by siding with the proponents of the network theory. His whole evolutionary scenario hinged on the adoption of this theory. If you are wrong – and Parker was – the construct flounders. As discussed earlier, Parker chose to adopt Bethe's view of nerve net continuity and Sherrington's notion that synaptic integration was not yet within the grasp of coelenterates. To Parker's credit he was not alone in being swayed away from the discontinuity theory, despite Schäfer's solid work on jellyfish. We have

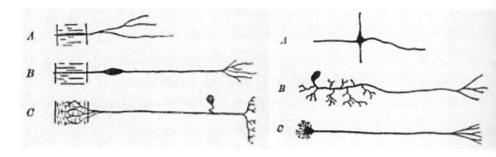

8.8 · (*Left*) Stages in the evolution of the sensory cells, from coelenterate (A) to mollusc (B) to vertebrate (C). (*Right*) Stages in the evolution of the motor nerve cell, from coelenterate (A) to the earthworm (B) to a vertebrate (C). From Parker (1919a), Figs. 51 and 52.

seen that several of the rookie neurohistologists portrayed in chapter 7 leaned in the same direction. In the mood of the time, Parker's bet seemed a sensible one.

Parker devoted the last few pages of his book to outlining his ideas on the evolution of sensory, motor, and interneuronal cells. To him the most primitive nerve cells that the synaptic nervous system inherited from the nerve net are sensory cells. He followed their gradation from the "sensory protoneurone" of coelenterates to the sensory neuron of higher invertebrates, and to the primary sensory neuron of vertebrates (Fig. 8.8, left, A–C). The striking trends here are the internal displacement of the soma and the emergence of the sensory dendritic tree in the epithelium of vertebrates. Similarly, he traced the transitions of the next innovative element, the neuron, from the "protoneurone" of the coelenterate nerve net to the motoneuron of higher invertebrates and finally to the primary motoneuron of the vertebrates (Fig. 8.8, right, A–C). Here one notices the polarization of the post-coelenterate cell and the diverging position of the soma between the invertebrate and vertebrate models, with direct consequences to the dendritic arrangement. Parker identifies what he calls the internuncial neuron – equivalent to today's interneuron – as a third type of nerve cell that arose in the course of the development of central nervous systems. Although sparse in the earthworm level of neural organization, their numbers and importance, according to Parker, rose considerably in arthropods and molluscs, up to the level of vertebrates, in which "the internuncial type composes the chief mass of the central organs, a feature that reflects the enormous development of associative operations carried out by the vertebrate as compared with the invertebrate."

Reflecting on the evolutionary voyage of the nervous system, Parker's finishing coda meditates on "the system that arose secondarily around an independent effector muscle, [and] has in the end gained such supremacy as to take to itself a number of independent effectors, any one of which might in the beginning have served as the nucleus around which the first nervous tissue could have taken origin." None of Parker's predecessors had invested as much of their scientific career in this problem and none had achieved the scope of investigation and depth of understanding that Parker displayed, even taking warts and all into account.

Swan Song with a Glow

After the publication of *The Elementary Nervous System*, George and Louise spent the summer of 1919 in La Jolla in Southern California, where Parker conducted his last original research on the neurophysiology of a coelenterate. The species he now scrutinized was the sea pansy *Renilla amethystina* (now *Renilla köllikeri*). It is a member of the sea pen family and consequently a colonial anthozoan. We saw in chapter 7 that a few young investigators had taken an interest in the way nervous systems are organized in anthozoan colonies, including sea pens. It appears that Parker undertook his investigations unaware of Niedermeyer's histological work on the sea pen *Veretillum* (1914) or the related works on alcyonarians by Hickson's laboratory (1895–1904) and Kassianow (1908). In an unusual departure from Parker's methodical scholarship, none of these contributions were cited in the publications resulting from his summer stint at the Scripps Institution.

Parker was quite candid about his motive for the new research venture (Parker, 1919b): "The curious sea-pen *Renilla* is a most favorable form in which to study colonial organization, for the relatively large size of its autozooids and its complete and natural freedom from attachment make it an unusually satisfactory organism for experimental study." The colony is organized around the colonial mass, made of the rachis, shaped loosely like the pansy flower, and the peduncle, which extends from the lower surface of the rachis, the whole effect recalling the shape of a kidney. The upper surface harbours the individual polyps forming the colony, called zooids. There are two types of zooids, the autozooids (feeding and reproductive polyps) and the siphonozooids (tiny polyps involved in water circulation).

Fluid circulation, Parker found, is an important colony-wide activity in which the peduncle plays a central role. Water is pumped in by ciliary activity of the myriad inhalent siphonozooids, resulting in an inflated colony; it is expelled by a single, larger, exhalent siphonozooid at the base of the peduncle, resulting in a flattened colony. The peduncle undergoes periodic peristaltic movements which spread to the rachis, thus generating internal water currents. Parker believed that the autozooids acted independently of each other, implying that no nervous coordination of muscle-based activities exists between them. The only integration centre he sees is purely mechanical and concerns the peristaltic activity of the "organ of inflation," the peduncle.

But Parker could not resist the conclusion that another colony-wide activity, bioluminescence, was under nervous control. The phenomenon, first noticed seventy years earlier by Louis Agassiz, is manifested by the display of a wave of bright blue-green light that crosses the upper surface of the rachis upon tactile stimulation in the dark. Parker observed that light emission originates in "microscopic white granulations" of the siphonozooids and around the base of the autozooids. Upon repetitive electrical stimulation, Parker saw that the luminescent "waves succeed one another at such a rapid rate that the whole superior surface seems to be covered with a rippling glow emanating from the region of stimulation." Any naïve observer is in awe at such a light display, but Parker was particularly struck by its physiological implications: "It is difficult to understand how these successive activities are induced unless it is assumed that the luminous points are all controlled by a nerve-net whose form of transmission is reflected in the outward moving circles of light."

Parker's encounter with the sea pansy unleashed a train of thoughts on the nature of colonialism in these animals. While he viewed the tentacles of autozooids as organs that are part of the makeup of an individual, itself part of a colony, he could not bring himself to use the epithet "organ" for the peduncle, which is not part of an individual but "a structure that serves the whole colony of individuals." He therefore calls the peduncle a superorgan. "In Renilla," he adds, "they are represented not only by the peduncle as a structure concerned with the inflation of the colony as a whole, but by the nerve-net that controls colonial luminosity. Superorgans give a unity to a colony that is often unexpressed in the individuals of which it is composed." Thus Parker was able to push his reflection on the existence of a colonial nerve net much further than the embryonic attempts of Kassianow and Niedermeyer before him, despite his apparent ignorance of the contributions of these precursors.

Parker expanded on his introductory paper, which was based on his first exposure to sea pansies in 1916, by publishing comprehensive analyses of the circulation of water in the colony (1920a) and of muscle and bioluminescent activities (1920b), based on the more extensive work accomplished in 1919. The first of these papers had no bearing on the issue of the involvement of the nervous system, but the second tackled it without reservation. Parker relied heavily on elaborate cutting experiments to investigate peristalsis, withdrawal contractile responses, and the luminescent waves. In that way he found that peristalsis proceeded from the proximal part of the peduncle – the location of the pacemaker – and travelled by splitting in two waves, on the one hand to the distal end of the peduncle, and on the other to the distal edge of the rachis. The slow progression of the wave (1.1 mm per second) and the analogy of this peristalsis with the functioning of the vertebrate heart led Parker to conclude that the peristaltic wave is myogenic in origin, even though magnesium sulfate anaesthesia, which he usually employed to reveal nervous system involvement, interrupted the waves.

Parker noted that while siphonozooid ciliary beats were responsible for water entry and colony distension, muscle contraction was involved in the reverse process, that is, expulsion of water through the exhalent siphonozooid and collapse of the rachis (withdrawal). Similarly, the bioluminescent wave is stimulated by tactile and electrical stimulation and is suppressed by magnesium sulfate anaesthesia. He observed that stimulation of the peduncle or rachis led to a wave of autozooid withdrawals as well as contraction of the colonial mass, but that stimulating an autozooid was ineffective in recruiting adjacent autozooids to withdraw. Parker concludes from these and other experiments that, contrary to peristalsis, these activities are neurogenic:

> This is especially striking in their rapidity of transmission, some sixty or sixty-five times that of the peristaltic waves. Apparently they are the product of an unpolarized nerve-net, which serves not only phosphorescence, but also the general contraction of the colony as a whole and the combined withdrawal of the autozooids. These activities, though they involve the autozooids, are strictly colonial, for they excite the withdrawal of these zooids all together and not as individuals and, though they can be readily induced by stimulating almost any part of the surface of the peduncle or the rachis, it is remarkable that they cannot be called forth by stimulating individual autozooids. Phosphorescence, general contraction, and the withdrawal of the autozooids, then, are also colonial activities, probably dependent upon a nerve-net and certainly not involving the organization of the zooid. (Parker, 1920b)

In other words, colonial activities largely overshadow individual activities of the polyps, and the colony functions as an integrated unit, thanks to the colonial nerve net's exercise of command over the localized nerve nets of individual polyps. As Parker put it, "such a nervous organization is social rather than individual." Thus ended Parker's twenty-five-year infatuation with the elementary nervous system.

The Afterglow

A year after his last original papers on the cnidarian nervous system were published, Parker took the place of his old mentor E.L. Mark as head of the Zoological Laboratories. The new administrative duties signalled a break in Parker's research orientation, but in no way impeded his research output. If anything, his research interests increased in diversity and eclectism, from the excretion of carbon dioxide by frog nerves to "the time of submergence necessary to drown alligators and turtles." However, Parker maintained a focus on the sensory systems of fishes and the comparative physiology of animal colour changes, which culminated in monographs establishing his leadership in these fields. He also published books on subjects of more general interest to a wider audience, about biological evolution, heredity, and social issues as seen from a biological perspective. He had a remarkable talent for making science accessible.

On the personal side, his marital life continued its serene course. Early in the 1910s the Parkers moved to 16 Berkeley Street in Cambridge, where they lived for the rest of their lives. The location was conveniently within short cycling distance from his workplace. His promotion to full professor in 1906 had made the large cottage house, which suited their tastes, affordable. Built in 1905, it was described as having "some of the same design elements one sees in the Prairie Style – the blocky geometric form, the horizontal quality, strongly reinforced by low, lidlike, and wide-projecting roofs" (Shand-Tucci, 1978).

Parker's demeanour was said to be at times stern and severe, and he was known occasionally to lose his temper. One such occurrence took place in 1920 – when his sea pansy papers were being published – and involved his neighbour on Berkeley Street, who came to Parker's door to use his telephone because hers was out of order. Parker may have been absorbed in professional ruminations that tolerated no disturbance – as was his wont – because he broke into a tempestuous rage. The neighbour took it in good stride – and humour – and composed "An humble and abject APOLOGY to neighbor G.H. Parker from neighbor E.Q. Bumstead on this 23rd day of June 1920." It continued thus:

> Whereas, said neighbor Parker, having a telephone that could be used; said neighbor Bumstead trespassed most grievously on said neighbor Parker's willing (?????) hospitality in using said telephone that could be used, whilom said neighbor Parker ramped and champed and tore his hair and acted in a manner quite unbecoming of the dignity of a Professor of ye ancient seat of learning called HARVARD UNIVERSITY: the reason therefore being that said neighbor Parker had upon the instant decided that he himself desired to use said telephone at the self same moment that said neighbor Bumstead so desired herself to use it. IN TOKEN OF DEEPEST PENITENCE said neighbor Bumstead

desires to offer this crimson ROSE of the hue of said HARVARD UNIVERSITY colors to said NEIGHBOUR PARKER. (Parker Papers at Harvard University Library Archives)

The fact that Parker preserved this little bit of memorabilia suggests that he found the humour endearing. He also had a fun-loving side (Romer, 1964). In his autobiography he recalls episodes of mischievous behaviour and banter. One involved him and his friend William Ritter – of Schripps Institution fame – performing a tap dance to a group of Italian immigrant miners in the West. Another was associated with his "refuge from the undue pressure of American life" – great food and wine. He was a regular of the restaurant Capri in Boston's Little Italy, where he had struck up a friendship with Aristeo, the owner. As Parker testified, "I frequented the Capri during our period of national prohibition and I was treated by Aristeo to a portion of homemade red wine served in a coffee cup."

Parker officially retired in 1935 at the age of seventy, but he went on working at his usual brisk pace. He carried on until he was almost ninety, and only then was he stopped by serious health problems. Louise died in 1954, and Parker followed the next year. In a letter to John H. Welsh Jr of the Harvard Biological Laboratories, dated 19 January 1956, Alfred Romer, director of the Museum of Comparative Zoology, wrote in remembrance: "(1) the way he kept on working after retirement; it took a prostate and a fractured skull to slow him down; (2) except for Harrison, he was the last of the grand old men in zoology that ruled the roost for nearly half a century (including Morgan, Wilson, Lillie et al.)" (Parker Papers at Harvard University Library Archives).

An American Crop

This grand old man had trained directly, or participated in the training of, many graduate students – forty-six according to an estimate by his student John H. Welsh – but no "Harvard School" developed from Parker's body of work on the elementary nervous system. None of his students, it seems, took up work on sponges or coelenterates, with the exception of Welsh, who graduated in 1929 but picked up coelenterate research only in 1956. Welsh was an invertebrate specialist who was appointed professor at the newly established Harvard Biological Laboratories – an initiative in which Parker participated – which he chaired by the time he initiated research on sea anemones aimed at identifying substances of pharmacological interest, such as serotonin.

Another of his graduate students attracts interest not because he worked on coelenterates –which he did not – but because he became linked to one of Ramón y Cajal's works. Raoul Michel May was born in 1900 in Sonora, Mexico, of French parents. He moved to the United States in his teens, first to California where he attended Stanford University, then to Johns Hopkins. He enrolled in graduate studies at Harvard's Zoological Laboratory under Parker's supervision, earning his PhD in 1924. His thesis topic was the role of innervation in taste bud regeneration in catfish, one of Parker's research interests that had superseded the coelenterate nervous system. The work (May, 1925) supported Ramón y Cajal's hypothesis of the existence of neurotrophic factors – which May called hormones – in neural development. In 1924 he was awarded a travel fellowship by Harvard to further his studies at the Laboratoire d'anatomie comparée of the Sorbonne in Paris, after which he travelled to Madrid and learned techniques in Ramón y Cajal's laboratory. There he took advantage of his mastery of Spanish to translate Cajal's *Degeneration and Regeneration of the Nervous System*, which was published by Oxford University Press in 1928. He returned to Paris, where he remained for the rest of his life, rising to the chair of the Laboratoire de biologie animale at the Orsay campus of the University of Paris. In his autobiography Parker cited among other works a book by May (1945) in which Parker's views on the origin of the nervous system received special treatment.

If no Harvard students pursued Parker's work on sponges and coelenterates, others picked up the pursuit from the vantage point of various American academic institutions. Among the more notable were Alfred Goldsborough Mayer (1868–1922), Lewis Robinson Cary (1881–1956), Robert Mearns Yerkes (1876–1956), and Carl Hiram McConnell (1896–1939). Yerkes, who was to become known for his trail-blazing work on primate behaviour, was the first to follow in Parker's footsteps. Born on a Pennsylvania farm, he struggled to pursue academic interests that ran contrary to his father's wish that he take on farm work (Hilgard, 1965). A loan of money allowed him to enroll at Harvard, where he earned a bachelor's degree in 1898. Yerkes's potential must have been recognized early on, because he was immediately wooed by several members of the Zoological Laboratory staff, including Mark, Parker, and. Davenport. Under their influence and especially that of Harvard psychologists of the William James persuasion, Yerkes embarked on doctoral studies of jellyfish neurobiology and behaviour. The species he chose was the hydromedusa *Gonionemus murbachii*, which was abundant in the waters around Woods Hole, and the same species whose neuroanatomy Ida Hyde was studying in the same time period (see chapter 7).

Two papers emerged from his doctoral dissertation. In the first (Yerkes, 1902a) Yerkes explored the sensory capabilities of jellyfish and in the second (Yerkes, 1902b), the functional role played by the nervous system in motor behaviour. Yerkes had a very personal and direct style of stating the problem as he saw it, by formulating specific questions for which he was seeking answers through experimental observation. For the first paper he enumerated the following: "1. Has *Gonionemus* a sense of taste (i.e., a chemical sense) distinguishable from the tactual sense? 2. If there is such a sense, where is it located? 3. Are all parts of the body equally sensitive to all forms of stimuli, and if not, what is the localization? 4. Do different qualities of stimuli call forth different kinds of reactions, or is intensity of stimulus alone significant? 5. Do any stimuli have a directive influence upon the movements of the medusa?"

In seeking answers to these questions Yerkes found that *Gononemus* possesses a strong chemoreceptive as well as tactile (mechanoreceptive) and photosensitive abilities. The tentacle tips were the most sensitive, but the lips, manubrium, and subumbrella were also receptive. For chemoreception Yerkes distinguished feeding responses to "nutriment" chemicals from locomotor responses to harmful substances. For photosensitivity he found that the jellyfish is phototactic during swimming but seeks shade when settling. Finally, he observed that "when chemical, mechanical, or photic stimuli affect symmetrical points of the body unequally, they have a directive influence upon the movements of the organism."

For the second paper Yerkes formulated these pointed questions: "1. Do the special reactions of the tentacles, manubrium, and bell depend upon the activity of the central nervous system? 2. What is meant by irritability and spontaneity, and in what relation do they stand to the central nervous system? 3. Does coordination depend upon the functional activity of the central nervous system? 4. Is there any evidence of the existence of special nerve centres of spontaneity and coordination? 5. Finally, what are the functions of the central nervous system?" Yerkes could not use the neuroanatomy of *Gonionemus* as a reference point for answering these questions because he was unaware of Hyde's work, which appeared the same year as his own. So he used the Hertwigs' monograph on jellyfishes instead, believing that their neurohistological description was representative of what existed in *Gonionemus*.

Yerkes revisited some of the cutting experiments Romanes had performed on the scyphomedusa *Aurelia* and added more to gain as accurate an understanding of the situation in a hydromedusa as he could at the time. From such experiments he concluded first, as did Parker with regard to sea anemones, that appendages such as the tentacles display autonomous motor responses quite independent of a nervous centre. Contrary to Romanes, he found that spontaneous movements as represented by the rhythmical pulsations associated with swimming are dependent on nervous centres (marginal bodies and nerve rings) only in the sense that those centres may have a low sensitivity threshold for initiating pulsations. Removal of these centres caused the relative paralysis of the bell only because strong stimuli were now necessary to induce responses. The pacemaker that sets the pace of the pulsations is not located at a specific and stable site, but is found wherever the most sensitive part of the bell margin happens to be. From there it overrides the lower rates of pulsation present in other parts of the jellyfish by transmitting impulses it initiated. In this manner, coordination of contraction waves is achieved and harmonious swimming strokes ensue.

In answering his third and fourth questions, Yerkes offers a fresh insight into the mode of motor coordination in hydromedusae. He states that "coordination is not dependent on the functioning of the nerve ring or of any special nerve centres, but upon the rapid transmission of an impulse, which is either nervous or muscular, probably the latter." Chun and Jickeli had previously alluded to the possibility that impulse transmission was effected through the mediation of the layer of epitheliomuscular cells where a nerve plexus cannot account for it, but Yerkes was the first to provide some experimental evidence, however indirect, to support such a conclusion. He does not speculate on mechanisms by which such non-nervous transmission can be achieved, but we now know that communication of impulses between hydromedusan epithelial cells occurs through gap junctions, which are membrane protein channels that align and dock themselves in the contact zone between adjoining cells, thereby creating a cytoplasmic continuity through which action potentials and small molecules can travel. This is called epithelial conduction (Anderson, 1980). Although Yerkes's insight on epithelial conduction is impressive, his "either/or" approach to the question was proven unfounded. Today it is known that both nerve nets and epithelia can transmit impulses through gap junctions in hydromedusae.

Yerkes's jellyfish work led him to a research career wholly focused on the comparative psychology of animals. If he soon abandoned research on coelenterates, the field remained open for other budding American zoologists. Alfred Mayer was one of them. At the urging of his father, a university physics professor, Mayer tried unsuccessfully to study and pursue a career in physics before turning to biology, a topic more to his taste (Davenport, 1927). He became assistant to Alexander Agassiz in 1892, accompanying his boss to scientific expeditions in the Bahamas, the Pacific South Seas, and Australia between 1893 and 1898. After earning his PhD at Harvard in 1895 Mayer was appointed assistant curator for "radiate" animals at the Museum of Comparative Zoology, a post he held until 1900. Between 1900 and 1904 he held curator jobs at the newly created museums of the Brooklyn Institute of Arts and Sciences. In 1904 he was appointed director of the Department of Marine Biology of the Carnegie Institution of Washington, which was based in the Tortugas, the most western of the Florida Keys. While Mayer was interested in butterflies, he became a specialist of jellyfish under the influence of Alexander Agassiz, and in 1910 he published a highly influential three-volume treatise, *Medusae of the World*, which heralded an unrelenting interest in jellyfish neurophysiology.

Between 1906 and 1920 Mayer published ten papers dealing with the neural basis of bell pulsations and factors determining nerve conduction in scyphomedusae. Thus he conducted his jellyfish studies at roughly the same time as Parker conducted his own on sponges and sea anemones. Strangely there is no evidence of professional intercourse between the two. Mayer never cited Parker's work and Parker, in his 1919 book, cited only two of Mayer's papers. Both Mayer and Parker were friends of Davenport from their graduate days at Harvard in the early 1890s, but their respective connections never developed into a triangular friendship. One is tempted to speculate, for lack of evidence to the contrary, that an inimical relationship arose between Mayer and Parker because of a jealous rivalry. Mayer's drive for the establishment of the Tortugas laboratory may have been one fertile ground for hostility. There is epistolary evidence suggesting that while Davenport, Agassiz, and others supported Mayer's campaign, the prominent zoologists affiliated with Woods Hole, of whom Parker was part, were either cool to the idea or frankly against it (Stephens, 2006). The reasons for their objection were the forbidding remoteness of the Tortugas and, of course, their own push to consolidate the Marine Biological Laboratory at Woods Hole as the Mecca of American biological stations, accessible to all and an ideal meeting point for exchanges among scientists. To Parker, Edwin Conklin, Thomas Morgan, and others, funding a Tortugas laboratory was a wasteful deployment of resources. Although Parker wrote at length about seaside laboratories in his autobiography, he never mentioned the Tortugas laboratory. Nor does Mayer's name ever come up in Parker's autobiography.

Mayer was every bit the experimentalist that Parker exemplified. However, jellyfish being a delicate, pelagic animal, it is usually less amenable to experimentation than sea anemones or sea pansies. Mayer was fortunate in selecting a scyphomedusa that proved hardy and could survive for weeks in laboratory aquaria. The species was *Cassiopea xamachana* and it was abundantly available in Tortugas waters. In his first papers (Mayer, 1906, 1908) Mayer took Romanes' experimental approach but went further in the analysis, especially with regard to the effects of the chemical environment on the neuromuscular performance associated with pulsatile swimming. Mayer produced surgically what he called a disc – basically a jellyfish bell without the marginal sensory organs. He found that discs in which rings of subumbrellar muscle layers were removed could be induced by various means to break out "into rapid, rhythmic pulsation so regular and sustained as to recall the movement of clockwork." Thus sensory input was not necessary for this pulsation, and regeneration and other experiments strongly suggested that the "stimulus which causes pulsation is transmitted by the diffuse nervous or epithelial elements of the subumbrella." Despite this cautious statement, Mayer's detailed discussion makes it clear that he favours the involvement of the subumbrellar nerve net.

The outstanding findings of Mayer's 1906 paper concern the facilitatory effects of sodium, potassium, and calcium on the pulsations of the disc. Magnesium salts tended to suppress the pulsations, but "*calcium in the sea water assists* the NaCl to resist the retarding effect of magnesium." Unwittingly, Mayer had hit not only upon the role of sodium and potassium in the excitability level of nerve cells but also upon the depolarizing effect of potassium and the role of calcium in synaptic excitability. The 1908 paper set out "to correct certain errors in the previous report, and to announce some new results." Here Mayer is no longer ambiguous about the role of the nerve net, stating that "the pulsation-stimulus is conducted by the diffuse nervous network of the subumbrella, and is independent of the muscles which may or may not respond to its presence by contraction." In a bizarre statement betraying the influence of Jacques Loeb's physicochemical theories, Mayer concludes that sodium oxalate is excitatory to the marginal sense organs by way of its precipitation of calcium salts in the endoderm.

Mayer's last significant papers on the neurophysiology of *Cassiopea* (Mayer, 1916, 1917) concern the nature and rate of propagation of nerve net excitability associated with pulsations. Using the ring of subumbrellar tissue preparation that he developed, through which a "neurogenic contraction wave" travels continuously in a circular fashion, Mayer this time used a kymograph borrowed from a Princeton colleague to record these waves in real time and thereby "obtain an accurate determination of the rate of nerve-conduction" under various experimental conditions (Fig. 8.9). He concluded that "nerve-conduction is due to a chemical reaction involving the cations of sodium, calcium and potassium," and "some proteid element." Although Mayer saw the chemical reaction as an adsorption of ions on colloids (proteins), the statement was anticipatory in the sense that neuronal excitability is now known to be determined by the permeability of transmembrane protein channels to these

8.9 · (*Opposite, above*) Kymograph used by Mayer, showing the lever attached to the preparation by a thread and to a stylet pen grating on the blackened paper of a rotating drum. (*Opposite, below*) A kymograph recording with time marker and hand-written notations, showing the effect of calcium on the contraction waves. From Mayer (1917), Figs. 13 and 15. Courtesy of the Carnegie Institution for Science.

sea water   25°.8 C.

98 NaCl 0.59 nv..+ 2 Ca Cl₂   26°.3 C.

90 (NaCl + CaCl₂) + 10   of 0.97 × 10⁻⁶ H₂O

ions (Hille, 2001). On the other hand, Mayer thought that his experiments refuted the "local action" theory of Ralph S. Lillie (1875–1952) – not to be confused with his brother Frank R. Lillie, the second director of the Marine Biological Laboratory at Woods Hole.

Ralph Lillie had published three papers between 1914 and 1916 in which he formulated hypotheses on the mechanism of nerve conduction (Lillie, 1914, 1915, 1916). Building on Julius Bernstein's seminal membrane theory of bioelectric potentials in excitable cells (Bernstein, 1902), Lillie proposed that nerve cell stimulation makes the cell membrane more permeable to cations, thus inducing a depolarization of charges. This local depolarization is in turn associated with a transmembrane current flow that moves from the active to adjacent inactive sites through successive local circuits, thus propelling the nerve impulse along the nerve fibre. Lillie specifically cited Mayer's findings in support of his theory while criticizing Mayer's adsorption theory. Although Lillie had it right as far as applying his hypothesis to the individual nerve fibre was concerned – the biophysicist and Nobel Prize laureate Alan Hodgkin was inspired by Lillie's ideas when he designed his critical experiments to demonstrate the mechanism of action potential propagation along the squid giant axon – his scheme was hardly applicable to the jellyfish nerve net. In the nerve net, impulse propagation has as much to do with crossing numerous synapses as with travelling along the relatively short neuronal processes between synapses.

In his neurophysiological studies, Mayer had a close collaborator who was affiliated with Princeton, where he himself held a position as honorary lecturer. Lewis Cary, who was appointed assistant professor of biology at Princeton in 1910, attached himself to Mayer for many years until the latter's death from tuberculosis in 1922. He accompanied Mayer in his scientific expeditions to the South Pacific and Australia, where they studied the formation of coral reefs. Between 1915 and 1917 Cary published five papers dealing with work he had conducted in the Tortugas laboratory on the neurophysiology of Mayer's pet animal, the jellyfish *Cassiopea*. It can be said of Cary that he was the kind of personality that subordinates to others, who finds purpose only through the purposes of a figure he greatly admires. As a result Cary never developed his own independent research program, and he suffered the academic consequence, remaining assistant professor at Princeton until his retirement in 1948. Cary summarized his original findings on the regeneration of ablated elements of the jellyfish nervous system in his extensive final paper (Cary, 1917).

Regeneration as a research topic had not before been considered in coelenterates, so Cary's experiments in this regard proved of great importance. Because one cannot sever a discrete nerve in coelenterates in order to assess the role of the nervous system in regenerative capacity, Cary resorted to removing the sensory rhopalia of *Cassiopea*. He found that removal of the sensory nervous tissues caused a slowing of the early stages of regeneration, whereas maintaining high levels of neuromuscular activity accelerated the early stages of regeneration. He concluded that "some sort of trophic influence is exerted in general metabolic activities by the sense-organs." Cary's work was the first to uncover neurotrophic influences in a coelenterate. By the same token it suggested that neurotrophic factors would have emerged at the same time as nervous systems in evolution, but Cary himself never alluded to this possibility.

The last of the four Americans who made significant contributions to coelenterate neurobiology in the wake of Parker's work was Carl McConnell. Born in Tennessee in 1896 to parents from the Old South, Carl showed interest in following his father's vocation as a college professor. After earning a bachelor's degree at Lynchburg College, he obtained his master's degree at the University of Virginia. He then volunteered to serve in the First World War and was stationed in France. After the war McConnell pursued graduate studies in British Guiana, and earned a doctorate from the University of Virginia. In 1930 he married Rio Loretta Tucker, a school teacher from West Virginia, with whom he had two sons. Between 1929 and 1933, while holding teaching positions at the University of Virginia and Hartwick College in New York State, he produced several papers on *Hydra*, of which only one pertained to the nervous system (McConnell, 1932). Having received a research fellowship, he went to Ljubljana, Slovenia, to work under Jovan Hadži, who, we may recall, had produced a groundbreaking study on the nervous system of Hydra in 1914. In the Slovenian capital, McConnell undertook an examination of the genesis of the nervous system in the developing stages of *Chlorohydra viridissima* and *Pelmatohydra oligactis*.

McConnell found that the staining methods used by his predecessors, including his host Hadži, were useless when applied to budding

Hydras. By adapting the reduced methylene blue (Rongalit white) method introduced by Paul Gerson Unna (Unna, 1916), however, he obtained good results. He showed that the nervous system of new buds derived not from the existing nervous system of the adult generating the bud, but rather by proliferation and differentiation of interstitial (stem) cells in the bud. First, he describes the development of the sensory cells in the head region:

> The development of the sensory cells takes place along with the development of the ganglion cells. One first finds the interstitial cells, which are being elaborated into sensory cells, on the area around the mouth and on the developing tentacles; they are found, in the majority of cases, protruding through the surface of the epithelio-muscular cells. Occasionally they are found between the epithelio-muscular cells. Regardless of their location, they reach to the ectodermal surface and from their proximal ends send out typical processes which eventually find and coalesce with the processes of the ganglion cells. (McConnell, 1932)

From the above citation it is clear that the coalescence of the processes is meant to depict the formation of a syncytium, and further into the article McConnell reiterates that he firmly believes the nerve net is continuous. As to the dynamics of the formation of the nerve net itself, here is what he has to say:

> The formation of the nerve net takes place by the growing together of the processes from the ganglion cells: these processes advance between the epithelio-muscular cells until they meet other processes from ganglion cells with which they fuse. Other processes from the ganglion cells grow out in various directions and end among the muscular processes of the epithelio-muscular cells. These nerve endings on the muscular prolongations of the ecto-epithelio-muscular cells are entirely undifferentiated as far as can be ascertained microscopically, and it appears that the enervation of these muscular processes is only through contact with the processes from the ganglion cells. The processes which form the nerve net are usually longer than the free-ending processes. (McConnell, 1932)

McConnell observed a "wave of development which moves downwards from the head pole towards the foot pole." As the wave progresses the connections between sensory and ganglion cells are formed before those between ganglion cells and epithelio-muscular cells. Once the nervous system is fully in place in the newly formed adult hydra, interstitial cells continue to co-exist with other cells in the basal ectoderm. Although McConnell failed to see the discontinuity of the nerve cells, his contribution was seminal for the future understanding of neurodevelopment in coelenterates.

What legacy did his four successors leave? Of the investigators who followed immediately in Parker's wake, only Mayer and McConnell founded their research careers on coelenterates, devoting themselves in part to the nervous system in one form or another. This might have changed had they lived longer, but Mayer's life was prematurely cut short at fifty-three by tuberculosis and McConnell's from unknown causes at the age of forty-three.

What is particularly striking about Mayer, Cary, Yerkes, and McConnell is that none of their papers addresses the issue closest to Parker's heart: the origin of the nervous system. There is not the least passing remark on the possible significance of their findings with regard to the problem of how nervous systems arose. It is as if they could not bring themselves to intrude in what they may have construed as Parker's intellectual property. Yerkes narrows his discussions to behavioural issues, Mayer and Cary to mechanistic or physico-chemical considerations, and McConnell to descriptions of neural development. The latter especially would have been expected to suggest from the developmental narrative a scenario for the evolutionary formation of the nervous system, without necessarily implying a Haeckel-style recapitulation event, but nothing of the sort was penned. They clung to their reductionist approaches.

CHAPTER NINE

Emil Bozler and the Rise of the Comparative Physiologists

Soon it turned out that our previous views of jellyfish locomotion were largely unfounded; it was therefore necessary to
pursue the investigation on a broader basis.

Emil Bozler, 1926

As we have seen, among the few individuals who took up the cause of
generating ideas on the origin of nerve cells and nervous systems, only
George Parker made it a personal, forceful, and sustained agenda in his
career. We have also seen that the continuity/discontinuity controversy
continued to fester among coelenterate neurobiologists several decades after Schäfer had made his critical observations on a jellyfish and
Waldeyer successfully championed the discontinuity theory. Some of
the coelenterate investigators observed continuity between nerve cells,
but their observations were in some cases due to inadequate methodology; in others to prejudice against Waldeyer's theory and Ramón y
Cajal's body of work, and in still others, like Parker, to an evolutionary outlook that predisposed them to assume that the syncytial nerve
net represented the basal condition from which the synaptic nerve net
would arise in higher forms.

Several coelenterate investigators, especially in Europe, saw their
career stopped or altered by the Great War. After the years of slaugh-
ter a new generation of zoologists was slowly reconstituted in Europe.
Among them the young coelenterate zoologists were eager to make
their mark by observing their material with fresh eyes. One who made
an indelible imprint, especially by addressing the thorny issue of the
nature of the nerve net, was Emil Bozler (1901–1995).

A New Wunderkind and a Fresh Outlook

Emil Bozler (1901–1995) was born in Steingebronn in Germany's Baden-Wurttemberg state. No document has emerged about his family background, childhood, and early training. We catch up with him only when he earned his doctorate at the Zoological Institute of the Ludwig-Maximilians-Universität of Munich in 1923 (Rall, 1995). The feat of earning a doctorate at twenty-two suggests that young Bozler was a precocious scholar of exceptional ability. The topic of his dissertation is unknown, but his first papers, published in 1924 and 1925, dealt with the protozoan *Paramecium* and with insects. In the ensuing years Bozler remained based in Munich, where he was employed as instructor of physiology at the Zoological Institute, but he also spent time in A.V. Hill's laboratory at University College, London, where he began research on muscle physiology, a field in which he was to become a leader. Obviously full of energy and endowed with a prodigious capacity for work, Bozler sought out research projects to calm his restlessness and fill whatever leisure time his job as an instructor left him. Presumably this is how he took up the jellyfish projects.

It is not clear if Richard Hertwig had a hand in Bozler's choice. After all, Hertwig was the grand elderly fixture of the Zoological Institute at the time, although he had recently retired (1925). However, Hertwig's name is never mentioned in Bozler's papers except in the bibliographies. The only help Bozler acknowledged was from Anton Dohrn, director of the Naples Zoological Station where Bozler did his fieldwork, and from Paul Hoffmann (1884–1962), a neurophysiologist at the University of Freiburg, who provided the expertise and advice Bozler needed. If Hertwig had any input, it could only have been in the initial conception of the project. Besides, at twenty-four Bozler showed every sign of having reached sufficient maturity as a scientist to be able to conceive and execute his research projects quite independently.

The inspiration for Bozler's project on the neurophysiology of jellyfish seems to have come from a paper by a student of his age at the Zoological Institute. Gottfried Samuel Fränkel (1901–1984), a Munich native, had gone to Naples for his research project on leeches parasitizing on fish, but he was soon mesmerized by watching the balancing acts of the local jellyfish in the water column. So he switched projects and ended up demonstrating experimentally that jellyfish statocysts function as gravity receptors. In his seminal paper (Fränkel, 1925) Fränkel noted that the body conformation keeps the jellyfish upright during swimming, but when the animal is tilted the uppermost coronal muscles go into a state of tonic contraction and do not fully relax, thus lessening their effectiveness in comparison to the rest of the swimming muscles and allowing the animal to resume its normal position. Bozler wanted to understand how the jellyfish coordinates its muscles to compensate for tilting movements, so he followed Fränkel's lead and conducted his own research at the Naples Zoological Station to address these issues.

In the two papers that resulted (Bozler, 1926a, 1926b), Bozler brought fresh experimental evidence that allowed new, incisive insights into the behavioural workings of jellyfishes. These he owed to a superior experimental dexterity and observational acumen matched only by his predecessor Romanes. He analysed three behavioural activities in *Rhizostoma*, *Pelagia*, and *Cotylorhiza*: swimming movements, slow contractions, and compensatory movements. By means of surgical attrition Bozler determined that only the ganglion inside the tentaculocyst (rhopalium) was responsible for swimming movements; removing the sensory epithelium only slowed the frequency of pulses transiently. He concluded that the rhythm of swimming movements had its roots in impulses generated rhythmically by ganglion cells, which therefore act as pacemakers. The rhythmic impulses are next propagated through a fast nerve net, travelling at 24 cm per second. This conduction speed, and the fact that the tentaculocysts around the margin regulate each other, ensures a synchronous contraction pulse of the bell.

Slow contractions, Bozler noticed, travel radially over the bell at only 1 cm per second, and only in one direction, from the margin toward the apex of the bell. As the slow contractions are mediated by at least some of the muscles used for swimming, Bozler deduced that the nerve net associated with the control of slow contractions must be different and separate from the nerve net associated with the faster contractions for swimming. In addition, the directionality of the slow contractions suggested to Bozler that the putative synapses of the nerve net must be polarized; that is, be able to transmit impulses between nerve cells only in one direction and not the reverse. This, in his mind, contrasted with the fast nerve net used for swimming, in which synapses are considered unpolarized. Furthermore, the slow contraction wave tended to spread decrementally; that is, with attenuation after it travelled a certain dis-

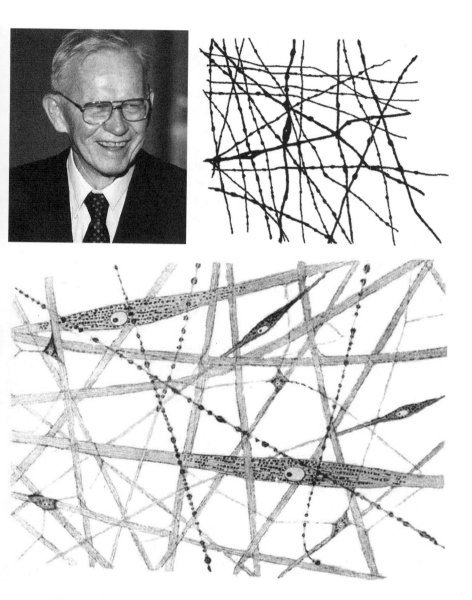

9.1 · (*Above, left*) Emil Bozler late in his life. Courtesy Ohio State University College of Medicine. (*Below*) Depiction of the subumbrellar ectodermal nerve net of *Rhizostoma*. (*Above, right*) Depiction of the endodermal nerve net of *Rhizostoma*. From Bozler (1927a), Figs. 12 and 13. Courtesy Springer-Verlag.

tance, whereas the swimming contractions showed no decrement. Thus Bozler uncovered the physiological properties that distinguished the two presumptive nerve nets.

Bozler concurred with Fränkel that the tentaculocyst, in addition to having a pacemaker role for the swimming movements, acts as a statocyst for compensatory movements. Roswitha Marx explains the experimental basis in her doctoral thesis: "When he [Bozler] kept an animal in a fixed position, bending the statocyst downwards (towards the subumbrella) with a needle led to an increased beating frequency. The compensatory movements consisted of synchronous beats, but the circular muscle closest to the manipulated statocyst did not fully relax. Bozler proposed that a third nerve net was responsible that could slow down the muscle contractions locally" (Marx, 1997).

Now, all these experimental results left Bozler perplexed. How could his results be reconciled with the then (ca. 1926) prevailing view of the syncytial nervous system in coelenterates? We may recall that when Schäfer claimed that nerve cells were discontinuous in the jellyfish nerve net, other coelenterate investigators disregarded the claim, either because their observations could not settle the issue one way or the other, or because they were swayed by Apathy's and Bethe's strongly held views that the nerve net is composed of nerve elements anastomosed together by neurofibrils. But, while the concept of anastomoses of nerve cells traps one into admitting the existence of one all-encompassing, syncytial nerve net, Bozler deduced the existence of three nerve nets working independently of each other. In addition, Bozler's observations of decremental and polarized conduction appeared incompatible with the functional nature of a syncytial nerve net, in which impulses are expected to be through-conducted and in any direction.

Bozler's assessment of the neurohistology of his predecessors was rather negative. He faulted many, and particularly Apathy and Bethe, for their allegedly poor histological preparations and inaccurate observations. He resolved, therefore to rely on no one and to settle the question of the state of connectivity between nerve cells for himself. For his purpose he found that the specimens of *Rhizostoma* captured in Naples were the most suitable, if only because the subumbrella affords areas where the epithelium is thinnest and least muscular, thereby allowing ganglion cells and nerve fibres to be viewed in live material with adequate microscopic settings. But this procedure had its technical limits, as the finer nerve fibres could not be followed. So he chose the Rongalit white method for staining the finer fibres. Bozler obtained good and reliable results only if a drop of hydrochloric acid was added to the methylene blue solution, a key constituent of Rongalit white. It is fair to say that Bozler's publication of his histological results (Bozler 1927a) had the effect of reversing the wave of approval for the syncytial theory.

He started by describing the types of ganglion cells present in the nervous system, discriminated on the basis of the number of processes: bipolar and multipolar. In the ectodermal nerve net of the subumbrella Bozler saw a network of intertwining nerve cell processes (Fig. 9.1, below). There he found three types of bipolar neurons, based on size. All bipolar cells have unbranched processes and assume different spatial orientations except for those emerging from the tentaculocyst, where the processes run parallel to each other. Multipolar neurons are all small, and their fine and varicose processes branch repeatedly. Bipolar cells cross each other at contact zones, thus forming a synaptic nerve net. While observing that the multipolar cells also contact each other, Bozler did not identify them as forming a separate nerve net because he reported the presence of contacts between multipolar and some bipolar cell processes. But there is one point about which Bozler was emphatic: the fine processes can be followed to their final destination without any evidence of fusion, and they remain separate even at their zones of intimate contact with each other. The evidence he provided was persuasive in this regard.

The existence of an endodermal nerve net had been a point of contention among coelenterate neurohistologists for a long time, but Bozler put the issue to rest by providing careful observations of a dense net constituted solely of small bipolar neurons with long, varicose processes (Fig. 9.1, above right). The cells cross each other profusely at different parts of the soma or processes, but as for the ectodermal nerve net Bozler found no evidence of fusion between processes. With regard to muscle innervation, he had little to say except to note that bipolar cell processes in both ectoderm and endoderm are associated with muscles. Finally, he came to a far-reaching conclusion: that, contrary to Bethe's contention, there was "no evidence for the presence of a true, continuous nerve net, and that therefore the structure of the nervous system of coelenterates could not be used as the main argument against the neuron theory" (Marx, 1997). Unfortunately, however, like too many of his predecessors except Parker, Bozler failed to see the evolutionary significance of his discovery. His observations, building on those of Schäfer, should have suggested that nervous systems emerged from the start as assemblages of individual, separate nerve cells, and that this feature had remained unchanged through the course of evolution.

While Bozler found in his histological observations confirmatory evidence for nerve cells working as individual units and nerve nets functioning independently, the correlations between the neuroanatomy and physiology were less than satisfactory. He had anticipated the presence of three nerve nets, but his neurohistology came up short. Bozler failed to fully realize, as noted by Theodor Bullock and Adrian Horridge (1965), that the multipolar neurons in the ectoderm formed a nerve net quite separate from the nerve net formed by bipolar cells, and that he may have been misled in thinking that synaptic-like contacts existed between multipolar and bipolar cells. Bozler attributed physiological significance to the presence of three types of bipolar cells in the ectodermal nervous system, but it is not clear whether he meant that they acted as cellular substrates for the three physiological activities he described or whether he thought each of the three types formed a separate nerve net. This muddiness was the only flaw of the paper.

Bozler devoted a second paper to debunking the notion that neurofibrils play a physiological role in the jellyfish nervous system (Bozler, 1927b). He observed what he describes as a "neurofibrillary tangle" in live, large bipolar neurons, but failed to observe Bethe's fibrillary lattice in the cell body. By manipulating his live material, he found that the neurofibrillary tangle had a soft consistency and lacked resilience, and that its role in shaping the form of the cell was therefore highly unlikely. And of course he found no evidence that neurofibrils created bridges that allowed continuity between nerve cells. This concluded his investigations on the jellyfish nervous system, a topic he never returned to in subsequent years.

Like-minded Zoologists of German-speaking Europe

If Bozler was the most talented coelenterate neurobiologist of his generation and the most articulate champion of new approaches to investigating the earliest nervous systems, he was not alone in his part of the world to embrace the new path. Adolf Portmann, Victor Bauer, Ernst Horstmann, and Karl Heider shared his willingness to challenge old ways of addressing problems and his readiness to ask questions appropriate to the new era. The question of how the first nervous systems were organized and how they served the functional needs of these organisms was not the only zoological pursuit that was jettisoning the old approaches and heralding innovative experimental designs. It was the pervasive influence of this new generation of young PhDs and *Habilitation* fellows who flourished in German-speaking Europe after the First World War that produced the "modern" thinking. Many of these researchers found their intellectual banner in newly founded scientific journals where their approaches received sympathetic editorial treatment.

The most important of the new German journals was the *Zeitschrift für vergleichende Physiologie*, to which young zoologists with a taste for ingenious experimentation and a keen interest in understanding how animals work from a comparative perspective flocked to publish their results. The journal was founded in April 1924 by two researchers then sharing interests in animal orientation in space, and bee colour perception in particular: Alfred Kühn (1885–1968) and Karl von Frisch (1886–1982). In 1924 Kühn was professor of zoology at the University of Göttingen and his research was directed toward understanding the links between development, inheritance, and evolution, using insects as experimental models. Today he is considered the precursor of the field of molecular developmental genetics (Laublichler and Rheinberger, 2004). Von Frisch, on the other hand, had been a doctoral student of Richard Hertwig and in 1924 he occupied the Chair of zoology at Breslau University. He had already made significant progress toward understanding the "psychology" of honeybees based on sensory and behavioural studies. These and his future observations on the "bee dance" earned him the Nobel Prize with fellow pioneering ethologists Konrad Lorenz and Nikolaas Tinbergen. Such were the off-the-beaten-track research orientations of the co-founders of the journal.

A similar philosophy presided over the inauguration of a new scientific journal in England the year before: the *British Journal of Experimental Biology*. It was put together by Francis A.E. Crew, professor of physiology and animal genetics at Edinburgh University, with the assistance of fellow affiliates on campus. One of the latter, Lancelot Hogben, who was soon to spend a few years as a professor at McGill University, recalls the birth of the journal:

> Facilities for publication in comparative physiology were sparse in that time [1923]. Before coming to Edinburgh, I had sounded H.G. Wells, whose son (now a fellow of the [Royal] Society) had been one of my students, about costing the production of a journal to meet the need and to become a focus for detaching academic zoology from its preoccupation with phylogenetic speculation. In my first summer at High School Yards, I raised the issue on one of our late night sessions with Crew, [Julian] Huxley and [J.S.] Haldane. All three responded with enthusiasm. Crew himself offered to underwrite some of the initial cost from his army gratuity and to invoke the good offices of a local publishing firm with loose affiliations to the University. The response justified his optimism, and we constituted ourselves as the Editorial Board of the *British Journal of Experimental Biology* with Crew as editor and Chairman of the Board. (Hogben, 1974)

The British journal soon experienced financial troubles and was rescued by the newly instituted, not-for-profit Company of Biologists in Cambridge, which dropped the "British" from the journal's name. Although the German and British journals shared a common mission, the articles published in their first issues betray a difference of emphasis in their choice of topics. The *Journal of Experimental Biology*, of which more will be said in the next chapter, emphasized endocrinology, visceral physiology, and physiological genetics, whereas the *Zeitschrift für vergleichende Physiologie* concentrated on sensory physiology and neurophysiology. Only two years after the latter's birth Bozler was among the young "moderns" who published his physiological work on jellyfishes in the *Zeitschrift*. And in the same year a contribution appeared in the German journal that dealt with the neuromuscular organization that subtends the creeping behaviour of a sea anemone (Portmann, 1926).

Adolf Portmann (1897–1982) was one of those zoologists, of whom we have seen several examples in this book, who in their time were destined to make intellectual contributions of universal importance. Born in Basel, Switzerland, he did all his university studies there and in 1921 completed his doctoral dissertation on the dragonflies of Central

Europe. The limited world of inland fauna soon wore on him and he sought diversity in the marine fauna available at several of the seaside laboratories of Europe (Banyuls-sur-Mer, Roscoff, and Villefranche-sur-Mer in France, and Helgoland in Germany). While still an assistant at the Zoological Institute of Basel University, Portmann travelled to Banyuls-sur-Mer to conduct an investigation of the creeping locomotion of sea anemones. Sea anemones move along the sea bottom by expanding their adhesive foot (pedal disc) in the direction of movement while their body column shortens to reduce drag. This is followed by a contractile retraction of the foot to the new location along with the elongation of the body column, and so on for another cycle.

Portmann noticed that foot expansion was associated with contraction of the circular muscle of the body column, which also contributed to body column shortening with the assistance of a weak contraction of the longitudinal muscle. Although there was no clear sensory cue for the behaviour, the sea anemones tended to start crawling at nightfall, and crawling activity displayed a day-night rhythm. To Portmann it suggested the involvement of something akin to an autonomous nervous system and, by removing the nerve-rich mouth region and observing that the crawling behaviour went on as usual, he discarded the role of a centralized or condensed nervous system. Only simple, diffuse nerve nets associated with the circular and longitudinal muscles could play a role in the crawling movements.

Portmann's sea anemone study was just one of many he conducted on the broad variety of animals that attracted his hypercharged curiosity about morphology and behaviour. His industry helped raise him to full professor of zoology at Basel University in 1931. In a series of books dating from 1948 to the 1970s Portmann constructed a philosophical concept, based on his cumulated observations of the animal world, in which he "considered the outer surface of living organisms as a specific organ that serves a self-representational role" (Kleisner, 2008). In this view the animal surface is not so much a functional projection of, or a protective shield for the internal organs, as it is an organ of display for biosemiotic purposes. Portmann's essays on the subject gained him prestige not only within the biological community but also among the intelligentsia at large.

The year after Bozler's and Portmann's contributions appeared in the *Zeitschrift für vergleichende Physiologie*, another paper, this time on jellyfish swimming, appeared in the same journal (Bauer, 1927). Victor

Bauer (1881–1927) had an eclectic and rather difficult life. He was born in Samarang in today's Indonesia but went to school in Germany. He earned his doctorate in 1903 under the mentorship of the famous biologist August Weissmann in Freiburg. His dissertation topic was the metamorphosis of the central nervous system of insects during development (Reif, 1987). Probably thanks to inherited family wealth Bauer did not pursue a tenure-track academic position but instead joined the staff of the Physiology Group at the Naples Zoological Station as an assistant between 1903 and 1910. In 1910 he built his own private laboratory in Positano, near Naples, where he conducted research on the sensory physiology and ecology of marine animals. The outbreak of war put an end to Bauer's idyllic life in Naples; he spent wartime in the militia and then in the Foreign Office, containing locust swarms in Asia Minor, and overseeing the management of fisheries in Turkey. By the time he returned to Germany, Bauer had lost all his wealth and was reduced to accepting a low-paying job at the Institute of Lake Research by Lake Constance between 1919 and 1922, followed by an assistantship at the Institute of Physiology of Bonn University until 1926. In that year he managed to obtain a late *Habilitation* and took up a professorship in zoology and comparative anatomy in Bonn. It was shortly before this new, hard-earned academic stability that he conducted a groundbreaking piece of research on jellyfish.

Bauer conducted his research in the summer of 1925 at the Naples Zoological Station, his old haunt of pre-war days. He had a long-standing interest in the question of nerve-based reflexes associated with swimming movements, dating back to the years 1908–10, when he published physiological studies on mysid shrimps, comb-jellies, and other planktonic animals. This time he focused on the swimming beat of the jellyfish bell, asking himself whether the rhythm of the bell contractions is reflex or automatic. Incredible as it may seem, Bauer conducted his research unaware that Bozler was completing a research project that overlapped with his own, in the very same Neapolitan location. The issue at hand goes back to Romanes's time when, we may recall, the jellyfish swimming pulsations presented an analogy with the heartbeat. The question of whether the rhythmic oscillation of the heartbeat was generated by cardiac muscle or nerves had been transferred to the jellyfish conundrum, and although it had been suggested that in jellyfish the pacemaker was nerve-based, no irrefutable proof had been forthcoming. Although Bozler provided in his 1926 papers further, more persuasive evidence that the pacemaker was seated in the ganglia of the marginal bodies (rhopalia), he could not solve the riddle of the reflexive versus endogenous nature of the rhythmic activity. Bauer endeavoured to settle the matter.

Initially, Bauer's project was broader, encompassing in fact the very physiological questions Bozler was asking. However, as Bauer was writing up his paper, he became aware of Bozler's first 1926 paper. Realizing that some of his conclusions overlapped with Bozler's own, he communicated with Bozler and learned that Bozler had a second paper forthcoming. Bozler was generous enough to send Bauer the manuscript or proof of his second paper, so that Bauer could adjust the gist of his own forthcoming paper accordingly.

At the root of the debate was Jakob von Uexküll's view on the matter. Uexküll (1864–1944) was an impoverished, Estonia-born aristocrat who studied in the Baltic city of Dorpat and in Heidelberg, but never obtained a regular doctorate. The closest he came was an honorary doctorate from the University of Heidelberg in 1907, when he was forty-two years old and had garnered fame for his biological theories. These he encapsulated in his book *Umwelt und Innenwelt der Tiere*, published in 1909 (Buchanan, 2008). In reaction against Jacques Loeb's representation of animals as "physico-chemical machines," Uexküll proposed the concept that animals are defined by their surrounding world (*Umwelt*) and respond to their environment by 'harmonizing" behaviours. A key feature of his thought is the notion that the entire animal world is coordinated by a circular feedback, whereby the environment broadcasts "carriers of significance," sensory cues that are picked up by the sensory receptors of the animal and given meaning, after having been processed in the animal's inner world (*Innenwelt*), through the effector responses organized as behaviour. Uexküll's idea of environmental cues carrying potential meaning led to the foundation of the discipline of biosemiotics, to which, as we just saw above, Portmann adhered. Portmann was just one of many who were deeply influenced by Uexküll's views, including intellectual luminaries such as Martin Heidegger, Maurice Merleau-Ponty, and Gilles Deleuze (Buchanan, 2008).

Uexküll's insistence on the representational power of the environment being channelled through a reflex circuit of sorts permeated his many biological investigations at the Naples Zoological Station. In 1901 he had produced a paper on the swimming pulsations of jellyfish (Uexküll, 1901) in which he concluded that each executed swimming pulsation is a reflex response to the sensory input from the gravity/balance receptors (statoliths), themselves influenced by the physical impact of the previous pulsation. This reflex theory of the swimming rhythm was exactly what Bauer set out to test experimentally, employing surgical cuts and electrical stimulation on healthy specimens. Bauer's experiments led him to conclude that the genesis of the rhythm was not dependent on a feedback loop. He found at best that sensory input only modulated the tone and frequency of the ongoing rhythmic pulsations. Far from agreeing with Uexküll that the origin of the rhythm lay in the swimming muscles, Bauer concurred with Bozler that the rhythm originated in the ganglia of the marginal bodies. He went further than Bozler, however, by establishing that rhythmicity is an intrinsic property of the ganglia, in no way dictated by sensory input. He further suggested that this "autonomous rhythmic excitation" of the ganglion cells was due to "a periodic release from a refractory state" by analogy with the prevailing view of the heartbeat. Bauer's definitive demonstration "meant that the jellyfish was the simplest animal with a true nervous pacemaker" (Bullock and Horridge, 1965), although Bauer himself failed to see the evolutionary ramification of his discovery.

Unfortunately Bauer was not given a chance to reflect on this scientific achievement. On the night of 18 October 1927, eight months after his jellyfish paper appeared in published form, he died of pneumonia at the premature age of forty-six (Reif, 1987).

The challenge posed by the issue of the neurophysiology of swimming in jellyfish continued even after Bozler's and Bauer's momentous contributions. Seven years later a young compatriot added two papers on the subject, casting them in a comparative physiological framework (Horstmann, 1934a, b). Ernst Horstmann (1901–1973) was born into the family of an architect in a town on the Rhine near the French border (Fleischhauer, 1973). He studied biology and medicine in Munich and Kiel, where he took up graduate studies. He earned his doctorate in 1933 under the supervision of Baron Wolfgang von Buddenbrock (1884–1964), who occupied the Chair of zoology at Kiel University and distinguished himself, beyond his inherited aristocratic title, by his research on the respiratory physiology of insects and the sensory physiology of molluscs. As one of the early practitioners of the new discipline of comparative physiology, Buddenbrock encouraged Horstmann to adopt the comparative outlook in his doctoral project on jellyfish.

For his studies Horstmann selected two scyphomedusae, which had received little attention of late and therefore suited his comparative approach. The species, *Aurelia aurita* and *Cyanea capillata*, were readily available in the Kiel Fjord. That Horstmann's papers are organically linked with those of Bozler, Fränkel, and Bauer is made forcefully clear by the frequent comparisons of his results with theirs. For example, Horstmann concluded in his first paper that *Aurelia* and *Cyanea* differ from the scyphomedusae studied by Bozler in that their rhythmic motor outputs do not arise automatically in the ganglia of the tentaculocysts, but instead that sensory inputs in the tentaculocysts contribute to the generation in the ganglion cells of the rhythmic swimming impulses around the bell margin. Although Horstmann cited Bauer as well as Bozler, he failed to gauge the significance of Bauer's results and to realize that alternative interpretations could be proposed for his own experimental observations. However, in other respects Horstmann made interesting discoveries.

He found that low light levels can reduce, and high light levels increase, the frequency of swimming pulses in *Aurelia*, which possesses ocelli in the marginal bodies. He also found that "the marginal body ganglia are necessary only in older animals to maintain the rhythmic beat under normal environmental conditions. In younger animals any piece of subumbrella can pulsate by itself" (Horstmann, 1934a; my translation).

In the second paper (Horstmann, 1934b) he paid special attention to the neuromuscular physiology associated with swimming. For this purpose Horstmann built a humidity chamber of his own design inside which strips of the subumbrella of *Cyanea* were suspended and attached to a kymograph for recording of muscle activity. Silver wire electrodes were inserted in the chamber and inserted in the preparation to stimulate muscle (direct stimulation) or nerve net (indirect stimulation) activity. Horstmann was particularly interested in understanding the role of muscle tonus in the contractility associated with swimming. (Tonus may be defined as a residual, continuous muscle tension at rest or as

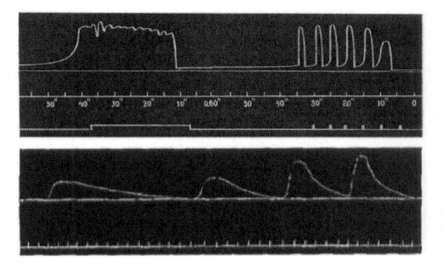

9.2 · Two sets of kymograph recordings of muscle contraction made by
Horstmann (1934b) on the scyphomedusa *Cyanea*. The lower traces in the two
sets are time markers. See explanations in text. Courtesy Springer-Verlag.

the result of passive muscle resistance to stretch.) He found that both
muscle tonus and rhythmic swimming pulses are eliminated by the removal of the marginal body ganglia, thereby suggesting that both phenomena have a central nervous origin. However, he also found that the
muscle tonus of each subumbrellar sector associated with a marginal
body can be independently regulated.

Finally, Horstmann examined the relationship between muscles and
nerves in bringing about rhythmic contractions. Direct faradic stimulation, which bypasses the nerves to excite muscles directly, induces
rhythmic contractions transiently, followed by twitching or tonic episodes, in other words uncoordinated muscle activities (Fig. 9.2, top,
left trace). In contrast, indirect faradic stimulation causes all muscles
to contract and slacken synchronously to maintain precise rhythmic
contractions, as a result of fast impulse transmission in the subumbrellar nerve net innervating the muscles (Fig. 9.2, top, right trace). Horstmann also revisited the phenomena of the "staircase effect" and the
refractory period first noted by Romanes and Eimer (see chapter 5).

The staircase effect is in fact a misnomer. As seen in the trace of Fig.
9.2 (bottom), the contractions do not remain at a raised level until the

next stimulated contraction unfolds, in contrast to the real staircase associated with temporal summation of consecutive contractions. The kymograph recording illustrates instead the phenomenon of facilitation, as explained in chapter 5. Facilitation is now known to result from an increased release of neurotransmitters acting on muscle receptors at successive stimulations, but Horstmann's interpretation of it departed farther from reality than Romanes's actually did. He believed that the mounting contractile responses, with their characteristically faster rising phases, are due to physico-chemical changes in the muscles, such as a lowering of viscosity or friction after the first contractions. From his recordings Horstmann concluded that the refractory period, that is, the period of time during which stimulations fail to arouse a response because the preparation is currently in an excited state, is longer in the nerve than in the muscle cells. To him it explained why stimulating the muscles directly led to tonic contractions peaking to a tetanus-like plateau as shown in Fig. 9.2 (top).

Coelenterate investigators of the postwar German school manifested a heightened intensity of scientific curiosity in experimental matters. They also showed a willingness to embrace the new field of comparative physiology and use it to good effect for the elucidation of questions, particularly related to jellyfish swimming, which had lingered since the days of Romanes. The rewards for their efforts came largely from the functional insights they gained, despite some accompanying misconceptions and flawed interpretations. And, even though these investigators had but a dim awareness of their contributions on that front and did not prioritize that theme, they did move forward in their understanding of evolutionary issues. Bozler put the final nail in the coffin of the continuity theory of the coelenterate nerve net, thus refuting Parker's vision of the first nervous systems arising as networks of nerve cells in protoplasmic continuity with each other. Bauer's discovery of the jellyfish neuronal pacemaker meant that rhythmic firing of impulses was already an intrinsic property of at least some of the first evolving neurons.

So far, the field of coelenterate neurobiology was heavily lopsided toward the Cnidaria. Knowledge on the neurophysiology and neuroanatomy of ctenophores had progressed poorly since the works of Eimer, Chun, Hertwig, Schneider, and Samassa in the nineteenth century. A few years before the Great War, Victor Bauer, whose jellyfish work we have just discussed, undertook in Naples a neurophysiological study of comb-jelly ciliary locomotion, taking an approach that closely resembled that later taken by Bauer for jellyfish swimming (Bauer, 1910). As mentioned early in chapter 7, swimming in ctenophores is accomplished by the metachronous beating of ciliary comb-plates. These comb-plates are stacked in several rows that run radially from the apical (aboral) end to the mouth. The comb-plates beat toward the aboral end, much as the paddling movements of a canoeist do, such that the animal moves in the water mouth first. At the time the prevailing view was that signal transmission creating the wave of ciliary motion occurs between the cilia-bearing epithelial cells. A champion of this view up until the publication of *The Elementary Nervous System* was George Parker himself, who regarded the phenomenon as an example of neuroid conduction.

As his experimental model Bauer selected a ctenophore species found in abundance in Naples: *Beroe ovata*. He used indirect experimental evidence to suggest that nervous regulation was involved. The tentativeness of his conclusion is reflected in the cautiously worded title of his paper, in which the phrase "apparent nervous regulation" appears. In essence, Bauer found that a light touching on the mouth momentarily interrupted movements of the comb-plate, whereas stronger stimulation caused an increase in comb-plate activity. Removing the sensory centre of the apical pole, which includes the statocyst and was then believed to coordinate the activity of the rows of comb-plates, had no effect on the responses to tactile stimulation. Bauer concluded from these observations that the diffuse, subepithelial nerve net of these animals controls the comb-plate beating movements as well as muscle activity. However, he had not performed neuroanatomical investigations to support his conclusion.

Karl Heider (1856–1935) made his contribution, seventeen years later, by revisiting the neuroanatomy of ctenophores, which had not progressed at all since the 1890s. Heider was born in Vienna, son of Dr Moritz Heider, the founder of scientific dentistry in Austria, and of Baroness Marie von Thinnfeld, who came from an aristocratic line of early industrialists and ministers to Emperor Franz Joseph I. After starting medical studies Heider moved to natural history and in 1879 he completed his doctorate under Carl Claus at the Institute of Zoology of the University of Vienna. He returned to medical school and

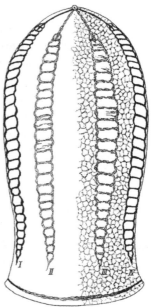

9.3 · (*Left*) Karl Heider in 1932. Entry in "Embryo Project Encyclopedia" (1932): ISSN1940-5030. (*Right*) Schema of the topography of the nervous system of *Beroe ovata*, in which are seen four nerves coursing down from the apical body and merging with the scaffolds of tissue strands along which run nerve cells. The subectodermal nerve net is seen in the right half of the schema, and the nerve ring around the mouth is depicted at the bottom. From Heider (1927), Fig. 20. Courtesy Springer-Verlag.

graduated in 1883. After completing his military service in 1884, Heider was granted his *Habilitation* and took a professorship at the University of Innsbruck – and even became its rector in 1904. In 1917 he was appointed to the Zoological Institute of the Friedrich-Wilhelms-Universität in Berlin, of which he became director in 1924. In the course of his career he became famous for his studies on the comparative embryology and phylogeny of invertebrates. So he turned his attention to the nervous system of ctenophores in the twilight of his academic life, bringing to his subject the ripe experience of histological work.

Heider did his research at the Naples Zoological Station in the winter of 1926–27, using the species previously favoured by Bauer, *Beroe ovata*. Using Unna's vital methylene blue staining method for the most part, he distinguished three components of the nervous system. First, he confirmed Richard Hertwig's identification of a cnidarian-like subectodermal nerve plexus, diffusely distributed throughout the body but more densely as a ring around the mouth. Second, he described eight nerves issuing from the apical sensory body; and third, he observed a scaffold of tissue strands around "vascular tubes" coming down toward the mouth, along which nerve cells course from the apical nerves (Fig. 9.3, right). He saw these eight nerves and associated scaffolds as the likely substrates for the nervous regulation of the comb-plate rows envisaged by Bauer. He could not make out the mesogleal nerve net that Hertwig had spotted.

Heider concluded that the level of organization of the ctenophore nervous system is very similar to that of cnidarians, especially with regard to the nerve net and its condensation as a ring around the mouth. He regarded the organization of the nervous tissue associated with the comb-plate rows as an evolutionary step toward the development of the nervous system of deuterostomes (echinoderms, protochordates, and vertebrates). Although he thought the nervous system of higher metazoans evolved by detaching itself from the ectodermal epithelium and forming larger ganglionic masses deeper in the body, with nerve fibres of higher conduction speed than in coelenterates, he nevertheless believed that the nerve-rich tissue tracts on the vascular tubes along the comb-plate rows were passed along for the development of the innervation of the vascular system of deuterostomes, especially echinoderms.

Heider was the only member of the German school of coelenterate neurobiologists of the era to consciously bring forward the evolutionary implications of his findings. The others refrained from openly discussing evolutionary issues, either because the impetus of their work leaned entirely toward function, or because they demurred from discussing evolutionary perspectives out of distaste for speculation or general unease about dealing with the topic. However great their misgivings may have been, the following generation of coelenterate neurobiologists saw clearly the evolutionary importance of the body of work that came out of the German school, and they did not shy away from highlighting it in their own work.

CHAPTER TEN

Carl Pantin and the Well-Tempered Nerve Net

The right way to look at the lower animals is to think of them as essentially simpler engineering solutions of the same physiological problems that beset us all.

Carl Pantin, 1952

Once it was established by Romanes for jellyfish and by Parker for sea anemones that the nerve net is the first evolutionary manifestation of an organized nervous system, investigations began springing up in the attempt to understand the physiology of such systems and how their functioning differs from that of higher animals. Romanes and Parker themselves, of course, looked beyond the pregnant evolutionary implications of their discoveries and pioneered such investigations as we have seen in chapter 9, stimulating others to emulate them. While it can be said that their contributions, piecemeal as they came, helped throw light on some features of nerve net activity, a satisfying comprehension of the overall properties of nerve nets had by the mid-1930s still not been reached. One scholar entered the stage who was destined to devote, in the way Parker had done in a previous generation, the largest part of his research career to reach this comprehensive understanding. It can truly be said that this man was the legitimate heir to Romanes and Parker.

Building on the Foundations

British zoologist Carl Frederick Abel Pantin (1899–1967) was born in Blackheath, a suburb of London, to a wealthy family that had traded

in timber and other materials since 1788, when his great-great-grand-
father, Thomas H. Pantin, established the business (Russell, 1968).
Carl's father, Herbert, led the prosperous Pantin Company and his
mother, Emilie Juanita Abel, was a descendant of the German musician
Karl Friedrich Abel (1727–1785), who became attached to the English
Court and was a close friend of Johann Christian Bach. Carl's early
interest in natural history is documented in an account by his younger
brother William:

> At the sea shore he would collect fossils and dissect sea ur-
> chins (whereas I would be trying to carve medieval sculptures
> in chalk); I remember him reading Sir Ray Lankester's *Science
> from an easy chair*, I think this would be about 1910–1911, soon
> after it came out, and this would fit in I suppose with his inter-
> est in zoology; and about that time I think he was also reading
> H.G. Wells's science fiction books, and browsing in the *Children's
> Encyclopaedia*. About 1909–1910 he taught me (then a small boy)
> how to draw plans of buildings; I remember this well and have
> been eternally grateful to him for it. (Quoted in Russell, 1968)

Pantin spent his high school years at Tonbridge (Kent), a major
independent boarding school for boys, between 1913 and 1917. After
graduation he was soon enlisted as an officer cadet of the Royal En-
gineers and in March 1918 was shipped to France, where he served in
various engineering support tasks until the end of the war. He enrolled
at Cambridge in the spring of 1919, first with the intention of reading
physics. But his academic tastes quickly drifted toward physiology and
zoology, thanks to the influence of inspiring teachers at the Zoology
Department, especially the physiologist James Gray, only eight years
his senior.

James Gray, one of the founders of the field of experimental cytology and a student of the reflex control of movement, was one among
many distinguished experimentalists boasted by Cambridge. The promotion of physiology by pioneer Michael Foster (see chapter 5) had
borne many fruits over the years, not least the creation of a physiology
department. There, in the years during which Pantin received his university education and eventually joined the faculty, many luminaries
thrived, as explained by Adrian Horridge in his memoirs:

In physiology in Cambridge, first-year lectures on the nervous system were given by Lord [Edgar] Adrian, who had been awarded the Nobel Prize in 1932 for his demonstration of nerve impulses and the discovery of the frequency code in single axons. In the practical class, activity in nerve fibres was made visible with an amplifier and cathode ray tube put together by Sir Bryan Matthews, who had perfected the string galvanometer that did the same job in the 1930s. Second-year lectures on the nervous system were given by Willie Rushton, who discovered the principle of univariance in vision, Andrew Huxley and Alan Hodgkin, who, with Jack Eccles, were awarded the Nobel Prize for medicine in 1963. At the time, they were working actively on the squid's giant axon membrane. (Horridge, 2009)

The Zoology Department was influenced by the entraining effect of the dynamic Physiology Department. It was, as Horridge put it, "stuffed with Fellows of the Royal Society and was committed to the experimental approach." Among them were Eric Smith, an expert on annelid worm and echinoderm neurobiology, and Gray. The latter acted as Pantin's mentor for his first foray into research by suggesting a topic to his own liking: amoeboid movement. This mode of locomotion, used by the protists known as amoebas, was considered a form of contractility that predated the muscle fibres of multicellular animals. This activity, it was believed at the time, allowed amoebas to crawl along a surface by pushing streams of protoplasm in the direction of movement. But when Pantin set out in 1922 to the laboratory of the Marine Biological Association of the United Kingdom (MBA) in Plymouth, it was not only to investigate a marine form of amoeba, but also to take on an appointment as member of the scientific staff of the laboratory. The MBA was founded in 1884 out of concern over the state of the fisheries industry in the kingdom and the need for rationalization through the establishment of a fisheries science. The prominent zoologists Thomas Huxley and E. Ray Lankester had lobbied for many years in favour of such an organization. Interestingly, the committee that recommended putting the MBA in place in 1883 counted among its members some of the protagonists in this story, such as Michael Foster, George Romanes, and John Burdon-Sanderson. The MBA chose Citadel Hill in Plymouth for its laboratory, and activities began there in 1888.

At the Plymouth laboratory Pantin, whose attachment to the MBA would lead him to become its president between 1960 and 1966, was assigned to the Department of General Physiology, then headed by W.R.G. Atkins. He started his research in such earnest that the following year, 1923, saw the publication of his first paper on the physiology of amoeboid movement, followed by a second in 1924. As F.S. Russell explained in his memoir:

These observations supported the hypothesis that in the movement of the *limax* type of *Amoeba* a pseudopodium is extended at the front end formed of fluid endoplasm resulting from local swelling and absorption of water extracted from the posterior end. As the pseudopodium advances new ectoplasm is being formed by gelation. As water is extracted from the posterior end of the ectoplasmic tube its contraction drives the fluid endoplasm forward. Continuous forward progress thus results from the constant trans-formation of the fluid endoplasm at the front into gelated ectoplasm and the transformation of the gelated ectoplasm in the rear into fluid endoplasm. (Russell, 1968)

Pantin's papers on amoeboid movement, as with many of his other investigations, stood the test of time and established him as a first-rate experimentalist. That his intellect and technical approach to research strikingly resembled Parker's is attested by his former student, W.G.M. Pryor (quoted in Russell, 1968), who remarked that Pantin's research topics "show a gradual progression of interests, each growing out of the last. In most of them he was something of a pioneer; his observations are simple and ingenious, and never involve any very complicated apparatus. The subjects he worked on all seem to have continued to develop at a high rate – he selected important problems, and he stimulated interest so that his early work is now mostly where early work should be, buried in the foundations."

Pantin showed himself a prolific author of scientific papers on amoeboid movement as well as other subjects, including crustacean muscle physiology, during his years in Plymouth. But his appointment there ended in 1929, when he returned to Cambridge the happy recipient of a Fellowship at Trinity College. In his capacity as reader, Pantin was assigned to teach courses on invertebrates and continued the line of research he had successfully conducted in Plymouth. But in 1933 he suddenly decided to favour the Naples Zoological Station for field laboratory work, and engaged in a new field of investigation. The relocation may have been related to his contracting tuberculosis in 1933; it is likely that he had been advised to seek a warmer climate for his condition.

Once in Naples he undertook an investigation of the nervous control of crustacean muscles, which he speedily completed, leaving him extra time to look around for another project (Pantin, 1968). Robert Josephson recounts how it happened (Josephson, 2004), "The previous occupant of the laboratory in which he was working had left some sea anemones in an aquarium, and Pantin chose to examine the neural control of muscular contraction in these." Serendipity and Pantin's health problems seem to have led to his thirty-four-year infatuation with the neurobiology of sea anemones and with Parker's notion of the elementary nervous system.

When he arrived at Naples in 1933, Pantin occupied the Cambridge University table at the *stazione*. Needing scientific equipment of his own design in a remote location, he obtained funds from the Grant Committee of the Royal Society. The remaining logistics were provided by Reinhard Dohrn (1880–1962), who had succeeded his father, Anton, as director of the *stazione*. Pantin soon made his choice of experimental animal: the sea anemone *Calliactis parasitica*, which had a habit of settling on mollusc shells. If we can trust the opening statement of his first paper, it seems clear that Pantin trained his focus, more directly than any of his predecessors, on the nerve net itself at the operational level:

> The most primitive form of nervous system in the Metazoa appears to be the nerve net, and in it undoubtedly many features of the nervous systems of the higher animals may be exhibited in a simplified form. But in comparison with such systems the nerve net seems to show certain far-reaching contrasts which suggest not merely a more primitive organization but a fundamentally different method of transmission of excitation. It is our object to investigate the nature of the special properties of the nerve net. (Pantin, 1935a)

The personal gifts that Pantin brought to the subject were twofold. First was his ability and inclination to envision physiological problem solving in the context of the whole organism and its environment. This, of course, he shared with many of his predecessors, such as Romanes and Parker, and also his fellow comparative physiologists of the 1920s and 1930s who were eager to distance themselves from paramedical physiology. However, Pantin seemed to possess a more heightened sense of the importance of this school of thought than most of his colleagues. Second, he combined an uncanny ability to assess all the implications of a scientific problem with the technical wizardry best fitted to address it.

Faced with the problem of how nerve nets work, Pantin was conscious that all his predecessors had failed to comprehend the whole picture. He seized on the phenomenon of summation/facilitation that had been periodically noticed since the days of Romanes (and which we have reported in several preceding chapters). Many of his predecessors had muddled the distinction to be made between simple summation of consecutive responses with facilitation as a factor of sensitization for subsequent stimuli. Pantin, just as Romanes before him, was clear on this point. But even when these investigators understood the distinction, they took a narrow view of the functional implications of facilitation, seeing only the process of facilitated responses to consecutive stimuli but not the result of its operation in the nerve net.

Pantin's first major contribution arose from finding that stimulating the body column of the sea anemone by touch or by applying a single pulse through the stimulating electrodes induced contractile responses that travelled all around the column. This he regarded as a clear case of through-conduction of excitation in the nerve net, a non-decremental spread of excitation that required no priming. The feature of the phenomenon that struck him most was not only that a single stimulating pulse could initiate a "wave of excitation" that passed over the entire nerve net, but also that the muscles that happened to be in the path of the wave of excitation were induced to contract "on the go," so to speak. "We are thus brought to the remarkable conclusion," Pantin exclaimed, "that the whole nerve net of the column acts as a conducting layer in its most simple form, directly transmitting excitation from the site of stimulus to the muscle. Such an arrangement is far simpler than the most elementary reflex arc, and indeed it is simpler than anything which the nerve net has previously been considered to exhibit" (Pantin, 1935a).

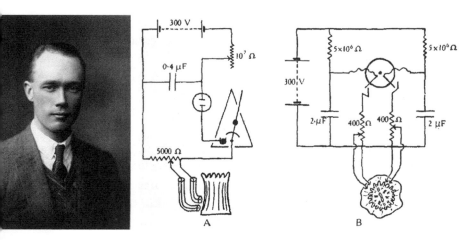

10.1 · (*Left*) Carl Pantin in his thirties. Courtesy of the Master and Fellows of Trinity College, Cambridge, UK. (*Right*) Pantin's circuit diagrams of the electrical stimulation protocols used in his 1935a paper. Courtesy of the Company of Biologists, Cambridge, UK.

But in the oral disc of the anemone the local nerve net behaved differently than in the column. There stimulation led to contractile responses that spread to a point and then stopped short of running the entire course open to them. This decremental spread of excitation had been observed before, notably by Bozler, but its significance for the mode of transmission of the nerve net had eluded him and other coelenterate neurophysiologists. It was Pantin's *eureka* that he saw facilitation as a critical tool in extending the spread of transmission in diffuse nerve nets displaying decremental conduction.

And his insights did not stop there. Pantin soon realized that it was methodological shortcomings that had led his predecessors astray. Toying with methods of stimulation led him to expose the differences between mechanical (tactile) and electrical stimulation with regard to what is actually stimulated. For a start, he found that increasing the intensity of tactile stimulation of the oral disc led to summation of responses, whereas varying the intensity of pulses conveyed by electrical stimulation of the same region had no effect on the strength of the contractile responses. Pantin reasoned that the touch probe excited sensory endings, which in turn transmit their signal either directly to the muscles or indirectly via the nerve net. With increased pressure of the probe on the disc, more sensory units are recruited and contractile responses are thus summated. By contrast, with electrical stimulation, the silver electrodes he used were inserted into the wall of the disc or body column such that the sensory cells were bypassed and the nerve net could be excited directly once threshold was reached, thereby revealing its all-or-nothing mode of response.

To reveal how facilitation plays a crucial role in spreading excitation through the nerve net, Pantin designed and built a stimulation apparatus containing circuits that allowed him to precisely control the frequency at which multiple stimuli are delivered to the preparation, and to reasonably estimate the refractory period during which stimuli may fail as an excitation wave washes over the nerve net (Fig. 10.1, right). This procedure empowered him to make critical observations that had been out of reach to his predecessors. He succinctly stated his main findings for the disc in the following passage:

> In the disc, therefore, a single stimulus produces only a slight response or none at all. All the more complex responses require a battery of stimuli, and it is the time interval between stimuli which governs the nature of the response, and not their strength. It seems reasonable to suppose, therefore, that the various responses of the disc are called up in succession through facilitation.
>
> The second special feature of the disc responses is the apparent existence of a decrement of conduction. This is apparent in the "edge-raising" reaction. The stronger the mechanical stimulus, the more widely is the reaction propagated round the disc. The reaction is easily evoked electrically, but only by a battery of several stimuli. Here again above the threshold, the strength of the stimulus is of no importance; but as the frequency of the stimuli in the battery increases there is a great increase in the size of the response and also it is propagated further. Its propagation is easily studied with low-frequency stimuli. There is not a single contraction wave propagated outwards with an intensity that decreases as it gets further from the point of stimulation; though this might have been expected were the conducted excitation to undergo a decrement. Actually each stimulus of the series calls up

a contraction, but with each successive stimulus the contraction extends by stages further and further round the disc, and ceases to extend the moment the battery of stimuli comes to an end. While, therefore, the muscles under the electrodes respond with contractions to the first few stimuli, those some distance away round the disc do not begin to respond till several stimuli have already passed, after which they respond in a normal manner. One cannot interpret these effects by supposing that a stimulus is conducted with a decrement. Evidently, in a battery of successive stimuli, each member paves the way for the propagation of its successors into fresh sectors of the disc. The most simple explanation of this is that the conducting path extends by progressive facilitation between the conducting units of the adjoining sectors; that is, the response of a sector on the outskirts of the contracting area differs from the response immediately under the site of the electrodes only in that several stimuli are required before a conducting path is established to it. It will not therefore begin to respond to each stimulus until several stimuli have taken place. This implies that facilitation is taking place between different sections of the nerve net itself. (Pantin, 1935a)

This newly discovered phenomenon Pantin called "interneural facilitation." In his view, interneural facilitation was the key method by which obstacles to the spread of impulses in a synaptic nerve net, such as the one in the oral disc of sea anemones, are overcome.

But there are upper limits to frequencies of stimulation, above which no further increase of responses will occur. Pantin acknowledged this when he wrote: "The refractory period sets a limit to the frequency of effective stimulation of the nerve net." Once again, he devised an ingenious stimulation protocol to measure this refractory period. He placed two pairs of electrodes four centimetres apart at the base of the body column (Fig. 10.1, right) and applied condenser discharges at increasing time delays between the two pairs. In this way, as Pantin points out, "the intensity from the leading condenser [first fired] is maintained just in excess of the threshold value, while the strength required to give a facilitated response following the discharge of the second condenser is determined for various time intervals." From these experiments Pantin calculated a relative refractory period of 0.5 second, meaning that a frequency of around ten pulses per second is the upper limit to elicit facilitated responses in the nerve net of sea anemones.

In short, as Pantin concludes in his first paper: "The true characteristics of the nerve net are diffuse conduction and the extreme development of facilitation. Diffuse conduction may be total, as in the column, or restricted, as in the disc of *Calliactis*. Facilitation may be between the nerve net and the muscles, or between parts of the nerve net" (Pantin, 1935a). This was a remarkable achievement for a beginner in the field of coelenterate neurophysiology. More understanding of the workings of the "primitive" nerve net was gained in that single paper than with the accumulated results of all his predecessors since Parker. It opened a new era for the field to be seized upon by the next generation, as we shall see.

After Pantin had shown in broad strokes how the nerve net works, he immediately set about revealing the underlying design of the nerve net – what he calls the plan of the nerve net. In this we see the mind of a frustrated engineer at work. Pantin's biography leaves little doubt that had he not pursued zoological studies he would have taken up a career in engineering. Here again, his inquiring mind and technical abilities converged to yield stunning insights. In his second paper (1935b), Pantin establishes that the greater part of the nerve net in the column behaves as a single conducting unit, the through-conducting system. But the different muscles in the path of this nerve net will respond according to their specific responsiveness to facilitatory pulses; that is, the initiation of their response is determined by their pulse frequency threshold (Fig. 10.2). Thus, a large number of pulses of low such frequencies will elicit contraction of parietal muscles, and pulses at a higher frequency will recruit the faster longitudinal muscles of the mesenteries. A smaller number of pulses at still higher frequencies will fire up both the longitudinal and sphincter muscles, and short bursts of pulses will activate only the marginal sphincters of *Calliactis*. In this way a simple nerve net can coordinate various and relatively complex behaviours through patterning of neuromuscular facilitation.

Pantin also distinguishes between vertical and lateral conduction in the nerve net. He finds that through-conduction runs vertically and involves the mesentery and the sphincter; this is the through-conduction unit that exhibits the highest conduction speed and the least facilitation. In lateral conduction designed to spread the wave of excitation

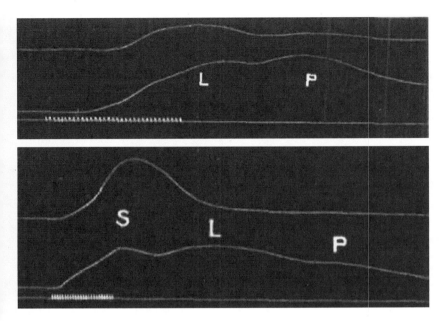

10.2 · Two strips of Pantin's kymograph recordings of muscle contraction of sea anemones, showing the different dynamics of facilitation responses between sphincter (s), parietal (p) and longitudinal (l) muscles in response to trains of electrical stimuli shown in the lower trace. From Pantin (1935b). Courtesy of the Company of Biologists, Cambridge, UK.

around the column, by contrast, the nerve net is sparser and involves numerous synaptic obstacles, necessitating many facilitatory impulses, to ensure transmission; lateral conduction is therefore much slower. Pantin's inspection of the neuroanatomical monograph of the Hertwig brothers on actinians (1879) supports his conclusions based on experiments. The numerous nerve cells with long, relatively thick processes running vertically in the mesenteries are fitting candidates for the faster through-conducting system, whereas the sparse distribution of shorter nerve cells in the column wall can only serve for the slow conduction system. The complex layout and orientation of nerve cells in the tentacles and oral disk, Pantin assumes, provides ample ground for interneural facilitation.

Finally, Pantin turns to the functional and evolutionary thoughts inspired by his findings. While he refrains from idle speculation on the origin of nervous systems, insofar as such speculation would not improve on Parker's ideas expressed in *The Elementary Nervous System*, Pantin is willing to submit his thoughts on the evolving nervous system of sea anemones and cnidarians in general. He presents his view that the system based on interneural facilitation is the basic design of the nerve net, whereas through-conducting systems, whether in the form of the mesenteric-sphinter nerve net of sea anemones or the nerve ring of jellyfish, are to be considered specializations that arose from the basic plan. In fact, he sees the transitional stages of this specialization within the actinian group, from species that possess but a stump of a through-conduction system to others in which it is better developed, and finally to species like *Calliactis*, in which it reached its highest development. As for functional implications, here is Pantin's view: "The effect of this specialization [through-conduction] is to divide the neuromuscular system of *Calliactis* into two distinct parts, the disc, where feeding reactions predominate, and the column; this, with its specialised through conduction and its successive muscular responses of parietals, mesenteries and sphincter, gives a graded series of protective reactions."

After Pantin, the emphasis on the two main behaviours of coelenterates besides locomotion – that is, feeding and protection – would continue to pervade the field of coelenterate neurobiology. Feeding, even in coelenterates, is a complex, stepwise, flexible activity toward which the nervous system must accommodate itself. The variety of nerve cells, both in morphology and orientation, found in the tentacles and the oral disc, as well as the predominance of interneural facilitation, serve feeding behaviour well. On the other hand, protective behaviour against predators or environmental disturbances (such as cocooning or escape responses) must be swift to be effective; hence the role of more rapid, through-conducting nervous pathways. The latter, as specializations, Pantin found to have emerged repeatedly in other invertebrates, the prime example being the giant axons of annelid worms and arthropods (crustaceans and insects), which allow very rapid evasive or escape responses. Pantin is careful to draw the comparison strictly as analogy, not committing himself to the possibility that through-conducting pathways arose independently by parallel or convergent evolution.

Pantin next asked himself whether there were other properties of the nerve net, besides interneural facilitation, that are shared with central nervous systems of higher animals. He tried to answer this in his third paper of the bumper-crop year 1935 by looking at polarity of nerve impulse transmission and the phenomenon of after-discharge; that is, unsolicited excitation following an initial response to stimulation (Pantin, 1935c). Thus far, as we have seen in previous chapters, the consensus was that nerve nets by their very nature are unpolarized. But Pantin, building on a few suggestive observations by Parker and others, challenged this view, aided by his keen sense of observation and outstanding analytical powers. Like Parker, he noted the tendency of excitation of longitudinal muscle in sea anemone tentacles to spread from the point of stimulation inward toward the mouth; that is, preferential centripetal conduction. He found support for this in Pavel Grošelj's neuroanatomical work (1909), in which sensory cells extending loose bundles of long processes toward the oral disc were observed. However, he soon became aware that the reverse polarity – that is, centrifugal conduction – applied to the spread of excitation associated with circular muscle in the tentacles. No polarity of conduction was detected in the column. Pantin concluded that "all the data so far presented seem to be consistent with the centrifugal physiological polarity of interneural conduction combined with an anatomical polarity in the opposite sense." This checkered manifestation of nerve polarity in sea anemones suggested to Pantin that its evolutionary origin must be sought in "the gradual development of differential rates of facilitation between units of the nerve net."

After-discharges are defined narrowly by Pantin for sea anemones as one or more additional contractions that arbitrarily appear after the response of the muscle (marginal sphincter) to a given stimulus (at a point on the column). These are conditions that activate the through-conducting nerve net. Pantin goes to great lengths to discuss possible causes for this phenomenon, but its "fickle" occurrence and other characteristics forced him to dismiss these causes: "At present there is no satisfying explanation for these supernumerary contractions. While the phenomenon resembles superficially the after-discharge of the higher types of central nervous system, its origin, and in particular its relation to the synaptic properties of the nerve net is uncertain."

Notwithstanding the uncertainties, these experimental observations provided satisfying answers to his queries and led Pantin to offer yet new insights on the status of the coelenterate nerve net in relation to the development of the central nervous system.

Whatever view is taken of the supernumerary contractions, their existence together with that of polarity and of facilitation shows that some characteristics of the nervous systems of the higher animals are already foreshadowed in the coelenterate nerve net. It is often said that the nerve net possesses central nervous properties not present in peripheral systems. This is due to the fact that the vertebrate skeletal neuromuscular system remains the standard by which all others are compared. Unfortunately this is very specialised, owing to the high development of continuity of conduction between motor nerves and muscle fibres. In several phyla, as in the coelenterates, neuromuscular facilitation is found. It is an inversion to say that the neuromuscular arrangements in all other phyla show central nervous properties because they do not possess the unique peripheral conduction mechanism of the Vertebrata.

The true relation of the nerve net to the central nervous system of more highly organised phyla is found in the morphological absence of centralisation, and the simple relation of response to stimulus; these are questions of degree of organisation. (Pantin, 1935c)

Having thus illuminated the workings of the nerve net as no one before him, having uncovered interneural and neuromuscular facilitation as central features of this nerve net, and having understood shrewdly where the coelenterate nerve net stood both as a functioning entity and in its phylogenetic relationship with the centralized nervous system of higher, bilateral animals, Pantin moved on to minor aspects of the physiology of the nerve net – what may be regarded as fine-tuning or rounding out the knowledge base.

One of these minor topics concerned neuromuscular facilitation shaping the "staircase effect." Pantin was drawn into it while reading one of Albrecht Bethe's papers, which had appeared while he was still labouring on *Calliactis* in Naples. In that paper Bethe (1935), expanding on his previous work on jellyfish, had traced what Pantin called "a far-reaching parallelism between [jellyfish] rhythmic contractions and those of the vertebrate heart" (Pantin, 1935d). Given Pantin's current findings on *Calliactis*, Bethe's revival of an analogy dating back to Romanes's days instantly aroused his scepticism. Both in the vertebrate heart and in jellyfish, stimulation-induced staircase effects are obtained after removing the pacemaker. For Pantin, however, the analogy stopped there. Whereas in the vertebrate heart "an intrinsic increase of contractility of the whole contractile tissue" accounts for the staircase effect, in both jellyfish and sea anemones, he adamantly insists, evidence shows that the staircase "depends upon neuromuscular facilitation by which each stimulus in a series reaches more and more individual muscle fibres" (Pantin, 1935d).

Pantin took a well-deserved rest after his incredibly productive Neapolitan sojourn, if only to nurse his tuberculosis. When he returned to sea anemone work, he was no longer using Naples as a logistical base, but revisited his lab in Plymouth, where he worked primarily on *Metridium senile* for many years to come. Pantin acquainted himself with *Metridium* as an experimental model by closely studying Parker's work on this genus. *Metridium* differs from *Calliactis* in that its marginal sphincter is much slower and plays no role in the protective retraction of the animal; its longitudinal mesenteric muscle, however, retracts faster than that of *Calliactis*. Increasingly burdened with administrative tasks, both at Cambridge and at the family business, Pantin now sought student assistance. With his first graduate student, D.M. Hall, he produced a paper that revealed the dramatic slowing of muscle contraction and relaxation as well as a pronounced prolongation of facilitation with a lowering of the temperature (Hall and Pantin, 1937). This finding strongly suggests the dependence of these activities on metabolic (enzymatic) processes, but Pantin shied away from any such interpretation.

Next, Pantin recruited another graduate student, Donald M. Ross (1914–1986), a Canadian who had graduated from Dalhousie University with a bachelor's degree and MSc and developed a lifelong research interest in the behaviour of sea anemones. In the work that earned him his doctorate under Pantin's supervision at Cambridge's Zoological Laboratory and in Plymouth, Ross examined the effects on sea anemones of ions known to affect vertebrate neuromuscular systems (Ross and Pantin, 1940). He and Pantin found that magnesium and carbon dioxide had anaesthetic effects on facilitated responses of *Metridium* and *Calliactis*, whereas potassium and calcium enhanced facilitation. High potassium even directly induced muscle contractions, as in the skeletal muscles of vertebrates. The modulation of facilitation by these ions was interpreted by the pair to suggest that neuromuscular transmission was the target. In the paper they came tantalizingly close to intuiting what we now understand as the mechanism of presynaptic facilitation, whereby successive impulses reaching the nerve terminal allow more and more calcium to enter it, thus increasing the number of calcium-dependent transmitter release events.

Switching Gears

In the next decade Pantin turned his primary attention away from the nerve net toward other physiological issues related to the life of sea anemones. He first tackled a topic that had spawned unsatisfactory and controversial papers for a long time: the control mechanism of the stinging cells (nematocytes). With his customary level-headedness Pantin rightly concluded that excitation of nematocyst release depends on contact with the prey but is aided by sensitization to chemicals released by the prey (Pantin, 1942). He was also right in concurring with Parker that nematocytes are independent effectors, but he overlooked the now established role of the nerve net in modulating the sensitivity of these effectors. After the Second World War, Pantin recruited a female student whose entry on the scene heralded a fruitful and highly productive period in his career: Elizabeth J. Batham. She started out with several original papers on sea anemone muscle anatomy (Batham and Pantin 1951), muscle physiology, and motor behaviour (Batham and Pantin 1950a, b, c; 1954), but Pantin soon enlisted her in a new quest for the exploration of the physical nerve net.

Elizabeth Joan Batham, born in Dunedin, New Zealand, on 2 December 1917, was encouraged in her interest in natural science by her mother, while her father tried to pull her toward his own profession, engineering (Jillett, 2001). In the end she went about her university studies conservatively, majoring in English and "home science" at the University of Otago in 1936. Only later did she switch to science, graduating at the same university with a master's degree in botany in 1940, and with Honours First Class in zoology in 1941. Although she was awarded a fellowship to study abroad, the war forced her to postpone taking it. Meanwhile she did some marine biology research at the fledgling Portobello Marine Station near Dunedin on the Portobello peninsula. Finally, late in 1945 the twenty-eight-year-old Betty Batham, as she was known, made her way to Cambridge, where she blossomed as an experimental biologist under Pantin's guidance. After earning her doctorate in 1948 she remained in Cambridge as Pantin's research assistant for the next two years. The series of papers on sea anemones derived from her thesis and assistantship was slow in coming out, appearing only between 1950 and 1954.

The first inklings of Pantin's interest in studying the neuroanatomy of sea anemones are found in his Croonian Lecture delivered in 1952, seventy-six years after Romanes delivered his own. Seeing himself as the intellectual heir of Parker, Pantin entitled his paper "The elementary nervous system" (Pantin, 1952). Although intended as a timely opportunity to review the progress of his research, Pantin's paper nevertheless devoted a fair amount of space to as yet unpublished neuroanatomical results. These results, according to Pantin, stemmed from a work in progress by Betty Batham. Pantin was not trained in neurohistology but he admired the skills of the Hertwig brothers in this field – he playfully invented the "Hertwig Rule," according to which the Hertwigs are always right (Russell, 1968). He found in Batham a resource person who had some knowledge of histological techniques and, more important, had the determination and patience to see her neuroanatomical project through.

Notwithstanding his great admiration for the Hertwigs, in assessing the literature on the neuroanatomy of sea anemones, Pantin found that they as well as Grošelj and Havet had fallen short of providing an entirely satisfying picture of the actinian nervous system. He was especially critical of the silver impregration used by Havet, which can produce precipitates that mimic the appearance of nerve cell processes. Not wanting to repeat the errors of their predecessors, Batham and Pantin assessed the problem painstakingly from both the theoretical and practical angles:

> To be certain that structures should be correctly identified as nerve, a variety of nerve stain techniques were used so that each could check the other. Moreover, a prolonged search was made for methods which gave good fixation, so that cellular structure could be compared with that of nerve cells in other animals. Finally, care was taken that, by comparison with preparations stained with Mallory's [background] triple stain, etc., the possibility that muscle fibres or fibrous structures in the mesoglea might be mistaken for nervous elements was excluded. That is, a body of circumstantial evidence was built up not only to prove that certain structures resembled nerves in various ways,

but also that they did not resemble that which was not nerve.
(Pantin, 1952)

Applying these guidelines to *Metridium senile*, Batham and Pantin
arrived, at long last, at a set of techniques that allowed them to achieve
their goals. Their best results were obtained thanks to a new silver impregnation technique introduced by histologist William Holmes of
Oxford University (Holmes, 1942, 1943). Holmes, although described
by George Mackie in his unpublished autobiography as effete and
laid-back, and by mid-life in the 1950s fallen from the highpoint of
his career, had been a brilliant zoologist. He had made improvements
on the old techniques of Golgi and Ramón y Cajal, and inspired new
and succeeding generations of neurobiologists to apply them. In addition, Batham and Pantin had some success with the reduced methylene
blue technique.

The Pantin-Batham duo was most keen to discover what stood
behind the physiologically characterized through-conducting system
suspected to reside in the mesentery. They were elated to discover a
dense plexus of bipolar and tripolar nerve cells on the side of the mesenteries where the retractor (longitudinal) muscle resided (Fig. 10.3,
right). The density and the relatively large size of the nerve fibres gave
this nerve net all the trappings of a through-conducting system. Across
the mesoglea, on the mesenteric side bearing the radial muscle, only
a loose nerve net was present. Unmistakable contacts *en passant* were
observed between bipolar cells, reflecting possible synaptic zones. On
this basis Batham and Pantin definitely rejected the syncytial theory,
confirming observations of more than a decade earlier in both sea
anemones and jellyfish that discontinuity prevailed between nerve cells
(Woollard and Harpman, 1939). They also observed contacts between
fibres of the nerve net and muscle cells which resembled the end-plates
of vertebrate neuro-muscular junctions. Their failure to observe unequivocal nerve fibres crossing the mesoglea contributed to closing a
contentious issue that harked back to the nineteenth century and was
still alive as recently as Havet's time.

Batham and Pantin also achieved a break with the past in the technical
field. So far, microscopic representations of coelenterate nervous structures in publications had principally been drawings. Although microphotography was introduced as early as the mid-nineteenth century, its
use was largely restricted to textbooks on histology or microbiology

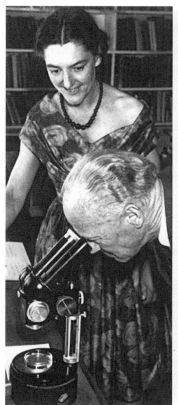
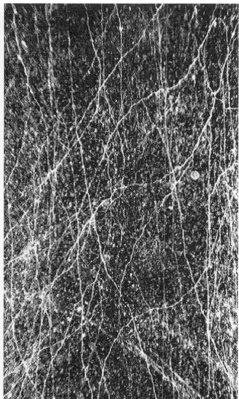

10.3 · (*Left*) Elizabeth Batham in 1962 in her Portobello laboratory with the chancellor of the University of Otago, Hubert Ryburn. From Jillett (2000). Courtesy of the Hocken Collections/Special Collections, University of Otago, Dunedin, New Zealand. (*Right*) A low-power microscopic view of the dense mesenteric nerve net of *Metridium senile*. From Pantin (1952). Courtesy Royal Society of London.

(Breidbach, 2002). By the early 1890s, however, microphotographs increasingly found their way into scientific journals, but the prohibitive costs of reproduction limited their affordability only to journals with a wide readership. For this reason medical and paramedical journals were the primary users, whereas zoological journals with smaller audiences held back until the 1930s when microphotography became cost-effective. The 1939 article by A.H. Woollard and J.A. Harpman appears to be the first in which microphotographs of coelenterate neurohistological material were inserted, although their technique and images were criticized as inadequate by Pantin and his co-workers (Batham et al., 1961). By the early 1950s, when Pantin's Croonian Lecture was published, photographic illustration of neurohistological observations had become the norm. With drawings there was no way to assess the quality of the techniques used by authors; microphotographs allowed readers to judge the reliability of the authors' interpretations by directly observing the quality of the histology. Pantin and Batham understood the limitations of photographic representation as well as those of histological techniques, and as a result their photographic images (see Fig. 10.3) went far in conveying a physical sense of the sea anemone nervous system vividly and convincingly.

The publication in full of Batham's neurohistological work – and with her as first author – came out only eight years after the preliminary descriptions in Pantin's Croonian Lecture (Batham et al., 1960). The delay was due to Batham's return to her homeland early in the 1950s to take charge of the Portobello Marine Station. The marine station had been championed by the Australian and future Nobel Prize neurophysiologist John C. Eccles (1903–1997), who was a professor at the University of Otago from 1945 until he was called back to Australia in 1952 to fill a position at the John Curtin School of Medical Research of the Australian National University. The tremendous workload entailed by the renovation of the station buildings, the acquisition of new infrastructure and equipment, and the accomplishment of routine administrative tasks left Batham no time for her own research activities in the years to come. In the end Pantin enlisted the help of Elaine Robson, one of his graduate students who had acquired histological experience in depicting the epithelio-muscular system of sea anemones (Robson, 1957), to round up Batham's neurohistological work and see it through to publication. In the years immediately following the 1960 paper, Robson added new observations on the actinian nervous system (Robson, 1961, 1963).

Batham was to meet a tragic end. As her biographer noted (Jillett, 2000), she never married and in her later years "she shared her leisure time, triumphs and frustrations with her widowed father." Early in 1974 she became ill and resigned her post as director of the marine station to take up residence in the New Zealand capital, Wellington. "In early July," Jillett continues, "her car was found abandoned near the seashore at Seatoun, Wellington. Mystery, rumour and conjecture surrounded her disappearance, and although there were a number of reported sightings, none were verified. Betty Batham was later presumed to have drowned off Seatoun on or about 8 July 1974." She was only fifty-six and had never returned to Cambridge.

For all his positioning as the intellectual legatee of Parker, Pantin seemed to have studiously avoided the very topic that was foremost in Parker's mind: the question of the origin of nerve cells and nervous systems. This seeming omission had everything to do with the way Pantin's mind worked. The recurrent theme of his sea anemone papers is the emphasis not only on function and behaviour as such, but in particular on the animal's workings as responses to engineering challenges posed by the environment and the "necessities of life." This vision was ably encapsulated by his favourite phrase, describing the sea anemone as a "behaviour machine." His Croonian Lecture of 1952 articulated this theme eloquently:

> The universality of physiological properties is a fundamental character of living organisms. It is not primarily a consequence of common evolutionary descent but of the unique limitations of the properties of matter and energy; so that for organisms, just as for engineers, any functional structure can only be built in a limited number of ways. But this does not mean that any functional machine must be built up in only one way. As in engineering practice, there are usually several ways in which a functional objective can be achieved. And it is precisely the discovery of these unexpected alternative mechanisms which has so often marked advance in biology. So that perhaps the value of a study of the elementary nervous system depends not only on the hope which its structural simplicity gives for a complete mechanical analysis, but also on the way in which it brings to light the variety of physiological phenomena available for the construction of even the simplest behaviour machines. (Pantin, 1952)

But Pantin made an exception to his reticence on evolutionary matters in an essay published in 1956, in which he took issue with Parker's grand scheme of the evolutionary development of the first nervous systems (Pantin, 1956). We may recall that Parker envisioned the emergence initially of independent effectors, such as the "myocytes" of sponges, to be followed later by sensory cells connecting to muscle cells differentiated from myocytes, and finally by the emergence of "protoneurons" interposed between sensory and muscle cells. Pantin objected to Parker's focus on cells rather than tissues whose functioning requires integrative processes, as his own observations led him to conclude; the self-contained time capsules of cell type emergence hypothesized by Parker would only produce unfit individuals and chaotic behaviour.

Pantin and Batham's observations of spontaneous muscle activity in sea anemones led Pantin to suggest that muscle sheets imbued with endogenous activity had preceded nervous tissue in evolution. Nerve nets, in his view, arose to support specialized effector functions such as that of the marginal sphincter of sea anemones, where through-conduction is necessary. Whereas muscle sheets had originally subserved spontaneous activity, Pantin thought, nerve nets had evolved to support integration of reflex actions.

Although Pantin mused on the propitious time in functional terms for the emergence of nerve nets, he remained silent on specifics of the origin of nerve cells: that is, from what pre-existing cell type did they evolve and how? This critical question may have fascinated Parker such that he strained to find an answer in his three-step theory, but Pantin's circumspect mind approached such questions cautiously, being more inclined to ask questions that an engineer can answer than be drawn into what he would consider speculation. Lacking relevant data from his observational experience with sea anemones, he may have felt forbidden to enter this slippery territory.

In his last review of the "coelentetrate behaviour machine," his swan song published just two years before his death, Pantin again confines himself to tractable evolutionary issues and discusses the adaptive constraints of coelenterate nerve nets shackled in diploblastic tissue organizations; that is, in the two-dimensional sheets of the ectoderm and endoderm, as opposed to the three-dimensional arrangement of triploblastic animals, in which the mesoderm is added. In the evolutionary move from a diploblastic to a triploplastic nervous system, Pantin envisioned the selective pressures to upgrade the sensory system as well as to reorganize the outlay and interconnectivity of the nerve cells:

> It seems probable that many actinians have at least [100,000] nerve-cells in their nerve nets. But their sources of information are far more restricted. There are no sensory instruments; even the eyes and otocysts of medusae do not seem designed for the

abstraction of complex information. Actinians must rely upon tactile, mechanical, and other such simple sensory information, at the surface of the body. The really important first step in the evolution of advanced behaviour is the replacement of simple stimuli or simple patterns of stimulation for the genesis of behaviour, by an abstracted model of objects in a real world – that same real world of objects with which our own naive realism endows the world. An ant reacts to stone and so do we, rather than reacting to the very different initial sensory inputs by which these are detected by ants and men ...

Nevertheless, the properties of the nerve net, as they are becoming elucidated, seem to be exactly the pre-adaptive features which, combined with exteroceptive sensory instruments and a three-dimensional development of the net, could give what is required. It is so significant that both [an insect] and an actinian may have about the same number of nerve-cells. What they can do with their predictor machinery depends upon how that number of nerve cells is organized in each case. (Pantin, 1965)

At the time of this review paper, based on a talk delivered at a symposium on coelenterates under the auspices of the American Society of Zoologists in which several of his past students also participated (Donald Ross, D.M. Chapman, Betty Batham, G.A. Horridge, Elaine Robson), Pantin had recently been diagnosed with leukemia (Russell, 1968). The following year he relinquished the chairmanship of the Zoology Department at Cambridge, which he had held since 1959. He died on 14 January 1967.

It can certainly be said that Pantin almost single-handedly revived the field of coelenterate neurobiology just as Parker had done a generation earlier. The investigators who had succeeded Parker in German-speaking Europe may have made significant contributions, as we saw in the previous chapter, but coelenterate neurobiology research was at best a stepping stone or a hiccup in the course of their respective careers. Pantin, on the other hand, made coelenterate biology the mainstay of his research career, and his lab developed a school of thought that carried over to the next generation. Indeed it belonged to the next generation to usher in the modern era.

CHAPTER ELEVEN

The Modern Era

Research for me is very much like the sort of aimless tinkering
that one does when young.

George O. Mackie

After the Second World War, new investigative techniques helped
breathe life into the field of coelenterate neurobiology, providing answers to questions that had defeated generations of researchers down
to Carl Pantin himself. Electron microscopy and electrophysiology
stand out as techniques that allowed the field to take giant leaps in
the next few decades. Carl Pantin's own students were the first to avail
themselves of the developing techniques: in 1953 Adrian Horridge produced the first extracellular recordings of action potentials in the jellyfish *Aurelia*, and in 1957 Elaine Robson used electron microscopy to
provide the first description of the fine structure of various *Hydra* cells.
An important offshoot of the new discoveries made possible by the
application of modern tools was a revised narrative of how nerve cells
and nervous systems may have arisen.

The Electrophysiologists: Wizards of the Trade

George Adrian Horridge was born in 1927 in Sheffield, England, to a
mechanically minded father who, according to the son's own recollections (Crompton, 2003), taught him to be entrepeneurial and to be
skillful with his hands. The chaotic war years, with bombardments that
seriously encroached on his high school studies, also taught him to
be resourceful. These qualities came in handy when in 1946 he won
a scholarship to Cambridge as one of the up-and-coming bright kids
of his generation. As an undergraduate he started out concentrating

in zoology, but developed interests in biochemistry and physiology. After being offered a PhD studentship in 1950, again at Cambridge, he became interested in Pantin's research, and the latter accepted to act as his supervisor. However, Horridge was pretty much on his own, as he explains:

> As a research student you had to find yourself a topic and also find somebody who would supervise you in it. Then you had to collect your apparatus. You built it yourself or scrounged it from somebody who had just finished or some other member of staff who had got some stuff in his cupboard, and collected things from various places in the lab such as the chemical room in the corridor, the workshop and the electronics workshop. So eventually, having assembled everything you thought you wanted, you might start some experiments. (Crompton, 2003)

It did not help that Pantin went on a research trip to Brazil for fifteen months, but Horridge was quickly learning to be his own man, relying on his bright mind and "street smartness." He spent time tinkering with various animals and ideas for experiments at the Plymouth Laboratory, where Pantin had come of age as a biologist. He set his sights on *Aurelia* as an experimental model and chose his research project in a very roundabout way:

> Then I went back to Cambridge, where I was able to use a phase-contrast microscope for the first time to look at transparent jellyfish, *Aurelia*, which I had collected by going on my motorbike to Brancaster Staithe, on the Norfolk coast. (The phase-contrast microscope was another recent invention at the time, and [Lord] Victor Rothschild had purchased some microscopy equipment out of his own private money.) The nerves of these jellyfish have a different refractive index, so although they are completely transparent you can see the nerves under phase. There they all were, living nerves. A very lucky break, I thought.
> I spent the whole of that first year learning electronics – building electronic equipment and playing about, recording from snails and odd things that had some nerves in them. I built first of all a power pack, then a multivibrator stimulator and a variety

of other stimulators with neon lamps in them, which flashed very slowly, and then a DC amplifier that gave me enormous trouble because it was totally unstable. After that I got a radar oscilloscope, which had a blue screen and fast time bases, and produced circles on the screen, but I changed the time base and the amplifiers inside so I had a new oscilloscope with a green screen that gave a long fluorescence. I had to learn a tremendous amount of stuff for the complete physical techniques for analysing nervous systems, but one result was that later whenever we had a problem in the lab I was able to solve it. Once you've made all the equipment and discovered the little details of exactly how to record nerve impulses, I suppose you become more confident. (Crompton, 2003)

Horridge soon had the opportunity to test the whole scope of his newly found self-confidence by spending the early summer of 1952 at the Marine Station of Millport in Scotland, where he found a ready supply of jellyfish. With his electrophysiological equipment transplanted to Millport, Horridge was able to pick up the larger nerve fibres of *Aurelia* with a platinum electrode and record individual nerve impulses. He found that each single nerve impulse was followed by a swimming contraction (Fig. 11.1), and that these neural events showed neither polarization nor interneural facilitation or conduction decrement. Thus the existence of neural activity similar to that of higher animals, suspected since Romanes's experiments but never formally proven, was finally demonstrated thanks to the savviness of the post-Pantin generation. The twenty-five-year-old's pioneer work was rewarded a few months later by its publication in the prestigious journal *Nature* (Horridge, 1953), and a complete account appeared soon afterward (Horridge, 1954).

This crucial breakthrough soon had an entraining effect on other coelenterate neurophysiologists. Some confirmed and added to Horridge's findings in other scyphomedusae (Yamashita, 1957; Passano, 1958; Passano and McCullough, 1960), while others explored hydrozoans (Josephson, 1961, on the hydroid polyp *Cordylophora*). Belatedly, reflecting the difficulty of handling the slimy sea anemones, recordings of muscle (Josephson, 1966) and nerve net impulses (Robson and Josephson, 1969) were achieved in Pantin's pet animals, helping to valid-

11.1 · Horridge's oscilloscope traces of his recording of the spike nerve potential (upper trace) followed by the swimming contraction (lower trace) in the jellyfish preparation. From Horridge (1953). Courtesy Nature Publishing Group.

ate much of the previous work of the Cambridge laboratory. But with few exceptions Horridge's followers were up-and-coming Americans. The new leaders in the field belonged to a country where electrophysiological recordings were pioneered and technically refined by leaps and bounds.

One landmark discovery that was to alter neurophysiological thinking significantly, however, originated from an Anglo-Canadian trained in electrophysiology by the Americans Leonard Passano and Robert Josephson. George O. Mackie was born in Lincolnshire, England, in 1929, but spent much of the Second World War as a child refugee in British Columbia. He studied at Oxford, where he earned his bachelor's degree in 1953 and his PhD in 1956 based on a thesis on siphonophores. Having fond memories of his childhood years in Canada and being faced with poor job prospects in postwar England, he took a position as a lecturer at the University of Alberta in Edmonton. While at Edmonton he undertook his first investigation of coelenterate neurobiology by revisiting the neurohistology of the siphonophore *Velella*, last studied by Carl Chun and Karl Schneider late in the previous century. Using the staining method of William Holmes, his thesis supervisor at Oxford (mentioned earlier), Mackie obtained such stunning visual results that the journal editor to whom he submitted his paper was a bit incredulous, as he explains:

The Editor, John Baker, wrote back to tell me I should insert a footnote saying that "ink had been added to the pictures," i.e. that I had retouched them to enhance the contrast! Instead of taking this as an unintentional compliment, I was outraged and hurt and promptly sent him my best slide to show him how wrong he was. I was hungry for recognition and for me the Velella work was my widow's mite, priced above rubies. Baker was impressed by the slide but then instead of returning it he put it up on a shelf and forgot about it, and was petulant and irritated at my repeated efforts to get him to look for it, but he eventually found it and sent it back. I may just add that Baker was fussy and irascible but he took an interest in all of us and spotted me as a microscopist before I knew it myself. (Mackie, *An Unstructured Affair*, electronic document at University of Victoria website)

The paper was published (Mackie, 1960), but to its author the process was a sobering introduction to the publishing experience in coelenterate neurobiology.

In 1964 Mackie took the opportunity of a sabbatical leave at the Villefranche marine station in France to try his hand at his friend Passano's electrophysiology set-up there, using local siphonophores as experimental animals. This foray led to Mackie's discovery of epithelial conduction, whereby impulses generated in epithelial cells are transmitted directly from cell to cell throughout a nerve-free epithelium. The paper that resulted from this work (Mackie, 1965) was the first to demonstrate this phenomenon in the animal world. Suspicions that non-nervous (neuroid) conduction participated in the behaviour of coelenterates had been aired in the past by investigators we have profiled in earlier chapters (Hadži, Parker, Pantin), and more recently by Horridge, Passano, and Mackie himself. But none had come up with clean experimental evidence until Mackie's work, his first as an electrophysiologist. Later, Mackie and Passano showed that epithelial conduction occurred widely in hydrozoans (Mackie and Passano, 1968), and others extended the range of animals exhibiting the phenomenon.

C. Ladd Prosser, a leading comparative physiologist at the time, used the word "revolutionary" to describe Mackie's discovery. And it was. From then on one had to accept that in some coelenterates behaviour was coordinated by two parallel conducting systems, one epithelial and the other nervous. Beyond the functional implications, however, Mackie soon realized that evolutionary implications could also be contemplated, and rewardingly so. In an essay paper published five years

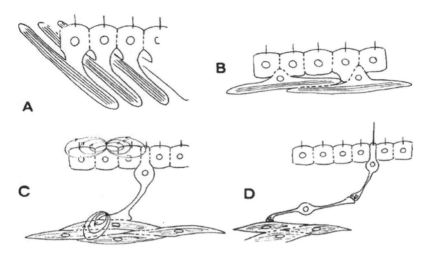

11.2 · Mackie's proposed steps for the early evolution of nerve and muscle cells. From Mackie (1970). Courtesy of the University of Chicago Press.

after his discovery (Mackie, 1970), Mackie framed the issue quite differently from past exponents: instead of asking how nerve cells arose in evolution, he purposedly asked: "How did conduction evolve in multicellular organisms?" Mackie's thoughts in answer to this question essentially amounted to a completely original construct of how nervous systems could have emerged.

Mackie was aware that Parker and Pantin were not the only coelenterologists to have addressed this issue squarely before him; his contemporaries among the comparative neurobiologists had done so as well. George Bishop (1956) and Harry Grundfest (1959), for example, both unfamiliar with coelenterates, regarded "primitive" animals as useful models to explore the gradations or hits-and-misses strewing the evolutionary trail to the formation of the first nervou systems. This sort of argument, Mackie felt, smacked of condescension, for, "It does not follow … that the mechanisms in so-called primitive forms represent tentative or incomplete attempts at the mechanisms found in higher animals" (Mackie, 1970). For Mackie, graded conduction is no more primitive than "self-propagating" conduction; rather, both represent different adaptive strategies to address different functional contexts. This view went a long way toward colouring his evolutionary hypothesis.

At the outset Mackie takes a relativist view of cell excitability and impulse conduction. Metabolites, including substances that may affect trophic responses of the plant, are transported from plant cell to plant cell. But in the few cases, where speedier responses are needed – prey capture by insectivorous plants, for instance – transmembrane ionic mechanisms have developed to make some cells excitable, and the bridges between cells that serve to transfer metabolites act also as paths for the conduction of electrical impulses from cell to cell. For Mackie this scenario repeated itself in animals. The starting point in a coelenterate prototype would be a specialized epithelium, as he explains:

> I envisage the starting point for a metazoan conducting tissue as something resembling a myoepithelial tissue sheet in a coelenterate, a sheet consisting of a single type of cell, and each cell having its own contractile myofibrils (Fig. 11.2, A). The cells would not resemble neurons morphologically. They would conduct graded or all-or-none events, contracting as they conducted, and their ability to do so would depend on the presence of pathways of communication between the cells, pathways which also serve for metabolic exchange. As in plants, the effective conducting unit would presumably be a group of cells rather than single cells.
>
> From such a primordial epithelium, whose cells were capable of reception, transmission, and contraction, specialized muscle tissues would have arisen by the sinking in and isolation of contractile cells (Fig. 11.2, B). Systems of neurons and neurosensory cells would have become segregated in the same way, losing their contractile component but providing for excitation of the evolving muscle tissues and for the coordination of their responses (Fig. 11.2, C–D). In some places, the primordial sheet would retain its conducting ability, either along with contractile ability or without it, giving us myoid and neuroid epithelia respectively, such as we find in the coelenterates. Elsewhere, conducting ability would be lost, as the presence of nerves would make neuroid conduction unnecessary. Neuroid conduction could, however, be reestablished in tissues secondarily. (Mackie, 1970)

This was as close to a plausible scenario of the evolutionary emergence of nervous systems as the knowledge base of 1970 would allow.

Mackie deduced from his own scenario that neuroid systems such as those based on epithelial or musculo-epithelial conduction are by nature no more specialized than undifferentiated nerve nets, in which all constituent neurons are similar and cumulate sensory, transmitting, and motor functions. Once cell specialization started to occur within neuroid systems, as seen in steps C and D (Fig. 11.2), and the cell modifications turned out cell units identifiable as neurons, the identity of neuroid systems as epithelial conductors faded. As this process went on, sensory cells, intereurons, and motor neurons became individualized in both their morphological and functional identities. Two years earlier, Horridge (1968) had alluded to a scenario of specialization in that vein.

As this scenario unfolded in coelenterates, Mackie noted, neuroid systems and nerve nets coexisted at first in the same animal, each fulfilling its functional role, as still exists for all to see in hydrozoans and ctenophores. However, neuroid systems largely disappeared in other forms, and in today's scyphozoans and anthozoans nerve nets have taken over full responsibility for initiating and transmitting signals meaningful for behavioural activities. But Mackie stressed that both neuroid systems and nerve nets likely survived or evolved independently in one form or another in animals with more elaborated nervous systems. This conjecture was based on scant information, but future investigations proved him right.

In the years that followed Mackie's landmark evolutionary formulation, electrophysiological tools became increasingly sophisticated, but his basic hypothesis was not shaken by these developments. So far electrophysiological recordings consisted of applying electrodes to the surface of coelenterate nerves and observing spike activity. Alan Hodgkin and Andrew Huxley had shown that, by inserting electrodes inside the giant axon of the squid, one obtained a record of the action potential in all its complexity, a finding that allowed them to unravel the ionic basis of these potentials. This was the effort that earned them the Nobel Prize in Physiology or Medicine. At the time they were able to accomplish this feat by selecting cells of inordinately large size, such as those of the the squid giant axon involved in fast escape reactions, so they could penetrate the cell stably without damaging it. Similarly, the first attempts to record intracellularly in coelenterates were made on large epithelial-like cells: Horridge (1965) claimed to have recorded action potentials in ctenophore comb-plate cells that were associated with ciliary beating, but these were later found to be artifacts (Tamm, 2014). Andrew Spencer, a student of Mackie, recorded, from the musculo-epithelial cells of a siphonophore subthreshold, graded potentials that failed to induce contractions, and above-threshold action potentials that were always followed by muscle contractions (Spencer, 1971). In all cases these represented examples of neuroid conduction where only all-or-nothing action potentials were conducted non-decrementally over the epithelial sheet.

The first successful intracellular recording of nerve cells in coelenterates followed the pattern set by the squid giant axon, in that it was accomplished also in a giant nerve fibre, that of the siphonophore *Nanomia*. The feat was claimed by Mackie himself (1973), who had made the original discovery of the existence of the unusually large neurons. He found that these action potentials possess all the trappings of nerve action potentials in higher animals except for the lack of sensitivity to tetradotoxin, which was known to block action potentials thanks to its blocking action on sodium entry into the cell. Mackie inferred that the action potentials of coelenterate neurons may be sodium-independent, but later work showed that sodium was usually required but that the transmembrane channels allowing sodium in are insensitive to tetradotoxin. This property sets coelenterate neurons apart from those of higher animals, so that it was reasonable to assume that tetradotoxin sensitivity was acquired later in neuronal evolution.

Advancement of knowledge of the electrophysiological properties of coelenterate neurons at the cellular level was a haltingly slow business, however, as it took four years for the next contribution to materialize. Peter Anderson, a postdoctoral student in Mackie's laboratory in Victoria, British Columbia, was able to confirm in the hydrozoan jellyfish *Polyorchis* that the intrinsic excitability of coelenterate neurons conformed to that of neurons from higher animals, including the reception of excitatory synaptic inputs from other neurons (Anderson and Mackie, 1977). But the intracellular recordings also yielded startling new information. The neurons impaled by the microelectrodes were located in the inner nerve ring and generated the swimming pulses in the form of bursts of action potentials. Although the jellyfish possesses eyes (ocelli), the neurons of the inner ring proved to be directly sensitive to light, and this property was linked to the increased pace of swimming when a shadow hovers on top of the jellyfish – an obvious escape reflex. But the most remarkable finding was that these neurons were electrically coupled to each other in a manner that presents a striking analogy with epithelial conduction. It suggested that Mackie's presumed evolutionary process, by which some epithelial cells sank and transformed into neurons, preserved in the newly formed neurons the original electrical coupling of the epithelial cells from which they were presumably derived.

Electron Microscopy and Immunohistochemistry: A Refinement of the Picture

Improvements in microscopic visualization technology and in tissue preparation led to higher resolution neuroanatomy and to the molecular typing of neurons. Electron microscopy promised to reveal the microstructure of objects at much higher amplification and resolution than then-current light microscopy. Electron microscope prototypes had emerged in Germany in the early 1930s, notably through the stewardship of the Siemens Company, and at the University of Toronto later in the 1930s, but their usefulness for biology remained unfulfilled. For neurobiology, the first breakthough came after the Second World War, thanks to MIT scientists Eduardo De Robertis and Francis Schmitt, who discovered the presence of microtubules in axons (Rasmussen, 1997). It was not until 1954 that the synapse, the existence of which had been suspected on physiological grounds by Charles Sherrington late in the nineteenth century, was spotted by electron microscopy as a tractable physical entity.

Although earlier attempts had met with disappointing results, a clear view of the synapse was achieved in the rat central nervous system (Palade and Palay, 1954) and in frog and earthworm neural elements (De Robertis and Bennett, 1955). It was seen to be composed of a presynaptic nerve ending in which synaptic vesicles reside and, across a small gap, a postsynaptic element belonging to another nerve cell. The apposed presynaptic and postsynaptic membranes showed a thickening relative to nonsynaptic zones, and there was a tendency for synaptic vesicles to gather at the thick presynaptic membrane. De Robertis and Bennett were quick to see in the synaptic vesicles the storage sites of neurotransmitters, such as acetylcholine and adrenaline, formed in the cell body and transported in vesicles along the axon to the nerve ending.

It is reasonable to assume that when Elaine Robson left Pantin's laboratory around 1957 to join colleagues in the United States for an electron microscopic examination of *Hydra*, she was looking forward to observing nerve structures. But the report of their investigation (Hess et al., 1957) made no mention of nerve elements, suggesting that their sparse distribution made it difficult to recognize them in the ultrathin tissue sections required for electron microscopy. In jellyfish some of the nerve tissue is conspicuously condensed in marginal ganglia or nerve rings, and thus less likely to be missed in electron microscopic surveys. Horridge took advantage of this peculiarity when he produced the first ultrastructural study of the nervous system of a coelenterate, with the assistance of a student who became proficient in the new technique during a sojourn at the electron microscopic facility of the University of Iowa (Horridge and Mackay, 1962). Again, after Robson, Horridge found that facilities in the United Kingdom were not yet available and the United States had taken the lead in electron microscopy as in many other technologies.

To Horridge and Mackay the most striking features of what they saw was that jellyfish neurons possess all the attributes of classical neurons except the possession of glial sheaths around the nerve processes, and that synaptic vesicles generally face each other on the pre- and postsynaptic sides (symmetrical synapses). Horridge and Mackay, strangely, did not address directly in what way their observations helped resolve once and for all the issue of continuity versus discontinuity of coelenterate nerve cells, although they may have felt that the depiction of synapses where extracellular space was clearly seen between the plasma membranes of adjoining neurons – the synaptic cleft – was self-explanatory in this regard.

Their electron micrographs distinctly show a crowding of vesicles, some of which harbour a dense core inside, at the synaptic membrane (Fig. 11.3, left), and they estimated that anywhere from five hundred to two thousand such vesicles are present at each synapse. The presence of symmetrical synapses (Fig. 11.3, left) was an important discovery, because it was consistent with the non-polarized nature of conduction in coelenterate neurons. However, perhaps echoing the current view of contemporary invertebrate neurobiologists, Horridge and Mackay

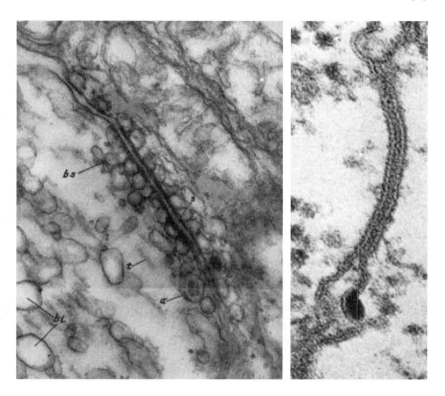

11.3 · Electron micrographs showing a chemical synapse in the scyphomedusa *Cyanea* (*left*, from Horridge and MacKay 1962), and a gap junction between two epithelial cells of *Hydra* (*right*, from Hand and Gobel 1972). Courtesy of the Company of Biologists, Cambridge, UK (left) and Rockefeller University Press (right).

failed to deduce the correct meaning of their discovery. They discarded the possibility of chemical transmission, which the presence of synaptic vesicles strongly suggested. They argued thus:

> Synapses so far described with vesicles on both sides have been shown to be electrically transmitting, i.e. the presynaptic action current is large enough to excite the postsynaptic fibre directly. Examples are the septate synapses in the earthworm and crayfish giant fibres, synapses between giant fibres and large motor axons in the crayfish (Hama, 1961); of these, each except the last mentioned transmit in either direction. Although a delayed chemical

transmission cannot be ruled out, the electrical transmission throws into doubt the possible function of synaptic vesicles as agents of the most rapid transmission of excitation across these synapses, and suggests that the vesicles have an additional function, possibly a trophic one in maintaining the existence of the synapses. (Horridge and Mackay, 1962)

Horridge and Mackay further argued that the synaptic cleft was narrower than that of chemical synapses, and therefore opened up the likelihood that the action current can flow directly from neuron to neuron at such synapses. But future investigations were to show that the scyphozoan jellyfish they were investigating – *Cyanea* – had no electrical synapse and that all synaptic transmission must be chemical in nature (Anderson, 1985; Anderson and Grünert, 1988).

Where epithelial conduction occurs, as in hydrozoans, electron microscopy provides the adequate resolution to spot physical evidence of intimate junctions between cells such as to allow electrical connectivity. Such contact zones between cells that were suspected of transmitting currents were first called nexuses; their ultrastructure was first observed in crayfish and squid nerves (Robertson, 1953), and their ability to translocate small molecules was elucidated in the mid-sixties (Loewenstein, 1966). It was found that an infinitely small gap space existed between the apposed membranes at the nexus, a mere two or three nanometres across. These patches of "gap junctions," as the low-resistance intercellular pathways became known, contain what were called subunits, and these subunits were later found to be micro-channels (called connexons) crossing the membranes, thus accounting for the very low resistance at these junctions. In 1972, Arthur Hand and Stephen Gobel, electron microscopists at the National Institute of Health in Bethesda, Maryland, were able to observe classical gap junctions between epithelial cells of the ectoderm and endoderm of *Hydra* (Hand and Gobel, 1972) (Fig. 11.3, right). In epithelio-muscular cells, they found gap junctions not only between the cell bodies but also between the muscular extensions of these cells. As the authors interpreted these junctions as the probable substrates for intercellular coupling, the physical conduit for epithelial conduction in hydrozoans was thereafter validated.

Jane Westfall, another electron microscopist, but one who was devoting her career to the study of coelenterate ultrastructure, next provided clear evidence that gap junctions existed also at neuromuscular junctions and between neurons in *Hydra* (Westfall et al., 1980). Later, hydrozoan jellyfish neurons were also confirmed to be interconnected by gap junctions. Because the connexons of the gap junctions form, as it were, tunnels giving access to the apposed neuron, it is conceivable that such interconnected neurons could be seen as syncytial. However, the connexons only allow the passage of small molecules and ions in addition to action currents; cytoplasm cannot flow through the connexons, and the continuity of cytoplasm between neurons had so far been a defining criterion for syncytial systems. One can therefore only regard neurons interconnected by gap junctions as being endowed with partial continuity.

This is an important point, in view of the often revisited controversy over syncytiality, continuity, or discontinuity in this historical narrative. The advent of electron microscopy in the second half of the twentieth century had the ironic consequence of revising the now entrenched paradigm of discontinuity between neurons, which had become accepted also by coelenterate neurobiologists. But observing static structures such as gap junctions is a poor substitute for actually seeing evidence of flow of material between the neurons. The year after Westfall provided the first evidence of gap junctions between hydrozoan neurons, Andrew Spencer exploited the use of a fluorescent dye, recently found to spread through gap junctions, to visualize a network of electrically interconnected neurons in the jellyfish *Polyorchis* (Spencer, 1981). This was compelling and graphic evidence that a continuity of sorts was at work in hydrozoan nerve nets. In contrast, the discontinuity theory continued to apply to the other classes of Cnidaria, scyphomedusae and anthozoans.

An interesting twist was added to this story. In Mackie's 1960 paper on the neuroanatomy of the nervous system of the siphonophore *Velella*, two superimposed nerve nets are depicted, an "open" nerve net in which the nerve cells have thin fibres and interconnect by contacts, and a "closed" nerve net in which the fibres are thicker and appear to form a syncytium. Mackie proposed a process by which such a syncytial system might arise, where an "adhesion bridge" of cytoplasm develops

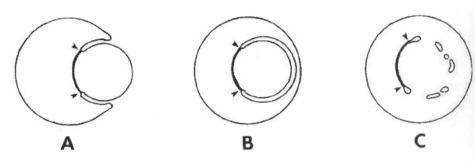

11.4 · Diagrams showing the reconstructed process by which a nerve fibre is engulfed by another as the gap junction keeps them attached to each other. From Mackie et al. (1988), Fig. 5. Courtesy John Wiley & Sons.

from a contact zone between two nerve fibres and stretches as the two fibres move apart from each other. To Mackie this was a coelenterate take on the syncytial process of some invertebrate neurons whereby growing axons fuse with each other to form a giant axon and multinuclear neuron, such as was known to exist in squids, crayfish, and annelid worms. The idea was that, in the absence of insulating sheaths, such fibres conducted impulses faster than small individual fibres and were better adapted to escape reactions. It was, in a sense, a jellyfish answer to this challenge, one that was dictated by the peculiar layout of nerve nets.

Mackie foresaw that the issue of syncytiality could only be resolved by electron microscopy; his light microscopic images, however exceptionally clear they appeared for 1960, were not decisive. It took him twenty-eight years to submit the *Velella* nervous system to ultrastructural analysis (Mackie et al., 1988). Electron microscopy revealed that Mackie's hypothesis based on adhesion bridges was invalid. Instead, he found that the formation of "giant" neurons in the closed nerve net was the result of engulfment of one nerve fibre by another (Fig. 11.4). This was made apparent by the presence of loose pieces of gap junctions inside the large fibres; ostensibly, the original nerve fibres had been connected by gap junctions, thus forming an electrically coupled and continuous nerve net; but the engulfing process, which resulted in a partly syncytial system, had the side effect of assimilating the now useless gap junctions, which were eventually digested. Mackie assumed that the giant neurons of the jellyfish *Polyorchis*, discovered by Spencer in 1979, were formed in a similar manner.

Investigations designed to find evidence for the presence of neurotransmitters in coelenterates by far preceded the discovery of classical chemical synapses in these animals. Already in 1956 John H. Welsh, who, as we saw in chapter 8, was one of Parker's students and also the latter's successor as professor at Harvard's Biological Laboratories, reported the presence of serotonin in sea anemones, on the basis of pharmacological and chromatography tests (Welsh, 1960). The following year, Donald Ross, who had left Pantin's lab to take a research position at University College, London, confirmed Welsh's finding with colleagues and found that serotonin was widely distributed in the sea anemone body (Mathias et al., 1957). These biochemical studies, by their very nature, could not determine whether serotonin was present in neurons, but a Swedish group in Lund was able to do so for an adrenaline-like substance (Dahl et al., 1963). They used their own newly discovered histofluorescent technique, which lit up neurons that contained biogenic amines such as adrenaline, dopamine, and serotonin. Their technical breakthrough inaugurated the era of neurochemical typing of neurons. The Swedes were quick to turn to coelenterates for trying out their technique, which led to the discovery of an "adrenergic" sensory nerve net in the tentacles and oral disc of sea anemones.

The specificity of the histofluorescent technique was far from ideal, but no alternative was available until immunohistochemistry came along in the early 1970s. Immunohistochemistry was predicated on the selective binding of antibodies to specific proteins in histological sections. By labelling the antibody with a conspicuously visible tag (fluorescent dye or enzyme-catalyzed precipitate), it became possible to locate important biological markers in identifiable cells. As the antibodies could only recognize large molecules such as proteins and enzymes, neurotransmitters were at first revealed in specific neurons only indirectly, by labelling neurons containing enzymes involved in the synthesis of the neurotransmitter of interest. Later, in the 1980s, it became possible to directly visualize neuropeptides, which are made of amino acid chains like proteins but smaller, and to visualize small transmitters like biogenic amines by conjugating them with proteins. In this way, a wider range of known neurotransmitters and neurohormones could be detected than with the histofluorescent technique, which targeted only biogenic amines.

Cornelius Grimmelikhuijzen, a Dutch expatriate working in Heidelberg, was the first to use the technique to visualize neuropeptide-containing neurons in coelenterates, and he dominated the field for two decades. He started swiftly in the early 1980s with a series of papers reporting the presence of immunoreactivity to vertebrate-like neuropeptides in neurons of *Hydra*, the animal on which he had started his research career thus far (Grimmelikhuijzen et al., 1981, 1982). But his claim to distinction in the field came when he charted the major components of coelenterate nervous systems by using an antibody against a neuropeptide belonging to a family in which representatives are found abundantly in coelenterates. Thus he discovered that even *Hydra*, the focus of the controversy over the very existence of a nervous system since the days of Nicolaus Kleinenberg (see chapter 3), not only possessed a well-developed nerve net but, more significantly, boasted a condensation of sensory nerve cells culminating in a nerve ring around the mouth region which could be compared with the jellyfish nerve ring (Grimmelikhuijzen, 1985). This finding alone brought home the pregnant point that the lack of adequate methods had led to gross underestimations of the grades of organization of coelenterate nervous systems, which historically led to misconceived ideas on the evolutionary origin of nervous systems.,

Grimmelikhuijzen's contribution did not stop here. He next teamed up with Jane Westfall, the electron microscopist mentioned previously in relation to the existence of gap junctions between hydrozoan neurons. In the wake of Horridge and Mackay's first description of synapses in a jellyfish, Westfall had accumulated confirmations of synapses between neurons and at neuromuscular junctions in a variety of coelenterates (Westfall, 1970, 1973). Now, in the early 1990s, she used Grimmelikhuijzen's neuropeptide antibodies to visualize neuropeptides in synapses at the electron microscopic level. Tagging the markers was more of a challenge in electron microscopic immunocytochemistry, for the tag needed to be of a size to fit the minute synaptic vesicles inside which the neuropeptide of interest was suspected to reside. For this purpose electron-dense gold particles had been designed to bring the resolution of the technique to a level that allowed the visualization of discrete immunoreactive spots over synaptic vesicles, and thereby demonstrate that the suspected neurotransmitter is intimately associated with chemical synapses. Westfall seized upon this method to effectively localize coelenterate neuropeptides in synaptic vesicles at interneuronal and neuromuscular synapses in sea anemones (Westfall and Grimmelikhuijzen, 1993; Westfall et al., 1995). These observations concluded the display of evidence over many decades that suggested how nervous systems evolved from a single organizational plan at the very start. If there were other designs, they must have vanished early in the evolution of multicellular animals, for they left no trace.

The Struggle Today: Current Trends

What is the status of the field today, and to what extent have the views expressed by pioneers of the past held up to what we know now? The summing-up published a few years ago in the magazine *Science* (Miller, 2009), gathering together the thoughts of the top contemporaries working in this field, is the best guidepost we currently have. What strikes the informed reader most on reading the article is that we stand hardly any closer to a satisfying and consensual understanding of the emergence of the first nervous systems than in the past. New discoveries bring about new intellectual challenges with their own intractability; or conceptual difficulties experienced by past biologists may remain as thorny today. One conceptual challenge that is still up for debate today is to define what a neuron is. It has not escaped the debaters – even going back to Romanes's time – that much of one's scenario for the early evolution of neurons depends on the way you define them.

To be sure, there is a wealth of new discoveries that were out of reach to past researchers for lack of modern tools of investigation. With the advent of cell membrane electrophysiology came a better grasp of cell excitability and the role played by the traffic of specific ions – potassium, sodium, calcium, chloride – in its complex management. Molecular biology tools not only allowed a sophisticated reading of a nerve cell's gene script for managing its machinery, but also uncovered gene signatures that are used in exploring the ancestral relationships between animal groups.

One lesson learned by more recent investigators is that excitability is not unique to nerve cells. Single-celled organisms such as *Paramecium* are electrically excitable and this excitability, involving voltage-gated pores in the cell membrane through which ions flow, plays a critical role in controlling the activity of their cilia which, in turn, control locomotion of these animals (Hinrichsen and Schultz, 1988). In the same way, microorganisms as well as single-celled animals possess genes that code for proteins typically associated with other neuronal functions besides excitability, such as synapses and neurotransmitters. Sponges lack nerve cells yet display excitability, contain genes associated with neurotransmission, and respond to classical neurotransmitters (Leys et al., 1999; Srivastava et al., 2010; Elliott and Leys, 2010). The message here is that neuronal attributes preceded the emergence of nerve cells. Possession of these attributes is not enough; the genes must be expressed together in a coordinated manner to obtain the nerve cells that characterize coelenterates. How this is accomplished remains to be discovered.

The zoologists chronicled in this book generally worked on the assumption that nerve cells arose once in some ancestor of coelenterates, and that nervous systems evolved and diversified from these cells. As explained by neurobiologist Leonid Moroz of the Whitney Laboratory for Marine Bioscience in Florida (in Miller, 2009), "If you look at any other organ or structure, people easily assume it could evolve multiple times ... but for some reason, people are stuck on [a single origin] of neurons." Already, Mackie's scheme for the emergence of neurons from an excitable and conducting epithelium (Mackie, 1970) should have alerted us to the possibility of multiple origins for nerve cells. Any such epithelium present in different animal groups, given that it displays several crucial attributes of nerve cells – sensory input, excitability, and impulse propagation from cell to cell – could have independently given rise to nerve cells through elongation of an axon-like process projecting to a muscle cell. Excitable epithelial cells electrically coupled to each other by gap junctions exist in the two traditional coelenterate groups, cnidarians and ctenophores. Could it be that nerve cells arose independently in these two groups?

This is precisely the hypothesis that Moroz put forward. He first presented arguments for the independent origins of neurons and nervous systems based on a broad and extensive analysis of morphological, physiological, and molecular evidence across phyla (Moroz, 2009). In a later article, in which the ctenophore genome was analysed for neural and related genes (Moroz et al., 2014), Moroz and his colleagues specifically addressed the relationship of ctenophores with other animal phyla. They came to the bold conclusion that ctenophores are ancestral to all other multicellular animals and therefore were the first to develop a nervous system that shares very few features with that of other animals. In later lines of descent, according to this scenario, animals such as sponges and the little-known placozoans lost the nervous system of their ancestors, whereas cnidarians and the more complex animals, which had preserved their "ancestral neuronal toolkit," were able to reconstitute a nervous system of their own, quite distinct from that of the ctenophores.

This new paradigm of neuronal evolution would never have entered the minds of past zoologists, who viewed ctenophores and cnidarians as related phyla under the umbrella of coelenterates. But Moroz's evolutionary musings extended even further. In a 2009 article he elaborated a theory about the evolutionary source of neurons and neurotransmitters. He proposed that the processes by which basal animals manage injury and regeneration – extensively studied in *Hydra* – were co-opted repeatedly in parallel evolutions of neurons and of their complement of neurotransmitters. To quote Moroz, "neurons might have evolved in ancestral metazoans as a result of development in the adaptive cellular regenerative response to localized injury and stress, leading to a coordinated (potentially defensive) reaction and behavior of the entire organism."

In the course of cellular stress or injury, several metabolites are released, and some of them are now part of the neurotransmitter arsenal of neurons, including adenosine triphosphate (ATP), nitric oxide (NO), glutamate, and small peptide signal molecules (Moroz, 2009). In summing up, Moroz suggests that "it is very likely that early secretory/peptidergic cells were evolutionary precursors of neurons and that the massive gene upregulation needed to repair an injury was co-opted to serve the needed neuronal integrative functions."

These new hypotheses may be destined to replace the old ones, now found inadequate, and they may most satisfactorily account for current evidence, but they remain hypotheses. At best, in Moroz's own words, they propose merely "a reconstruction of the dawn of neuronal organization." We are still far removed from the full understanding the pioneers of the past were struggling to achieve. The current approach, using genomics and molecular tools, is predicated on what Moroz calls "neurosystematics"; that is, phylogenetic reconstructions aimed at identifying "ancestral cellular lineages within nervous systems" and establishing "neuronal homologies or neuronal innovations across phyla." Unfortunately, it is an ever-reinvented pursuit. Reconstructions are as good as the phylogenies that support them. Larger databases, shifting reliabilities of molecular phylogenetic tools – all conspire to produce phylogenetic "flavours of the day" on which to construct the next theory, only to be debunked by the emerging, different phylogenetic tree that gets the spotlight and feeds an opposite theory. And so the search continues ...

CONCLUDING REMARKS

In recapping the historical narrative of this book, one is tempted to look for pivotal scientific breakthroughs that drove the inquiries into the first nervous system. The recognition of cells as the basic building blocks of biological organisms (cell theory) and acceptance of Darwinian evolution – both highlighted in the Introduction – were inescapable preludes to any musings on how nerve cells came into being. An awareness of the cellular structure of organisms and of what nerve cells look like in other animals allowed Louis Agassiz to identify nerve cells in jellyfish, but the theistic view of nature prevalent at the time – nine years before the publication of *On the Origin of Species* – precluded any discussion of nerve cell genealogy. All the subsequent actors in this story crafted their works after the Darwinian revolution, and while they overwhelmingly sided with some form of Darwinism, they most often shied away from committing to a theory of the evolutionary emergence of nerve cells.

The acknowledgment of Agassiz reminds us of the historic role of strong personalities in turning the paths of discovery in specific, sometimes fateful directions. The clash between Agassiz and Ernst Haeckel is a striking example. Agassiz dominated the field of natural history in America and used his dominion to frustrate the progress of evolutionary ideas in his adopted country. Haeckel, by contrast, not only heartily endorsed Darwinian ideas but also expounded at length on the subject and led the way for disciples such as the Hertwig brothers to examine coelenterate nervous systems in an evolutionary light. In England Romanes stood even closer to both Darwin the man and Darwinian ideas, and his emerging confidence led him to articulate his own ideas on how the first nerve cells arose. Consequently European currents of evolutionary dissemination and strong, self-willed German and British scientific personas ensured that Europe would outweigh America in the study of the first nervous systems. It took the emergence of George

Parker at the turn of the twentieth century to tip the scales toward the United States.

Two schools of scientific approach ran their course as they pitched battle over the best way to tackle the difficult problem of the emergence of the first nervous systems. Anatomists such as Haeckel, the Hertwig brothers, and a host of German-speaking followers proposed that direct observation of the organization of coelenterate nervous systems was the surest ticket to gaining insights on how the first nervous systems looked and arose. After all, phylogenetic trees were traditionally constructed from comparisons of morphological data among animal groups, so ancestry of nervous systems was best inferred that way. But this Haeckelian view was countered by experimentalists of Romanes's ilk who owed their bias to far-sighted men like the founder of the Cambridge School of Physiology, Michael Foster. To physiologists, the story of how coordination of sensory and locomotor activities came into being in multicellular animals is more relevant than focusing on the usual suspect, the nerve tissue itself. Neuroanatomy, for the physiologist, is useful insofar as it informs the neurophysiology; that is how Romanes saw Edward Schäfer's contribution.

One should not conclude, however, that Germans monopolized neuroanatomy and that Anglo-Saxons single-mindedly advanced physiology. After all, Foster established the Cambridge School partly to emulate the German physiology laboratories that had acquired such fame, and Theodor Eimer gave as much importance to physiology as to anatomy in his jellyfish work. But German physiologists of the late nineteenth century were medically oriented and showed little if any interest in zoologial problems. Only after Parker's time did German zoologists turn to physiology, just as the field of comparative physiology was created, and the likes of Emil Bozler addressed the question of the first nervous systems for relatively short periods. They were supporting players in the interlude between two leading Anglo-Saxon experimentalists, George Parker and Carl Pantin.

The protracted debate over the individuality of nerve cells – whether nerve cells existed only as constituents of a meshwork of unimpeded transmission lines (continuity theory) or existed as entities separate from neighbouring nerve cells and capable of distinctive activity (discontinuity theory) – played out largely beyond the small community of coelenterate biologists. But while the leading neurobiologists around

Golgi and Ramón y Cajal heated up the debate, the question of the individuality of nerve cells had an impact on the discussion of how the first nerve cells emerged in evolution. It is intriguing how little Schäfer's early discovery of the discontinuity of jellyfish nerve cells did to quell the debate, partly because his finding received such little notice, and partly because other findings, such as the description of neurofibrils, distracted the participants away from the discontinuity theory. The confusion led Parker to envisage that the first nervous systems were syncytial and that nervous systems later evolved toward discontinuity.

Even though later developments proved Parker wrong, it turned out that syncytial elements were part of the mix in coelenterate nervous systems. George Mackie's discoveries of epithelial conduction and nerve fibre engulfment showed that the nervous systems of at least some coelenterates (hydrozoans) function partly in a syncytial manner. And, as with Parker, this finding led Mackie to come up with a hypothesis of the evolutionary development of nerve cells that has as its starting point an epithelium of coupled cells endowed with sensory capability. The difference between Parker's and Mackie's hypotheses is that the latter was informed by data obtained by more reliable methods.

The importance of methods cannot be overemphasized in this story. The quality of the findings can only be as good as the quality of the methods. We saw how critical both tissue preparation and choice of microscope were in determining the path of significant and reliable discovery. Agassiz's investment in the optical quality of his microscope paid off with the first identification of nerve cells in a coelenterate. But identifying nerve cells had its own pitfalls. Who was to decide whether a cell was a nerve cell and not a connective fibre cell, with both sharing some morphological similarity? Taking as a reference point the well-studied nerve cells of higher invertebrates and vertebrates might help, but only on the assumption that such was the look of the earliest nerve cells. We saw how misguided the early reports of nerve cells in sponges were in this respect. The contradictory findings of some of the protagonists in this story can in many cases be attributed to neglect of the importance of methods or misjudgment as to proper methodology. Enlightened researchers such as the Hertwig brothers, Schäfer, Bozler, and Pantin were a minority, and a class unto themselves.

One puzzling finding in this story, which has hardly been addressed, is the dearth of developmental studies to assist in tracing the historical origin of nerve cells. This is all the more disconcerting in view of the fact that Haeckel, the great champion of the recapitulation theory, was a father figure, if not the mentor, to many of the protagonists in this book. Stated in neurological terms, Haeckelian recapitulation means that the temporal process by which nerve cells appear and differentiate in the embryo of a given species is a recapitulation of the way nerve cells appeared and diversified in the genealogical line leading to this species. Although the theory was discredited even by some of Haeckel's contemporaries, it held sway in some form or another with many of them. One can argue that Parker played to Haeckel's recapitulation tune in describing the presumptive four grades of neuromuscular organization in sea anemones (see chapter 8), but he performed no developmental study to confirm these presumptions. So the question is legitimate: why have coelenterate neuroanatomists not examined the neuroembryology of higher invertebrates for clues to the evolutionary process by which nerve cells arose first in metazoans and evolved in coelenterates?

Only four investigators looked at developmental stages for nerve cells: Karl Schneider (1890) and Carl McConnell (1932) in *Hydra*, Carl Jickeli in the hydroid *Eudendrium* (1883), and Robert von Lendenfeld in the jellyfish *Cyanea* (1882). Both Schneider and Jickeli observed or intimated that nerve and nettle cells develop at the same time, and Schneider went further to notice that they differentiated from "indifferent cells," today's interstitial cells. McConnell in *Hydra* draws attention to sensory cells differentiating ahead of ganglion cells and in the "head" region first (oral disc) before the wave of differentiation travels downward toward the foot; contacts between nerve cells are established before those between nerve cells and epithelio-muscular cells. None of these three attempted to draw any evolutionary inference from their observations; in recapitulation terms it would have meant inferring that nerve cells did not emerge alone in evolution, but in tandem with cells, such as nematocytes, that were also multifunctional. Also, recapitulation would dictate that making neuromuscular contacts was a late evolutionary step, occurring only after the nerve net was functionally in place.

In an early developmental stage of *Cyanea*, Lendenfeld observed that nerve (ganglion) cells were imbricated inside the ectoderm rather than lying underneath the ectoderm as in the mature form. Again, Lenden-feld refrained from any evolutionary interpretation, whether or not in a recapitulatory tone, even though his observation suggested an inward movement of nerve cells as they differentiated from "indifferent cells" in the ectoderm. There may have been two factors at play here: the logistical difficulty in obtaining young stages, except for cultured or sedentary forms such as hydra or other hydroid polyps, and a mixture of reluctance to speculate and lack of imaginative drive.

Theories on the evolutionary origin of nerve cells and emerging nerv-ous systems followed their own evolutionary paths. From Nicolaus Kleinenberg's hypothesis that the neuromuscular cell was the primor-dial multitasking cell from which nerve and muscle cells would have derived, to the Hertwigs' view that sensory, nerve, and muscle cells arose simultaneously from distinct precursor cells to function inter-dependently, a huge gap existed right from the start. While Romanes had no definite view on the matter except that the nerve plexus of jelly-fish is already fully formed and therefore throws little light on what the first nervous systems should look like, his fellow physiologist Theodor Eimer murkily proposed a compromise between Kleinenberg's and the Hertwigs' views. Parker did away with these schemes and presented a fresh view according to which a three-step evolutionary process took place: from independent effectors (musculo-epithelial cells) integrating sensory and motor functions, to a receptor-effector system comprising sensory-motor cells projecting a fibre to underlying muscle cells and between themselves, and finally to a fully developed nerve net where "protoneurons" are interposed between sensory and muscle cells. Such a nerve net, in its incipient form, would be syncytial.

Pantin rejected Parker's elaborate scenario in favour of a simple scheme that emphasized tissues rather than cells. Accordingly, Pantin saw musculo-epithelial sheets as the original conducting system sub-tending spontaneous activities, surmising that only later would nerve nets have arisen to mediate reflex actions and other responses. A somewhat similar but more articulated hypothesis was constructed by Mackie in the wake of his discovery of epithelial conduction. In Mackie's scheme the ancestral musculo-epithelium is composed of elec-trically coupled cells. The next step is the segregation of the muscle components into a separate muscle sheet that sinks inward, followed by the differentiation of epithelial protoneurons projecting a fibre to the new muscle cells. The last step is the further differentiation of interconnected sensory and nerve cells from the protoneurons.

After 150 years of debate, the question of the origin of nervous systems remains unresolved. Fascination by this topic impelled many of the protagonists in this narrative, and more players will enter new debates on the subject in the years to come. In keeping with the times, these debates are shifting to new lines of inquiry. As R. Lichtneckert and H. Reichert (2009) explained in a recent assessment, molecular tools will now enter the field: "The main question has been the identification of the primordial cell lineage from which nerve cells might have been derived. During the last decade, however, advances in molecular genetic techniques have focussed our interest on the genes that might have been involved in the generation of the first nervous system."

But the newly generated hypotheses will likely face the never-ending challenge of their validation, if only for the simple truth that the witnesses to the genesis of the first nervous system are all dead and have left no reliable record.

BIBLIOGRAPHY

Agassiz, E.C.C. (1885). *Louis Agassiz: His Life and Correspondence.* Boston and New York: Houghton Mifflin.

Agassiz, L. (1850). "Contributions to the natural history of acalephae of North America, Part 1: On the naked-eyed Medusae of the shores of Massachusetts, in their perfect state of development." *Memoirs of the American Academy of Arts and Science* 4(2): 221–316.

– (1862). *Contributions to the Natural History of the United States of America.* Boston and London: Little Trübner.

Akert, K. (1993). "August Forel: Cofounder of the neuron theory (1848–1931)." *Brain Pathology* 3: 425–30.

Allabach, L.F. (1905). "Some points regarding the behavior of *Metridium.*" *Biological Bulletin* (Woods Hole) 10: 35–43.

Anderson, C.G. and B. Anderson (1993). "Koelliker on Cajal: translated excerpts from Erinnerungen aus meinem Leben." *International Journal of Neuroscience* 70: 181–92.

Anderson, P.A.V. (1980). "Epithelial conduction: its properties and functions." *Progress in Neurobiology* 15: 161–203.

– (1985). "Physiology of a bidirectional, excitatory, chemical synapse." *Journal of Neurophysiology* 53: 821–35.

– and U. Grünert (1988). "Three dimensional structure of bidirectional, excitatory, chemical synapses in the jellyfish *Cyanea capillata.*" *Synapse* 2: 606–13.

– and G.O. Mackie (1977). "Electrically coupled, photosensitive neurons control swimming in a jellyfish." *Science* 197 (4299): 186–8.

Apathy, S. (1897). "Das leitende Element der Nervensystems und seine topographischen Beziehungen zu den Zellen." *Mittheilungen aus der Zoologischen Station zu Neapel* 12: 495–748.

Ariëns Kappers, C.U. (1929). *The Evolution of the Nervous System in Invertebrates, Vertebrates and Man.* Haarlem: De Erven F. Bohn.

– and A.B. Droogleever Fortuyn (1920). *Vergleichende Anatomie des Nervensystems. Erster Teil: Die Leitungsbahnen im Nervensystem der wirbellosen Tiere. Zweiter Teil: Die Vergleichende Anatomie des Nervensystems der Wirbeltiere und des Menschen.* Haarlem, Netherlands: De Erven F. Bohn.

– G.C. Huber, et al. (1936). *The Comparative Anatomy of the Nervous System of Vertebrates, Including Man*. New York: Macmillan.

Ashworth, J.H. (1899). "The structure of *Xenia hicksoni*, nov. sp., with some observations on Heteroxenia elizabethae, Kölliker." *Quarterly Journal of Microscopical Science* 42(3): 245–304.

Baker, J.R. (1948). "The cell theory: a restatement, history, and critique. Part I." *Quarterly Journal of Microscopical Science* 89(1): 103–25.

Barclay, O. (1977). *Whatever Happened to the Jesus Lane Lot?* Leicester: Inter-Varsity Press.

Barnes, E.J. (1998). "The Early Career of George J. Romanes, 1867–1878." UK: Newnham College, Cambridge University. Undergraduate thesis.

Batham, E.J. and C.F.A. Pantin (1950a). "Muscular and hydrostatic action in the sea anemone *Metridium senile* (L.)." *Journal of Experimental Biology* 27: 264–89.

– and C.F.A. Pantin (1950b). "Inherent activity in the sea anemone, *Metridium senile* (L.)." *Journal of Experimental Biology* 27: 290–301.

– and C.F.A. Pantin (1950c). "Phases of activity in the sea anemone, *Metridium senile* (L.), and their relation to external stimuli." *Journal of Experimental Biology* 27: 377–99.

– and C.F.A. Pantin (1951). "The organization of the muscular system of *Metridium*." *Quarterly Journal of Microscopical Science* 92: 27–54.

– and C.F.A. Pantin (1954). "Slow contraction and its relation to spontaneous activity in the sea anemone *Metridium senile* (L.)." *Journal of Experimental Biology* 31: 84–103.

–, C.F.A. Pantin, and E.A. Robson (1960). "The nerve-net of the sea-anemone *Metridium senile*: the mesenteries and the column." *Quarterly Journal of Microscopical Science* 101: 487–510.

–, C.F.A. Pantin, and E.A. Robson (1961). "The nerve-net of *Metridium senile*: artifacts and the nerve-net." *Quarterly Journal of Microscopical Science* 102: 143–56.

Bauer, V. (1910). "Über die anscheinend nervöse Regulierung der Flimmerbewegung bei den Rippenquallen." *Archiv für die gesammte Physiologie des Menschen und der Tiere*: 230–48.

– (1927). "Die Schwimmbewegungen der Quallen und ihre reflektorische Regulierung." *Zeitschrift für vergleichende Physiologie* 5: 37–69.

Beebe-Center, J.G. (1955). "George Howard Parker: 1864–1955." *American Journal of Psychology* 68(3): 492–4.

Bennett, M.R. (2001). *History of the Synapse*. London: Harwood Academic Publishers.

Berger, E.W. (1898). "The histological structure of the eyes of Cubomedusae." *Journal of Comparative Neurology* 8: 223–30.

Berger, E.W. (1900). "Physiology and histology of the Cubomeduse, including Dr. F.S. Conant's notes on the physiology." *Memoirs from the Biological Laboratory of the Johns Hopkins University* 4(4): 1–84.

Bernstein, J. (1902). "Untersuchungen zur Thermodynamik der bioelektrischen Ströme." *Pflügers Archiv* 92: 521–62.

Bethe, A. (1895). "Der subepitheliale Nervenplexus der Ctenophoren." *Biologisches Centralblatt* 15: 140–5.

– (1898). "Das Centralnervensystem von *Carcinus Maenas*. Ein anatomisch-physiologischer Versuch. II. Theil (3. Mittheilung)." *Archiv für mikroskopische Anatomie und Entwicklungsgeschichte* 51: 382–452.

– (1903). *Allgemeine Anatomie und Physiologie des Nervensystems*. Leipzig: Georg Thieme.

– (1909). "Die Bedeutung der Elektrolyten für die rhythmischen Bewegung der Medusen (II. Theil)." *Archiv für die gesammte Physiologie des Menschen und der Thiere* 127: 219–73.

– (1935). "Versuche an Medusen als Beispiel eines primitiven neuromuskulären Reaktionssystems." *Pflügers Archiv für gesammten Physiologie* 235: 288–315.

Bidder, G.P. (1898). "The skeleton and classification of calcareous sponges." *Proceedings of the Royal Society of London* 64: 61–76.

Bishop, G.M. (1956). "Natural history of the nerve impulse." *Physiological Reviews* 36: 376–99.

Bölsche, W. and J. McCabe (1906). *Haeckel, His Life and Work*. London: T.F. Unwin.

Borell, M. (1978). "Setting the standards for a new science: Edward Schäfer and endocrinology." *Medical History* 22: 282–90.

Bozler, E. (1926a). "Sinnes- und Nervenphysiologische Untersuchungen an Scyphomedusen." *Zeitschrift für vergleichende Physiologie* 4: 37–80.

– (1926b). "Weitere Untersuchungen zur Sinnes- und Nervenphysiologie der Medusen: Erregungsleitung, Funktion der Randkorper, Nahrungsaufnahme." *Zeitschrift für vergleichende Physiologie* 4: 797–817.

– (1927a). "Untersuchungen über das Nervensystem der Coelenteraten. I. Teil: Kontinuität oder Kontakt zwischen den Nervenzellen?" *Zeitschrift für Zellforschung* 5: 244–62.

– (1927b). "Untersuchungen über das Nervensystem der Coelenteraten. II. Teil: Über die Struktur der Ganglienzellen und die Funktion der Neurofibrillen nach Lebenduntersuchungen." *Zeitschrift für vergleichende Physiologie* 6: 255–63.

Breidbach, O. (2002). "Representation of the microcosm – the claim for objectivity in 19th century scientific microphotography." *Journal of the History of Biology* 35: 221–50.

Bridge, D., C.W. Cunningham, R. DeSalle, and L.W. Buss (1995). "Class-level relationships in the phylum Cnidaria: molecular and morphological evidence." *Molecular Biology and Evolution* 12: 679–89.

Buchanan, B. (2008). "Jakob von Uexküll's theories of life." Pages 7–27 in *Onto-Ethologies. The Animal Environments of Uexküll, Heidegger, Merleau-Ponty, and Deleuze*, edited by B. Buchanan. Albany, New York: SUNY Press.

Bullock, T.H. and G.A. Horridge (1965). *Structure and Function in the Nervous Systems of Invertebrates*. San Francisco: W.H. Freeman and Company.

Calder, D.R. (2013). "Harry Beale Torrey (1873–1970) of California, U.S.A., and his research on hydroids and other coelenterates." *Zootaxa* 6(4): 549–63.

Cary, L.R. (1917). "Studies on the physiology of the nervous system of *Cassiopea xamachana*." *Papers from the Tortugas Laboratory of the Carnegie Institution of Washington* 11(7): 121–70.

Castellucci, V., H. Pinsker, I. Kupfermann, and R.R. Kandel (1970). "Neuronal mechanisms of habituation and dishabituation of the gill-withdrawal reflex in *Aplysia*." *Science* 167: 1745–48.

Chernin, E. (1988). "The Harvard system: a mystery dispelled." *British Medical Journal* 297: 1062–3.

Chun, C. (1878). "Das Nervensystem und die Muskulatur der Rippenquallen." *Abhandlungen von der Senckenbergischen Naturforschenden Gesellschaft* 11: 181–230.

– (1881). "1. Das Nervensystem der Siphonophoren." *Zoologischer Anzeiger* 4: 107–11.

– (1882). "2. Die Gewebe der Siphonophoren." *Zoologischer Anzeiger* 5: 400–6.

Churchill, F.B. (2008) "Eimer, Theodor Gustav Heinrich." *Complete Dictionary of Scientific Biography*.

Clarke, E. and C.D. O'Malley (1968). *The Human Brain and Spinal Cord: A Historical Study Illustrated by Writings from Antiquity to the Twentieth Century*. Berkeley: University of California Press.

Conant, F.S. (1897). "The Cubomedusae." *Memoirs from the Biological Laboratory of the Johns Hopkins University* 4(1): 1–61.

Conn, H.J. (1948). "Professor Herbert William Conn and the founding of the Society." *Bacteriology Reviews* 12(4): 275–96.

Conn, H.W. and H.G. Beyer (1883). "The nervous system of Porpita." *Studies from the Biological Laboratory, Johns Hopkins University* 2: 433–45.

Creese, M.R.S. (1997). "Ida Henrietta Hyde (1857–1945)." Pages 246–53 in *Women in the Biological Sciences: A Bibliographic Sourcebook*, edited by C.A. Bierman and L.S. Grinstein. Santa Barbara, CA: Greenwood Publishing Group.

Crompton, B. (2002). "Interview with Professor Adrian Horridge." *Encyclopedia of Australian Science*. Australian Academy of Science.

Dahl, E., B. Falck, C. von Mecklenburg, and H. Myhrberg (1963). "An adrenergic nervous system in sea anemones." *Quarterly Journal of Microscopical Science* 104: 531–4.

Darwin, C. (1859). *On the Origin of Species by Means of Natural Selection*. London: J. Murray.

Darwin, C. and F. Darwin (1888). *The Life and Letters of Charles Darwin, Including an Autobiographical Chapter*. New York: D. Appleton and Company.

Davenport, C.B. (1927). "Biographical memoir, Alfred Goldsborough Mayor, 1868–1922." *Biographical Memoirs of the National Academy of Sciences* 21(8): 1–13.

De Robertis, E.D.P. and H.S. Bennett (1955). "Some features of the submicroscopic morphology of synapses in frog and earthworm." *Journal of Cell Biology* 1(1): 47–58.

Deiters, O.F.K. (1865). *Untersuchungen über Gehirn und Ruckenmark des Menschen und der Saugetiere*. Braunschweig: Vieweg.

Denmark, H.A. (1995). "The History of the Bureau of Entomology, Division of Plant Industry, Florida Department of Agriculture and Consumer Services." *Florida Entomologist* 78(1): 194–206.

Di Gregorio, M.A. (2005). *From Here to Eternity: Ernst Haeckel and Scientific Faith*. Göttingen: Vandenhoeck & Ruprecht.

Dutrochet, H. (1824). *Recherches anatomiques et physiologiques sur la structure intime des animaux et des végétaux et sur leur motilité*. Paris: J.-B. Baillière.

– (1837). *Mémoires pour servir à l'histoire anatomique et physiologique des Végétaux et des Animaux*. Paris: J.-B. Baillière.

Edwards, J.S. and R. Huntford (1998). "Fridtjof Nansen: from the neuron to the North Polar Sea." *Endeavour* 22(2): 76–80.

Ehrenberg, C.G. (1833). Nothwendigkeit einer feineren mechanischer Zerlegung des Gehirns und der Nerven vor der chemischen, dargestelt aus Beobachtungen von C.G. Ehrenberg." *Annalen der Physik* 28: 449–53.

– (1836). *Beobachtungen einer auffallenden bisher unbekannten Struktur des Seelenorgans bei Menschen und Thieren*. Berlin: Dummler.

Eimer, G.H.T. (1873). *Zoologische Studien auf Capri*. Leipzig: W. Engelmann.

Eimer, T. (1873). *Zoologische Studien auf Capri. I. Ueber* Beroe *ovatus. Ein Beitrag zur Anatomie der Rippenquallen*. Leipzig: Wilhelm Engelmann.

– (1874). "Ueber kunstliche Theilbarkeit von *Aurelia aurita* und *Cyanea capillata* in physiologische Individuen." *Verhandlungen der physikalisch-medizinischen Gesellschaft zu Würzburg* 5 (New Series): 137–61.

– (1878). *Die Medusen physiologisch und morphologisch auf ihr Nervensystem untersucht*. Tübingen, Verlag der H. Laupp'schen Buchhandlung.

– (1880). "Versuche über kunstliche Theilbarkeit von *Beroë ovatus*. Angestellt zum Zweck der Controle seiner morphologischen Befunde über das Nervensystem dieses Thieres." *Zeitschrift für Mikroskopische Anatomie* 17: 213–40.

‒ (1890). *Organic Evolution as the Result of the Inheritance of Acquired Characters According to the Laws of Organic Growth.* London & New York: Macmillan and Co.

Elliott, G.R.D. and S.P. Leys (2010). "Evidence of glutamate, GABA and NO in coordinating behaviour in the sponge, *Ephydatia muelleri* (Demospongiae, Spongillidae)." *Journal of Experimental Biollogy* 213: 2310‒21.

Fautin, D.G. and R.N. Mariscal (1991). "Cnidaria: Anthozoa." Pages 267‒358 in *Microscopic Anatomy of Invertebrates, Volume 2: Placozoa, Porifera, Cnidaria and Ctenophora,* edited by F.W. Harrison and J.A. Westfall. New York: Wiley-Liss.

Fleischhauer, K. (1973). "In memoriam Ernst Horstmann." *Anatomischer Anzeiger* 1335: 417‒30.

Florey, E. (1985). "The Zoological Station at Naples and the neuron: personalities and encounters in a unique institution." *Biological Bulletin* (Woods Hole) 168 (Supplement): 137‒52.

Fokin, S.I. (2000). "Professor W.T. Schewiakoff: life and science." *Protist* 151: 181‒9.

‒ (2008). "Russian biologists at Villafranca." *Proceedings of the California Academy of Sciences* 59(11): 169‒92.

Ford, B.J. (2009). "Charles Darwin and Robert Brown ‒ their microscopes and the microscopic image." *Infocus* 15: 18‒28.

Forel, A. (1887). "Einige hirnanatomische Betrachtungen und Ergebnisse." *Archiv für Psychiatrie und Nervenkrankheiten* 18(1): 162‒98.

Forel, A. (1937). *Out of My Life and Work.* New York: W.W. Norton & Company.

Foster, M. (1877). *A Textbook of Physiology.* London: Macmillan.

Foster, M. and C.S. Sherrington. (1897). *A Text Book of Physiology.* New York: Macmillan & Co. Ltd.

Fränkel, G.S. (1925). "Der statische Sinn der Medusen." *Zeitschrift für vergleichende Physiologie* 2: 658‒90.

French, R.D. (1970a). "Darwin and the physiologists, or the medusa and modern cardiology." *Journal of the History of Biology* 3(2): 253‒74.

‒ (1970b). "Some concepts of nerve structure and function in Britain, 1875‒1885: background to Sir Charles Sherrington and the synapse concept." *Medical History* 14(2): 154‒65.

Frixione, E. (2009). "Cajal's second great battle for the neuron doctrine: the nature and function of neurofibrils." *Brain Research Reviews* 59: 393‒409.

Gardiner, J.S. (1941). "Sydney John Hickson." *Obituary Notices of Fellows of the Royal Society* 3(9): 383‒94.

Geison, G.L. (1978). *Michael Foster and the Cambridge School of Physiology: The Scientific Enterprise in Late Victorian Society.* Princeton, N.J.: Princeton University Press.

Gerlach, J. von (1872). "Über die Struktur der grauen Substanz des menschlichen Grosshirns. Vorläufige Mittheilung." *Zentralblatt für die Medizinischen Wissenschaften* 10: 273–5.

Grimmelikhuijzen, C.J.P. (1985). "Antisera to Arg-Phe-amide visualize neuronal centralization in hydroid polyps." *Cell and Tissue Research* 241: 171–82.

–, A. Balfe, P.C. Emson, D. Powell, and F. Sundler (1981). "Substance P-like immunoreactivity in the nervous system of hydra." *Histochemistry* 71: 325–33.

–, K. Dierickx, and G.J. Boer (1982). "Oxytocin/vasopressin-like immunoreactivity in the nervous system of *Hydra*." *Neuroscience* 7: 3191–9.

Groeben, C. (1985). "Anton Dohrn – The Statesman of Darwinism." *Biological Bulletin* (Woods Hole) 168 (Supplement): 4–25.

Grošelj, P. (1909). "Untersuchungen über das Nervensystem der Aktinien." *Arbeiten aus den Zoologischen Instituten der Universität Wien* 17(3): 269–308.

Grundfest, H. (1959). "Evolution of conduction in the nervous system." Pages 43–86 in *Evolution of Nervous Control from Primitive Animals to Man*, edited by A.D. Bass. Washington, D.C.: American Association for the Advancement of Science, Publication No. 52.

Guillery, R.W. (2005). "Observations of synaptic structures: origins of the neuron doctrine and its current status." *Philosophical Transactions of the Royal Society of London*, B 360: 1281–307.

Hadži, J. (1909). "Über das Nervensystem von *Hydra*." *Arbeiten aus den Zoologischen Instituten der Universität Wien* 17(3): 225–68.

– (1963). *The Evolution of the Metazoa*. New York: Macmillan.

Haeckel, E. (1865a). *Beiträge zur Naturgeschichte der Hydromedusen. Erstes Heft: Die Familie der Russelquallen (Geryonida)*. Leipzig: Wilhelm Engelmann.

– (1865b). "On a new form of alternation of generations in the Medusae, and on the relationship of the Geryonidae and Aeginidae." *Annals and Magazine of Natural History* (Third Series) 90: 437–44.

– (1875). *Ziele und Wege der heutigen Entwickelungsgeschichte*. Jena: H. Dufft.

– (1897). *The Evolution of Man: A Popular Exposition of the Principal Points of Human Ontogeny and Phylogeny*. New York: D. Appleton and Company.

– and J. A. Thomson (1880). Report on the Siphonophoræ Collected by H.M.S. *Challenger* During the Years 1873–76. Edinburgh: Neill and Company.

Hall, D.M. and C.F.A. Pantin (1937). "The nerve net of the Actinozoa. V. Temperature and facilitation in *Metridium senile*." *Journal of Experimental Biology* 14: 71–8.

Hand, A.R. and S. Gobel (1972). "The structural organization of the septate and gap junctions of *Hydra*." *Journal of Cell Biology* 52: 397–408.

Hanström, B. (1928a). *Vergleichende Anatomie des Nervensystems der Wirbellosen Tiere*. Berlin: J. Springer.

– (1928b). "Some points on the phylogeny of nerve cells and of the central nervous system of invertebrates." *Journal of Comparative Neurology* 46(2): 475–93.

Harvey, E.N. (1912). "The question of nerve fatigue." Carnegie Institution of Washington, Year Book No. 10 (1911): 130–1.

Havet, J. (1899). "Note préliminaire sur le système nerveux des Limax (méthode de Golgi)." *Anatomischer Anzeiger* 16: 10–11.

– (1900). "Structure du système nerveux des Annélides: Nephelis, Clepsine, Hirudo, Lumbriculus, Lumbricus (Méthode de Golgi)." *La Cellule* 17: 65–137.

– (1901). "Contribution à l'étude du système nerveux des actinies." *La Cellule* 18: 385–419.

– (1922). "La structure du système nerveux des actinies. Leur mécanisme neuromusculaire." *Libro en honor de D.S. Ramón y Cajal.* Madrid: Publicaciones de la Junta para el Homenaje a Cajal. 1: 477–504.

Heider, K. (1927). "Vom Nervensystem der Ctenophoren." *Zeitschrift für Morphologie und Ökologie der Tiere* 9: 638–78.

Hertwig, O. and R. Hertwig (1877). "Ueber das Nervensystem und die Sinnesorgane der Medusen." *Jenaische Zeitschrift für Naturwissenschaft* 11(4): 355–74.

– (1878a). *Das Nervensystem und die Sinnesorgane der Medusen.* Leipzig: Verlag von F.C.W. Vogel.

– (1878b). *Der Organismus der Medusen und Seine Stellung zur Keimblättertheorie.* Jena: Verlag Gustav Fischer.

– (1879). *Die Actinien – anatomisch und histologisch mit besonderer Berucksichtigung des Nervenmuskelsystems untersucht.* Jena: Verlag Gustav Fischer.

– (1879–80). "Die Actinien – anatomisch und histologisch mit besonderer Berucksichtigung der Nervenmuskelsystems untersucht " *Jenaische Zeitschrift für Naturwissenschaft* 13: 457–640, 14: 39–89.

– (1882). "Die Coelomtheorie." *Jenaische Zeitschrift für Naturwissenschaft* 15: 1–150.

Hertwig, R. (1880). "Ueber den Bau der Ctenophoren." *Jenaische Zeitschrift für Naturwissenschaft* 14: 313–457.

Hess, A., A.I. Cohen, and E.A. Robson (1957). "Observations of the structure of *Hydra* as seen with the electron and light microscopes." *Quarterly Journal of Microscopical Science* 98: 315–26.

Hesse, R. (1895). "Über das Nervensystem und die Sinnesorgane von *Rhizostoma cuvieri.*" *Zeitschrift für wissenschaftliche Zoologie* 60: 411–57.

– (1924). *Tiergeographie auf Ökologischer Grundlage.* Jena: Gustav Fischer.

– and F. Doflein (1914). *Tierbau und Tierleben in ihrem Zusammenhang betrachtet.* Leipzig and Berlin: B.G. Teubner.

Hickson, S.J. (1895). "The anatomy of *Alcyonium digitatum.*" *Quarterly Journal of Microscopical Science* 37: 343–88.

Hilgard, E.R. (1965). "Robert Mearns Yerkes: 1876–1956." *Biographical Memoirs of the National Academy of Sciences* 38: 385–425.

Hill, L. (1935). "Sir Edward Albert Sharpey-Schafer. 1850–1935." *Obituary Notices of Fellows of the Royal Society* 1(4): 400–7.

Hille, B. (2001). *Ion Channels of Excitable Membranes*. Third Edition. Sunderland, MA: Sinauer Associates.

Hinrichsen, R.D. and J.E. Schultz (1988) "*Paramecium*: a model system for the study of excitable cells." *Trends in Neurosciences* 11(1): 27–32.

His, W. (1887). "Zur Geschichte des menschlichen Ruckenmarkes und der Nervenwurzeln." *Abhandlungen der königlisch Sächsischen Gesellschaft der Wissenschaften* 22: 477–514.

Hogben, L. (1974). "Francis Albert Eley Crew. 1886–1973." *Biographical Memoirs of Fellows of the Royal Society* 20: 135–53.

Holmes, W. (1942). "A new method for the impregnation of nerve axons in mounted paraffin sections." *Journal of Pathology and Bacteriology* 54: 132–6.

– (1943). "Silver staining of nerve axons in paraffin sections." *The Anatomical Record* 86: 157–87.

Horridge, G.A. (1953). "An action potential from the motor nerves of the jellyfish *Aurellia aurita* Lamarck." *Nature* 171: 400.

– (1954). "The nerves and muscles of medusae. I. Conduction in the nervous system of *Aurellia aurita* Lamarcq." *Journal of Experimental Biology* 31: 594–600.

– (1965). "Intracellular action potentials associated with the beating of the cilia in ctenophore comb plate cells." *Nature* 205: 602.

– (1968). *Interneurons: Their Origin, Action, Specificity, Growth, and Plasticity*. London and San Francisco: W.H. Freeman and Company.

– (2009). "What does the honeybee see and how do we know?: a critique of scientific reason." Canberra: Australian National University E-Press.

– and B. MacKay (1962). "Naked axons and symmetrical synapses in coelenterates." *Quarterly Journal of Microscopical Science* 103: 531–41.

Horstmann, E. (1934a). "Untersuchungen zur Physiologie der Schwimmbewegungen der Scyphomedusen." *Pflugers Archiv für gesammten Physiologie* 234: 406–20.

– (1934b). "Nerven- und muskelphysiologische Studien zur Schwimmbewegungen der Scyphomedusen." *Pflugers Archiv für gesammten Physiologie* 234: 421–31.

Hossfeld, U. (2003). "The road from Haeckel: the Jena tradition in evolutionary morphology and the origins of 'evo-devo.'" *Biology and Philosophy* 18: 285–307.

Huntford, R. (1997). *Nansen: The Explorer as Hero*. London: Duckworth.

Hyde, I.H. (1902). "The nervous system in *Gonionema murbachii*." *Biological Bulletin* (Woods Hole) 4: 40–5.

– (1938). "Before women were human beings: Adventures of an American fellow in German universities in the '90s." *AAUW Journal* 31(4): 226–36.

Hydén, H. (1967). *The Neuron*. Amsterdam, New York etc.: Elsevier Publishing Company.

Jacyna, S. (2009). "The most important of all the organs: Darwin on the brain." *Brain* 132: 3481–7.

Jarausch, K.H. (2004). "Graduation and careers." Pages 363–92 in *A History of the University in Europe, Vol. 3: Universities in the Nineteenth and Early Twentieth Centuries*, edited by W. Rüegg. Cambridge, UK: Cambridge University Press.

Jennings, H.S. (1905). "Modifiability in behavior. I. Behavior of sea anemones." *Journal of Experimental Zoology* 2: 447–72.

Jickeli, C.F. (1882). "Vorläufige Mittheilung über das Nervensystem der Hydroidpolypen." *Zoologischer Anzeiger* 5(102): 43–4.

– (1883a). "Der Bau der Hydroidpolypen. I. Über den histologischen Bau von *Eudendrium* Ehrbg. und *Hydra* L." *Morphologisches Jahrbuch* 8(3): 373–416.

– (1883b). "Der Bau der Hydroidpolypen. II. Über den histologischen Bau von Tubularia L., Cordylophora Allm., Cladonema Duj., Coryne Gärtn., Gemmaria M'Crady, Perigonimus Sars, Podocoryne Sars, Camponopsis Claus, Lafoëa Lam., Campanularia Lam., Obelia Pér., Anisocola Kirchenp., Isocola Kirchenp., Kirchenpaueria Jick." *Morphologisches Jahrbuch* 8(5): 580–680.

Jillett, J. (2000). "Batham, Elizabeth Joan, 1917–1974. Marine biologist, university lecturer." *Dictionary of New Zealand Biography*, vol. 5.

Jones, E.G. (1994). "The Neuron Doctrine 1891." *Journal of the History of Neuroscience* 3: 3–20.

Jones, W.C. (1962). "Is there a nervous system in sponges?" *Biological Reviews* 37: 1–50.

Josephson, R.K. (1961). "Repetitive potentials following brief electric stimuli in a hydroid." *Journal of Experimental Biology* 38: 579–93.

– (1966). "Neuromuscular transmission in a sea anemone." *Journal of Experimental Biology* 45: 305–19.

– (2004). "JEB classics: The neural control of behavior in sea anemones." *Journal of Experimental Biology* 207(14): 2371–2.

Kaas, J.H. (2009). *Evolutionary Neuroscience*. Oxford and San Diego: Academic Press (Elsevier).

Kassianow, N. (1901). *Studien über das Nervensystem der Lucernariden nebst sonstigen histologischen Beobachtungen über diese Gruppe*. Leipzig: Wilhelm Engelmann.

– (1903). "Ueber das Nervensystem der Alcyonarien." *Bergens Museums Aarbog* 1903(6): 1–5.

– (1908). "Untersuchungen über das Nervensystem der Alcyonaria." *Zeitschrift für wissenschaftliche Zoologie* 90: 478–535.

– (1908b). "Vergleich des Nervensystems der Octocorallia mit dem der Hexacorallia." *Zeitschrift für wissenschaftliche Zoologie* 90: 670–7.

– (1918). *La Sibérie et la poussée allemande vers l'Orient*. Berne, Switzerland: P. Haupt.

Kater, M.H. (1989). *Doctors Under Hitler.* Chapel Hill, NC: University of North Carolina Press.

Kleinenberg, N. (1872). *Hydra – Eine Anatomisch-Entwicklungsgeschichte Untersuchung*. Leipzig: Wilhelm Engelmann.

Kleisner, K. (2008). "The semantic morphology of Adolf Portmann: a starting point for the biosemiotics of organic form?" *Biosemiotics* 1: 207–19.

Kölliker, A.v. (1899). *Erinnerungen aus meinem Leben*. Leipzig: Wilhelm Engelmann.

Korotneff, A.A. (1884). "Zur Histologie der Siphonophoren." *Mittheilungen aus der Zoologischen Station zu Neapel* 5(2): 229–88.

– (1887). "Zur Anatomie und Histologie des Veretillum." *Zoologischer Anzeiger* 10: 387–90.

Kowalevsky, A. (1866). "Entwicklungsgeschichte der Rippenquallen." *Mémoires de l'Académie Impériale des Sciences de Saint-Pétersbourg* 10(4): 1–28.

Krasinska, S. (1914). "Beiträge zur Histologie der Medusen." *Zeitschrift für wissenschaftliche Zoologie* 109: 256–348.

Kühnelt, W. (1993). "Schneider Karl Camillo, Zoologe." *Österreichisches Biographisches Lexikon 1815–1950,* 10: 382–3.

Langton, G. (2006). *Armchair Mountaineering: A Bibliography of New Zealand Mountain Climbing*. Christchurch, New Zealand: New Zealand Alpine Club.

Lendenfeld, R. von (1882). "Über Coelenteraten der Südsee. I. Cyanea Annaskala nov. sp." *Zeitschrift für wissenschaftliche Zoologie* 37: 465–552.

Lendenfeld, R. von (1883). "Über das Nervensystem der Hydroidpolypen." *Zoologischer Anzeiger* 6(1): 69–71.

– (1883). "Zur Histologie der Actinien." *Zoologischer Anzeiger* 6(129): 189–92.

– (1885). "The histology and nervous system of calcareous sponges." *Proceedings of the Linnean Society of N.S.W.* 1(9): 977–83.

– (1889a). *A Monograph on the Horny Sponges*. London: Trübner and Co. (for the Royal Society of London).

– (1889b). "Experimentelle Untersuchungen über die Physiologie der Spongien." *Zeitschrift für wissenschaftliche Zoologie* 48: 406–700.

Leydig, F. (1857). *Lehrbuch der Histologie des Menschen und der Thiere*. Frankfurt: Verlag von Meidinger Sohn & Comp.

Leys, S.P. (2007). *Sponge coordination, tissues, and the evolution of gastrulation. Porifera research : biodiversity, innovation and sustainability*. Rio de Janeiro, Museu Nacional: M. Reis: 53–9.

–, G.O. Mackie, and R.W. Meech (1999). "Impulse conduction in a sponge." *Journal of Experimental Biology* 202: 1139–50.

Lichtneckert, R. and H. Reichert (2009). "Origin and evolution of the first nervous system." Pages 51–77 in *Evolutionary Neuroscience*, edited by J.H. Kaas. Oxford and San Diego: Academic Press (Elsevier).

Lillie, R.S. (1914). "The conditions determining the rate of conduction in irritable tissues and especially in nerve." *Americal Journal of Physiology* 34(4): 414–45.

– (1915). "The conditions of conduction of excitation of irritable cells and tissues and especially in nerve. II." *Americal Journal of Physiology* 37(2): 348–70.

– (1916). "The conditions pf physiological conduction in irritable tissues. III. Electrolytic local action as the basis of propagation of the excitation-wave." *Americal Journal of Physiology* 41(1): 126–36.

Linko, A.K. (1900). "Über den Bau der Augen bei den Hydromedusen." Mémoires de l'Académie Impériale des Sciences de Saint-Pétersbourg 10(3): 1–23.

Loeb, J. (1900). *Comparative Physiology of the Brain and Comparative Psychology.* New York: G.P. Putnam's Sons.

– (1918). *Forced Movements, Tropisms, and Animal Conduct.* Philadelphia and London: J.B. Lippincott.

Loewenstein, W.R. (1966). "Permeability of membrane junctions." *Annals of the New York Academy of Science* 137: 441–72.

López-Muñoz, F., J. Boya, and C. Alamo (2006). "Neuron theory, the cornerstone of neuroscience, on the centenary of the Nobel Prize award to Santiago Ramón y Cajal." *Brain Research Bulletin* 70: 391–405.

Louis, E.D. (2001). "Unraveling the neuron jungle: The 1879–1886 publications by Wilhelm His on the embryological development of the human brain." *Archives of Neurology* 58: 1932–5.

Lurie, E. (1960). *Louis Agassiz: A Life in Science.* Chicago: University of Chicago Press.

Mackie, G.O. (1960) "The structure of the nervous system in *Velella*." *Quarterly Journal of Microscopical Science* 101: 119–31.

– (1965). "Conduction in the nerve-free epithelia of siphonophores." *American Zoologist* 5: 439–53.

– (1970). "Neuroid conduction and the evolution of conducting tissues." *Quarterly Review of Biology* 45: 319–32.

– (1971). "Neurological complexity in medusae: A report of central nervous organization in *Sarsia*." Salamanca: *Actas del I Simposio Internacional de Zoofilogenia*, 269–80.

– (1973). "Report on giant nerve fibres in *Nanomia*." *Publications of the Seto Marine Biological Laboratory* 20: 745–56.

– (1989). "Louis Agassiz and the discovery of the coelenterate nervous system." *History and Philosophy of Science* 11: 71–81.

– (1990). "The elementary nervous system revisited." *American Zoologist* 30: 907–20.

– (2004). "The first description of nerves in a cnidarian: Louis Agassiz's account of 1850." *Hydrobiologia* 530/531: 27–32.

– and L.M. Passano (1968). "Epithelial conduction in hydromedusae." *Journal of General Physiology* 52: 600–21.

–, C.L. Singla, and S.A. Arkett (1988). "On the nervous system of *Velella* (Hydrozoa: Chondrophora)." *Journal of Morphology* 198: 15–23.

Martin, V.J. (2002). "Photoreceptors of cnidarians." *Canadian Journal of Zoology* 80: 1703–22.

Marx, R.M. (1997). "The development of the nervous system of *Aurelia aurita* (Scyphozoa, Coelenterata)." PhD Thesis, University of Victoria, BC.

Mathias, A.P., D.M. Ross, and M. Schachter (1957). "Identification and distribution of 5-hydroxytryptamine in a sea anemone." *Nature* 180: 658–9.

May, R.M. (1925). "The relation of nerves to degenerating and regenerating taste buds." *Journal of Experimental Zoology* 42(4): 371–410.

– (1945). *La formation du système nerveux*. Paris: Gallimard.

Mayer, A.G. (1906). "Rhythmical pulsation in animals. 1. Pulsation in jellyfishes, arms of *Lepas*, heart of *Salpa* and of loggerhead turtle." *Publications of the Carnegie Institution of Washington* 47: 1–62.

– (1908). "Rhythmical pulsation in Scyphomedusae. II." *Papers from the Tortugas Laboratory of the Carnegie Institution of Washington* 1: 113–31.

– (1916). "Nerve conduction and other reactions in *Cassiopea*." *Americal Journal of Physiology* 39: 375–93.

– (1917). "Nerve-conduction in *Cassiopea xamachana*." *Papers from the Tortugas Laboratory of the Carnegie Institution of Washington* 11: 1–20.

McConnell, C.H. (1932). "The development of the ectodermal nerve net in the buds of *Hydra*." *Quarterly Journal of Microscopical Science* 75: 495–509.

Menzel, R. (2004). Pages 455–7 in *The History of Neuroscience as Autobiography*, Vol. 4, edited by L.A. Squire. San Diego and London: Elsevier Academic Press.

Merejkowsky, C. (1878). "Etudes sur les éponges de la Mer Blanche." *Mémoires de l'Académie Impériale des Sciences de Saint-Pétersbourg* 26(7): 1–51.

Miller, G. (2009). "Origins. On the origin of the nervous system." *Science* 325(5936): 24–6.

Minchin, E.A. (1892). "The oscula and anatomy of *Leucosolenia clathrus*, O.S." *Quarterly Journal of Microscopical Science* 33: 477–95.

Morgan, T.H. (1919). *The Physical Basis of Heredity*. Philadelphia: J.B. Lippincott.

Moroz, L.L. (2009). "On the independent origins of complex brains and neurons." *Brain, Behavior, and Evolution* 74: 177–90.

– et al. (2014). "The ctenophore genome and the evolutionary origins of neural systems." *Nature* 510: 109–16.

Mossa, A. (1898). *Life of Man on the High Alps.* London: T. Fisher Unwin.

Müller, F. (1859). "Zwei neue Quallen von Santa Catharina. *Tamoya haplonema* und *quadrumana.*" *Abhandlungen der Naturforschenden Gesellschaft in Halle* 5: 1–12.

Nansen, F. (1886). "Preliminary communication on some investigations upon the histological structure of the central nervous system in the Ascidia and in *Myxine glutinosa.*" *Annals and Magazine of Natural History* (Fifth Series) 18: 209–26.

– (1887). *The Structure and Combination of the Histological Elements of the Central Nervous System.* Bergen: John Grieg.

Nezelof, C. (2003). "Henri Dutrochet (1776–1847): An unheralded discoverer of the cell." *Annals of Diagnostic Pathology* 7(4): 264–72.

Niedermeyer, A. (1914). "Beiträge zur Kenntnis des histologischen Baues von *Veretillum cynomorium* (Pall.)." *Zeitschrift für wissenschaftliche Zoologie* 109: 531–90.

Nyhart, L.K. (1995). *Biology Takes Form: Animal Morphology and the German Universities, 1800–1900.* Chicago: University of Chicago Press.

Oliver, G. and E.A. Schäfer (1894). "On the physiological action of extract of the suprarenal capsules." *Journal of Physiology* (London) 16: i–iv.

– (1895). "On the physiological effects of extracts of the suprarenal capsules." *Journal of Physiology* (London) 18: 230–76.

Osterhout, W.J.V. (1930). "Biographical memoir of Jacques Loeb, 1859–1924." *Biographical Memoirs of the National Academy of Sciences* 13(4): 316–401.

Palade, G.E. and S.L. Palay (1954). "Electron microscope observations of interneuronal and neuromuscular synapses." *Anatomical Record* 118: 335–6.

Pantin, C.F.A. (1935a). "The nerve net of the Actinozoa. I. Facilitation." *Journal of Experimental Biology* 12: 119–38.

– (1935b). "The nerve net of the Actinozoa. II. Plan of the nerve net." *Journal of Experimental Biology* 12: 139–55.

– (1935c). "The nerve net of the Actinozoa. III. Polarity and after-discharge." *Journal of Experimental Biology* 12: 156–64.

– (1935d). "The nerve net of the Actinozoa. IV. Facilitation and the 'staircase.'" *Journal of Experimental Biology* 12: 389–96.

– (1942). "The excitation of nematocysts." *Journal of Experimental Biology* 19: 294–310.

– (1952). "The elementary nervous system." *Proceedings of the Royal Society of London*: Series B, 140: 147–68.

– (1956). "The origin of the nervous system." *Pubblicazioni della Stazione Zoologica di Napoli* 28: 171–81.

– (1965). "Capabilities of the coelenterate behavior machine." *American Zoologist* 5: 581–9.

– (1968). *The Relations Between the Sciences*. Edited by A.M. Pantin and W.H. Thorpe. Cambridge, MA: Cambridge University Press.

Parent, A. (2003). "Auguste Forel on ants and neurology." *Canadian Journal of Neurological Sciences* 30: 284–91.

Parker, G.H. (1896). "The reactions of *Metridium* to food and other substances." *Bulletin of the Museum of Comparative Zoology of Harvard College* 29: 107–19.

– (1900). "The neurone theory in the light of recent discoveries." *American Naturalist* 34(402): 457–70.

– (1909). "The origin of the nervous system and its appropriation of effectors." *Popular Science Monthly* 75: 56–64, 137–46, 253–63, 338–45.

– (1910a). "The reactions of sponges, with a consideration of the origin of the nervous system." *Journal of Experimental Zoology* 8(1): 1–41.

– (1910b). "The phylogenetic origin of the nervous system." *Anatomical Record* 4: 51–8.

– (1911). "The origin and significance of the primitive nervous system." *Proceedings of the American Philosophical Society* 50: 217–25.

– (1914). "Experimentalism in zoology." *Science* 39: 381–5.

– (1916). "The effector systems of actinians." *Journal of Experimental Zoology* 21(4): 461–84.

– (1917a). "Nervous transmission in the actinians." *Journal of Experimental Zoology* 22(1): 87–94.

– (1917b). "The movements of the tentacles in actinians." *Journal of Experimental Zoology* 22(1): 95–110.

– (1917c). "Pedal locomotion in actinians." *Journal of Experimental Zoology* 22(1): 111–24.

– (1917d). "Actinian behavior." *Journal of Experimental Zoology* 22(2): 193–229.

– (1917e). "The activities of *Corymorpha*." *Journal of Experimental Zoology* 24(2): 303–31.

– (1918). "The rate of transmission in the nerve net of the coelenterates." *Journal of General Physiology* 1(2): 231–6.

– (1919a). *The Elementary Nervous System*. Philadelphia and London: J.B. Lippincott Company.

– (1919b). "The organization of *Renilla*." *Journal of Experimental Zoology* 27: 499–507.

– (1920a). "Activities of colonial animals. I. Circulation of water in *Renilla*." *Journal of Experimental Zoology* 31: 343–65.

- (1920b). "Activities of colonial animals. II. Neuromuscular movements and phosphorescence in *Renilla*." *Journal of Experimental Zoology* 31: 475–513.
- (1929). "The neurofibril hypothesis." *Quarterly Review of Biology* 4(2): 155–78.
- (1946). *The World Expands. Recollections of a Zoologist.* Cambridge, MA: Harvard University Press.
- and E.G. Titus (1916). "The structure of *Metridium* (*Actinoloba*) *marginatum* Milne-Edwards with special reference to its neuro-muscular mechanism." *Journal of Experimental Zoology* 21(4): 433–59.

Passano, L.M. (1958). "Intermittent conduction in scyphozoan nerve nets." *Anatomical Record* 132: 486.
- (1965). "Pacemakers and activity patterns in medusae: homage to Romanes." *American Zoologist* 5: 465–81.
- and C.B. McCullough (1960). "Nervous activity and spontaneous beating in scyphomedusae." *Anatomical Record* 137: 387.

Portmann, A. (1926). "Die Kriechbewegung von *Aiptasia cornea*. Ein Beitrag zur Kenntnis der neuromuskulären Organisation der Aktinien." *Zeitschrift für vergleichende Physiologie* 4: 659–67.

Pratt, E.M. (1904). "The mesogleal cells of *Alcyonium* (preliminary account)." *Zoologischer Anzeiger* 25(677): 545–8.

Purkyně, J.E. (1838). "Anatomisch-physiologische Verhandlungen." *Bericht über die Versammlung deutscher Naturforscher und Artze*: 177–80.

Rall, J.A. (1995). "Emil Bozler 1901–1995." *The Physiologist* 38(3): 103.

Ramón y Cajal, S. (1888). "Estructura de los centros nerviosos de las aves." *Revista Trimestrial de Histologia Normal y Patologica* 1(1): 1–10.
- (1894). "The Croonian Lecture: La fine structure des centres nerveux." *Proceedings of the Royal Society of London* 55: 444–68.
- (1937). *Recollections of My Life.* Philadelphia: American Philosophical Society.

Rasmussen, N. (1997). *Picture Control: The Electron Microscope and the Transformation of Biology in America.* Stanford, CA: Stanford University Press.

Reif, W.-E. (1987). "Victor Bauer (1881–1927), Sinnesphysiologie und Neo-Lamarckismus." *History and Philosophy of the Life Sciences* 9(1): 95–107.

Remak, R. (1836). "Verläufige Mittheilung mikroskopischer Beobachtungen über den innern Bau der Cerebrospinalnerven und über die Entwicklung ihrer Formelemente." *Archiv für Anatomie Physiologie und wissenschaftliche Medicin, Jahrgang* 1836: 145–61.
- (1837). "Weitere mikroskopische Beobachtungen über die primitivfasern des Nervensystems der Wirbelthiere." [Frorieps]: *Neue Notizen* 3: col. 35–40.

Richards, R.J. (2008). *The Tragic Sense of Life: Ernst Haeckel and the Struggle over Evolutionary Thought.* Chicago: University of Chicago Press.

Ringereide, M. (1979). "Romanes – Father and Son." *The Bulletin* (Committee on Archives and History of the United Church of Canada) 28: 35–46.

Robertson, J.D. (1953). "Ultrastructure of two invertebrate synapses." *Proceedings of the Society for Experimental Biology and Medicine* 82: 219–23.

Robson, E.A. (1957). "The structure and hydrodynamics of the musculo-epithelium in *Metridium*." *Quarterly Journal of Microscopical Science* 98: 265–78.

– (1961). "A comparison of the nervous system of two sea-anemones, *Calliactis parasitica* and *Metridium senile*." *Quarterly Journal of Microscopical Science* 102: 319–26.

– (1963). "The nerve-net of the swimming anemone, *Stomphia coccinea*." *Quarterly Journal of Microscopical Science* 104: 535–49.

– and R.K. Josephson (1969). "Neuromuscular properties of mesenteries from the sea-anemone *Metridium*." *Journal of Experimental Biology* 50: 151–68.

Romanes, G.J. (1874). "Locomotion of Medusidae." *Nature* 11(263): 29.

– (1876). "Preliminary observations on the locomotor system of medusae." *Philosophical Transactions of the Royal Society of London* 166: 269–313.

– (1877). "Further observations on the locomotor system of medusae." *Philosophical Transactions of the Royal Society of London* 167: 659–752.

– (1880). "Concluding observations on the locomotor system of medusae." *Philosophical Transactions of the Royal Society of London* 171: 161–202.

– (1885). *Jelly-fish, Star-fish, and Sea-urchins*. London: K. Paul, Trench & Company.

– and E.D. Romanes (1896). *The Life and Letters of George John Romanes*. London, New York, etc.: Longmans, Green and Company.

Romer, A.S. (1967). "George Howard Parker: 1864–1955." *Biographical Memoirs of the National Academy of Sciences* 1967: 357–90.

Ross, D.M. and C.F.A. Pantin (1940) "Factors influencing facilitation in Actinozoa. The action of certain ions." *Journal of Experimental Biology* 17: 61–73.

Rüegg, W. (2004). Themes. Pages 1–31 in *A History of the University in Europe. Vol. 3: Universities in the Nineteenth and Early Twentieth Centuries*, edited by W. Rüegg. Cambridge, UK: Cambridge University Press.

Russell, F.S. (1968). "Carl Frederick Abel Pantin 1899–1967." *Biographical Memoirs of Fellows of the Royal Society* 14: 417–34.

Samassa, P. (1892). "Zur Histologie der Ctenophoren." *Zeitschrift für Mikroskopische Anatomie* 40: 157–243.

Satterlie, R.A. (2002). "Neuronal control of swimming in jellyfish: a comparative study." *Canadian Journal of Zoology* 80: 1654–69.

Satterlie, R.A., P.A.V. Anderson, and J.F. Case (1980). "Colonial coordination in anthozoans: Pennatulacea." *Marine Behavior and Physiology* 7: 25–46.

Schaeppi, T. (1898). "Untersuchungen über das Nervensystem der Siphonophoren." *Jenaische Zeitschrift für Naturwissenschaft* 32 (N.S. 25): 483–550.

Schäfer, E.A. (1877). *A Course of Practical Histology: Being an Introduction to the Use of the Microscope*. London: Smith, Elde.

- (1878). "Observations on the nervous system of *Aurelia aurita*." *Philosophical Transactions of the Royal Society of London* 169: 563–75.
- (1893a). "The nerve cell considered as the basis of neurology." *Brain* 16: 134–69.
- (1893b). *The spinal cord and brain*. London: Longmans, Green & Company.
- , ed. (1900). *Text-Book of Physiology*. Edinburgh and London: Young J. Pentland.
- (1912) *Life: Its Nature, Origin, and Maintenance*. London: Longmans, Green and Co.

Schewiakoff, W. (1889). "Beiträge zur Kenntnis des Acalephenauges." *Morphologisches Jahrbuch* 15(1): 21–60.

Schmiedebach, H.P. (1990). "Robert Remak (1815–1865): A Jewish physician and researcher between recognition and rejection." *Zeitschrift für ärtzliche Fortbildung* 84(17): 889–94.

Schneider, K.C. (1890). "Histologie von *Hydra fusca* mit besonderer Berücksichtigung des Nervensystems der Hydropolypen." *Archiv für Mikrokopische Anatomie* 25: 321–79.

- (1892). "Einige histologische Befunde an Coelenteraten." *Jenaische Zeitschrift für Naturwissenschaft* 27 (N.F. 20): 379–462.

Schödl, G. (1988). "Samassa, Paul." *Österreichisches Biographisches Lexikon 1815–1950*, 9: 407–8.

Schulmann, R., A.J. Fox, M. Janssen, and J. Illy, eds. (1998). *The Collected Papers of Albert Einstein*, Volume 8, Part B: *The Berlin Years: Correspondence* 1918, p. 663. Princeton, NJ: Princeton University Press.

Schwann, T. (1839). *Mikroskopische Untersuchungen über die Uebereinstimmung in der Struktur und dem Wachsthum der Thiere und Pflanzen*. Berlin: G.F. Reimer.

Shand-Tucci, D. (1978). *Built in Boston: City and Suburb 1800–1950*. Amherst: University of Massachusetts Press.

Sharpey-Schafer, E. (1927). "History of the Physiological Society, 1876–1926." *Journal of Physiology* (London), Supplement: 1–76.

Shepherd, G.M. (1991). *Foundations of the Neuron Doctrine*. New York: Oxford University Press.

Sherrington, C.S. (1906). *The Integrative Action of the Nervous System*. New York: C. Scribner's Sons.

- (1935). "Sir Edward Sharpey-Schafer, 1850–1935." *Quarterly Journal of Experimental Physiology* 25(2): 99–104.

Siesser, W.G. (1981), "Christian Gottfried Ehrenberg: Founder of micropaleontology." *Centaurus* 25 (3): 166–88.

Smola, U. (2004). "Historical retrospect: Richard von Hertwig (1850–1937), a precursor of modern developmental biology." Pages xxi–xxiv in *Proceedings of the 11th International Echinoderm Conference, Munich, Germany*, edited by T. Heinzeller and J.H. Nebelsick. London: Taylor and Francis.

Sotelo, C. (2003). "Viewing the brain through the master hand of Ramón y Cajal." *Nature Reviews Neuroscience* 4: 71–7.

Sparrow, E.P. and S. Finger (2001). "Edward Albert Schäfer (Sharpey-Schafer) and his contributions to neuroscience: commemorating the 150th anniversary of his birth." *Journal of the History of the Neurosciences* 10(1): 41–57.

Spencer, A.N. (1971). "Myoid conduction in the siphonophore *Nanomia bijuga.*" *Nature* 233 (5320): 490–1.

– (1981). "The parameters and properties of a group of electrically coupled neurones in the central nervous system of a hydrozoan jellyfish." *Journal of Experimental Biology* 93: 33–50.

Srivastava, M. et al. (2010). "The *Amphimedon queenslandica* genome and the evolution of animal complexity." *Nature* 466: 720–7.

Stephens, L.D. and D.R. Calder (2006). *Seafaring Scientist: Alfred Goldsborough Mayor, Pioneer in Marine Biology*. Columbia, SC: University of South Carolina Press.

Striedter, G.F. (2009). "History of ideas on brain evolution." Pages 3–17 in *Evolutionary Neuroscience*, edited by J.H. Kaas. Oxford and San Diego: Academic Press (Elsevier).

Tamm, S.L. (2014). "Cilia and the life of ctenophores." *Invertebrate Biology* 133(1): 1–46.

Tansey, E.M. (1997). "Not committing barbarisms: Sherrington and the synapse, 1897." *Brain Research Bulletin* 44(3): 211–12.

Topsent, E. (1887). "Contribution à l'étude des Clionides." *Archives de zoologie expérimentale et générale*, Supplément 5 (Mémoire 4, Thèse à la Faculté des Sciences de Paris): 1–465.

Tucker, G.S. (1981). "Ida Henrietta Hyde: The first woman member of the Society." *The Physiologist* 24(6): 1–9.

Turner, F.M. (1974). *Between Science and Religion: Reactions to Scientific Naturalism in Late Victorian England*. New Haven and London: Yale University Press.

Uexküll, J. von (1901) "Die Schwimmbewegungen der *Rhizostoma pulmo.*" *Mittheilungen aus der Zoologischen Station zu Neapel* 14: 620–6.

– (1909) *Umwelt und Innenwelt der Tiere*. Berlin: Julius Springer.

Unna, P.G. (1916). "Die Sauerstofforte und Reductionsorte. Eine histochemische Studie." *Archiv für Mikrokopische Anatomie* 87: 96–150.

Vosmaer, G.C.J. and C.A. Pekelharing (1898). "Observations on sponges." *Verhandelingen der Koninklijke Akademie van Wetenschappen te Amsterdam* 6(2): 1–51.

Waldeyer, W. (1891). "Über einige neuere Forschungen im Gebiete der Anatomie des Centralnervensystems." *Deutsche Medizinische Wochenschrift* 17: 1213–18; 1244–6; 1267–9; 1287–9; 1331–2; 1352–6.

Weindling, P. (1991). *Darwinism and Social Darwinism in Imperial Germany: The Contribution of the Cell Biologist, Oscar Hertwig, 1849–1922*. Stuttgart, New York, Mainz: G. Fischer (Akademie der Wissenschaften und der Literatur).

Weissenberg, R. (1959). *Oscar Hertwig, 1849–1922: Leben und Werk eines deutschen Biologen.* Leipzig: Johann Ambrosius Barth Verlag.

Welsh, J.H. (1960). "5-Hydroxytryptamine in coelenterates." *Nature* 186: 811–12.

Westfall, J.A. (1970). "Ultrastructure of synapses in a primitive coelenterate." *Journal of Ultrastructure Research* 32: 237–46.

– (1973). "Ultrastructural evidence for neuromuscular systems in coelenterates." *American Zoologist* 13: 237–46.

– and C.J.P. Grimmelikhuijzen (1993). "Antho-RFamide immunoreactivity in neuronal synaptic and nonsynaptic vesicles of sea anemones." *Biological Bulletin* (Woods Hole), 185: 109–14.

–, J.C. Kinnamon and D.E. Sims (1980). "Neuro-epitheliomuscular cell and neuro-neuronal gap junctions in *Hydra*." *Journal of Neurocytology* 9: 725–32.

–, K.L. Sayyar, C.F. Elliott, and C.J.P. Grimmelikhuijzen (1995). "Ultrastructural localization of Antho-RWamides I and II at neuromuscular synapses in the gastrodermis and oral sphincter muscle of the sea anemone *Calliactis parasitica*." *Biological Bulletin* (Woods Hole), 189: 280–7.

Whiteley, W. (2006). "Pioneers in neurology: Fridtjof Nansen (1861–1930)." *Journal of Neurology* 253: 1653–4.

Willem, V. (1893). "L'absorption chez les Actinies et l'origine des filaments mésentériques." *Zoologischer Anzeiger* 16(409): 10–12.

Wilson, H.V. (1910). "A study of some epithelioid membranes in moaxonid sponges." *Journal of Experimental Zoology* 9(3): 537–77.

Winkelmann, A. (2007). "Wilhelm von Waldeyer-Hartz (1836–1921): an anatomist who left his mark." *Clinical Anatomy* 20: 231–4.

Winsor, M.P. (1991). *Reading the Shape of Nature: Comparative Zoology at the Agassiz Museum.* Chicago: University of Chicago Press.

Winter F.H. (1914). "Carl Chun." *Bericht der Senckenbergischen Naturforschenden Gesellschaft in Frankfurt am Main* 45 (3): 176–83.

Wolff, M. (1904). "Das Nervensystem der polypoiden Hydrozoa und Scyphozoa. Ein vergleichend-physiologischer und -anatomischer Beitrag zur Neuronlehre." *Zeitschrift für allgemeine Physiologie* 3: 191–281.

Woollard, H.H. and J.A. Harpman (1939). "Discontinuity in the nervous system of coelenterates." *Journal of Anatomy* (London) 73: 559–62.

Yamashita, T. (1957) "Das Aktionspotential des Sinneskörper (Randkörper) der Meduse *Aurelia aurita*." *Zeitschrift für Biologie* 109: 116–22.

Yerkes, R.M. (1902a). "A contribution to the physiology of the nervous system of the medusa *Gonionemus murbachii*. Part I. The sensory reactions of *Gonionemus*." *Americal Journal of Physiology* 6: 434–49.

– (1902b). "A contribution to the physiology of the nervous system of the medusa *Gonionema murbachii*. Part II. The physiology of the nervous system." *Americal Journal of Physiology* 7: 181–98.

Zoja, R. (1890a). "Alcune ricerche morfologische e fisiologische sull' *Hydra*."
Bolletino Scientifico 12(3): 65–92.

– (1890b). "Alcune ricerche morfologische e fisiologische sull' *Hydra*. Organi
della motilità." *Bolletino Scientifico* 12(4): 97–134.

– (1891). "Alcune ricerche morfologische e fisiologische sull' *Hydra*. Organi
della sensibilità." *Bolletino Scientifico* 13(1): 1–20.

INDEX